FATAL EXIT

IEEE Press
445 Hoes Lane
Piscataway, NJ 08854

FATAL EXIT
The Automotive Black Box Debate

THOMAS M. KOWALICK

IEEE PRESS

A JOHN WILEY & SONS, INC., PUBLICATION

Published by John Wiley & Sons, Inc., Hoboken, New Jersey
Published simultaneously in Canada.

For general information on our other products and services please contact our Customer Care Department within the U.S. at 877-762-2974, outside the U.S. at 317-572-3993 or fax 317-572-4002.

Wiley also publishes its books in a variety of electronic formats. Some content that appears in print, however, may not be available in electronic format.

Library of Congress Cataloging-in-Publication Data is available.

ISBN 0-471-69807-5

Printed in the United States of America.

10 9 8 7 6 5 4 3 2 1

Life feeds back truth to people in its own ways and time

To my beloved children: Ariel, Kassia and Michael

ACKNOWLEDGMENTS

One of the great pleasures of finishing a book is that it gives the author the opportunity to thank those who helped make the project a product.

I could not have generated the information, collected the photographs, or completed the project without the assistance and aid of others. I want to thank all of those who contributed to the successful completion of this book.

My first debt is, of course, to my three amazing children who encouraged me to discover my writing purpose and pursue it with intense passion and perseverance. I am grateful that they understood the world must know this story.

I am deeply grateful to my friend Mohsin Ali, former diplomatic editor for *Reuters*, who helped tremendously by serving as a guest lecturer to my college classes while I attended the many meetings. The hundreds of college students who participated in the research and surveys helped to get this book written. Unfortunately, some of these students perished in motor vehicle crashes. Many others were involved in crashes that caused them pain and injury. It was common to hear tragic stories on a daily basis. I constantly thank my college students and remind them that their involvement was important. What is written is never forgotten. I tell them that we learn more from our mistakes than from our successes. Hopefully, in this second century of motor vehicle travel, these students can experience the freedom to travel safely.

I received much support and advice, and sometimes the best advice came from the naysayers who told me I was wasting my time. Fortunately, I turned all their negatives into positives.

Many individuals are mentioned within, but this book is solely my project and all of the opinions expressed here (except for the direct quotations) are my own.

I do not speak or write for the automakers, government safety establishment, standards development organizations, or advocates—but I do include their own works in my book and also what others have commented about them. I do not bash any group for all are important and I am very careful to be factual. I express my grateful appreciation to those who gave permission to use news articles and extended quotations such as the National Academies of Sciences / Transportation Research Board (TRB), the *New York Times*, *Automotive News* and *EE Times*, the World Health Organization (WHO) and the World Bank.

Robert Kern, my literary agent in Chapel Hill, North Carolina and Cathy Faduska, my senior editor at IEEE Press / John Wiley & Sons, Inc. and Kay Ethier of Bright Path Solutions, Durham, North Carolina, helped craft the initial structure of my manuscript and greatly improved the book.

The entire point of an editor is to decide what is and what is not fit to print, and any book will have some selection criteria. Those criteria and the editor's judgment are its bias. The best that any author can do is make it as clear as possible why everything should be included so as to avoid "unbiased" reporting where no one can be portrayed as being wrong or opposing safety. No courage is required to publish a sanitized, non-critical version of events. To do otherwise requires a higher standard.

Vehicle and highway safety cannot be accomplished through the efforts of one person, a group, or a government agency. It is a shared responsibility among people who travel, the companies that provide transport, and the agencies that regulate travel. But, one person can make a difference toward the goal of safe travel.

Someday, when we are "actually all safer" while traveling in crash-proof vehicles on intelligent highways, I want to tell my children's children that I knew about this problem and did my best to erase it when I could.

I will tell them in life we have two choices, try or do nothing. To me it was impossible to witness the terrible pain and suffering and not get involved.

Road safety is no accident.

Silence is the ultimate weapon of power in vehicle and highway safety. This book will break that silence.

CONTENTS

On August 17,1896, the first human life was lost in a motor vehicle crash.

Bridget Driscoll, a 44-year-old mother of two, was fatally struck by an *auto-mobile* as she and her teenage daughter were on their way to see a dance performance at Crystal Palace in London. Witnesses said the car was going "at tremendous speed." (Back in those days, tremendous speed was eight miles per hour.) The driver, a young man giving free rides to demonstrate the new invention was, according to some, trying to impress a young female passenger. At the inquest, the coroner said, "This must never happen again."

Of course, it's happened since more times than we can bear.

According to World Health Organization (WHO) figures, road crashes throughout the world killed 1.8 million people and injured about 20 to 50 million more in 2002. Millions were hospitalized for days, weeks or months. Long term, perhaps 5 million were disabled for life. If current trends continue, the annual numbers of deaths and disabilities from road traffic injuries will, by the year 2020, have risen by more than 60%, placing motor vehicle crashes at number three on WHO's list of leading contributors to the global burden of disease and injury. As recently as 1990, car crashes ranked ninth on that list.

While the alarming statistics that point to the rapid escalation of this worldwide crisis numb our minds to the gravity of the situation, stark headlines occasionally grab our attention. Imagine for a moment the convulsions of grief that gripped the residents of small-town Millington, Tennessee, when they opened their Sunday papers on the morning of February 29, 2004, and were greeted with the headline, *Seven Teens Die in Car Wreck*. In the dark early hours of that Sunday morning, a car occupied by seven teenagers went airborne after speeding over a small hill, hitting a tree, and killing Michael, Samantha, Trey, Lauren, Jessica, Crystal and Eric.

Our impulse might be to say, "This must never happen again." But we know that it will. The key is, if we knew precisely why this happened, we might have the information necessary to see that it at least doesn't happen as often as we've come to expect.

To address the problem not exclusively as a transportation issue but as the public health global trauma that it is, we need to gather better crash data via Motor Vehicle Event Data Recorders (MVEDRs) to improve road safety. Quality data are vital to solving the mysteries of car crashes and working to improve the safety of our roads. Good science and good policy relies on good data. Without better data, the crisis will only continue to escalate, and the grief will only continue to mount.

FATAL EXIT: The Automotive Black Box Debate takes the reader inside the automotive industry and the government highway safety establishment. It provides all the information you need to understand the technology, consider the politics, and make an informed decision about the need for better data.

Your informed input will serve as a catalyst toward advancing the goal of making safe travel on the world's roads a reality, instead of the deadly gamble it has been for over a century.

In the midst of a late-night lightning storm some years ago, the unmistakable screech of tires and shattered glass woke me from my sleep. As I sat up in bed, sweating and shaking and unaware of my surroundings, I slowly regained my bearings and realized that what roused me was not anything in the waking world, but my own horrific dream. It was merely a dream, but the nightmare was rooted in the reality of my past, of car crashes that have shaped my life and altered my future.

And in this, I know I am not alone.

The odds are good that you are with me. Perhaps you've been in a car crash, or witnessed one, or suddenly lost someone dear to you on one of the world's roadways or highways. In all likelihood, you know someone whose life was turned upside down by a car crash. Maybe it was a friend of a friend, or the classmate of a daughter, or a former neighbor's child. If so, the illusion of automobile and highway safety has had a profound impact on you.

The truth is, though, that there is not a person alive in the civilized world who does not have a stake in the game of vehicle and highway safety. Statistically speaking, automobile crashes are the nation's largest public health hazard with over 3 million highway motor vehicle deaths in the United States since 1899. According to the National Highway Traffic Safety Administration (NHTSA), there were 6,327,252 crashes involving more than 24 million people in the United States in 2003. The victims suffered 2,889,000 injuries, and we mourned the loss of 42,643 men, women and children in those crashes. 193.3 million people drive, so for those of us behind the wheel, the stakes are obvious.

But even if you do not drive a vehicle, you surely ride in one. Your loved ones drive. You walk on the sides of roadways where cars and buses and trucks zip along at speeds that could take your life without a moment's notice. It is impossible not to be somehow impacted by the operation of 230,199,000 registered vehicles in our nation, all of which are potentially lethal weapons.

Perhaps you have never taken the time to ponder why it is that, as yet, there is no real solution to one of the most troublesome problems in contemporary civilization: motor vehicle injury and death. When you take the time to think about it, an obvious question is: why isn't someone doing something about all this? Of course, much is being done on many other issues besides MVEDRs. Or, another question might be: is there anything that CAN be done?

Now if you knew that the data currently culled from the scenes of automobile crashes were virtually useless, merely serve to create tragic statistics, and provide little information of practical, life-saving value. If you knew that simple, cost-effective technology existed that could record precisely what happens in car crashes so that the experts could work to create the conditions that would avoid them. If you knew that the technology existed that could make our roads, our vehicles, and our loved-ones much safer than they have ever been well, wouldn't you want to know why it's not being widely used?

Such technology does exist, but it is trapped in a quagmire of bureaucracy, politics, and debate. For over thirty years, the argument has raged behind closed doors among legislators, regulators, and captains of industry as to how to use it, when to use it, and whether to use it. As the arguments continue, sometimes over mundane things like which plastic connectors to use, the car crashes pile up.

So do the tragedies.

You need to know what is going on. Some say that MVEDRs are the silver bullet while others say they are not. You need to make your voice heard about the use of Motor Vehicle Event Data Recorders (MVEDRs), more commonly referred to as "black boxes." Until the public is aware of the debate, the necessary pressure will not be brought to bear on the powerful interests who hold the key to making a reality what to this point has only been a dream: the freedom to travel safely in an automobile.

Of course you know that the other modes of transportation—aircraft, for example—are outfitted with black boxes to help determine what's happened in the event that they crash. But if you're like most people, you might be unaware that motor vehicles also have black box technologies. If so, you also wouldn't know that there is a fierce debate about them among powerful individuals and special interest groups.

Although the vehicle safety problem is multi-facted and a shared responsibility among manufacturers, consumers, and governments, the fact that we've yet to arrive at solutions to the problem is compounded by the lack of objective data. We need objective data to enhance motor vehicle safety through the development and introduction of advances that provide real-world reductions of injuries and fatalities. Today, the challenge lies in combining technological breakthroughs in communication and vehicle design that will accomplish two tasks:

1. Enhance protection for motor vehicle occupants, and

2. Convey objective scientific data about what happens in a crash.

Most consumers are aware of automakers' efforts and advances in the development and actual safety features in motor vehicles. What we don't hear much about is the need for objective data to make cars safer and prevent crashes in the first place. During a century of rapid growth in transportation services, an "unknown" number of vehicles have crashed, injuring and killing an "unknown" number of people. Understanding what happens in a crash is also virtually an "unknown factor," because there have been no tools to objectively measure crash details precisely. Knowing exactly what happens in a crash is critically important to preventing carbon copies of similar events. A crash must be explained using scientific, objective data...not subjective analysis.

Only a handful of books have focused on the complex issues of motor vehicle crashes. Those with the requisite knowledge to write and publish such books have mostly been industry insiders. As such, their work betrays inherent biases that reflect the overall economic interests of the automakers who've generally sponsored the research. This is why the solution to this public health epidemic will never come from inside the establishment of vehicle and highway safety.

It will take an outsider with insider information to open the door for public involvement.

Currently, I am that outsider.

I am not employed within the auto industry, and I'm not a member of the political establishment. I am a professor with a passionate interest in seeing our vehicles and roadways become safer places. I have been involved with the players from the highest level of industry and government for a seven-year period (1997–2004). These players include: National Academy of Sciences (NAC), Transportation Research Board (TRB), National Transportation Safety Board (NTSB), National Highway Traffic Safety Administration (NHTSA), Federal Motor Carrier Safety Administration (FMCSA), National Safety Council (NSC), United Nations (UN), World Health Organization (WHO), Surgeon General of the United States, National Association of EMS Physicians (NAEMS), Truck Manufacturers Association (TMA), School Bus Manufacturers Technical Council (SBMTC), Insurance Institute of Highway Safety (IIHS), Owner-Operator Independents Drivers Association (OOIDA),Federal Highway Administration (FHWA), American Trucking Association (ATA), Automotive Occupant Restraints Council (AORC), American Insurance Association (AIA), National Association of Independent Insurers (NAII), American Bus Association (ABA), American Automobile Association (AAA), Mitsubishi Motors Corporation (MMC), Alliance of Automobile Manufacturers (AAM), Advocates for Highway and Auto Safety (Advocates), Association of International Automobile Manufacturers (AIAM), Society of Automotive Engineers (SAE), The Institute of Electrical and Electronic Engineers Standards Association (IEEE-SA), American College of Emergency Physicians (ACEP), and the University of Miami–William Lehman Injury Research Center.

This is a who's who list of the major players involved in vehicle and highway safety. If I were an employee of the many groups mentioned, this book would have been difficult, if not impossible, to publish. This book reflects the arguments on all sides of the issue while clearly presenting the public record regarding vehicle and highway safety. Admittedly, my personal observations are included, but I am careful to express my views within the boundaries of fair comment and criticism. The right of fair comment has been summarized in *Hoeppner vs. Dunkirk Pr. Co., 1930,* as:

> Everyone has a right to comment on matters of public
> interest and concern, provided they do so fairly and
> with an honest purpose. Such comments or criticism

are not libelous, however severe in their terms, unless they are written maliciously. Thus it has been held that books, prints and statuary publicly exhibited, and the architecture of public buildings, and actors and exhibitors are all legitimate subjects of newspapers' criticism, and such criticism fairly and honestly made is not libelous, however strong the terms of the censure may be.

There are many facets to vehicle and highway safety, and this book includes the majority of these issues within three points of view. On one side, there are the advocates of safety who view automotive black box technology as the technology of hope that will have a major impact on future vehicle and highway safety. On the other side, there are the advocates of privacy who envision a world of Big Brother surveillance. In between the two is the ongoing tension—the struggle for real safety checked by a respect for privacy, and the cunning dance that takes place between the various groups.

By including all sides of the debate in this book the end result is a non-bias account. Rather than censor constructive criticism the ideal is to permit anyone who claims bias to respond to the debate.

Any and all material in this book that mentions the IEEE-SA does so under the following guideline:

> At lectures, symposia, seminars, or educational courses, an individual presenting information on IEEE standards shall make it clear that his or her views should be considered the personal views of that individual rather than the formal position, explanation, or interpretation of the IEEE.

Furthermore, the IEEE Patent Policy notes the following:

> IEEE standards may include the known use of essential patents and patent applications provided the IEEE receives assurance from the patent holder or applicant with respect to patents whose infringement is, or in the case of patent applications, potential future infringement the applicant asserts will be, unavoidable in a compliant implementation of either mandatory or optional portions of the standard [essential patents]. This assurance shall be provided without coercion and prior to approval of the standard (or reaffirmation when a patent or patent application becomes known after initial approval of the standard). This assurance shall be a letter that is in the form of either: a) A general disclaimer to the effect that the patentee will not enforce any of its present or future patent(s) whose use would be required to implement either mandatory or optional portions of the proposed

IEEE standard against any person or entity complying with the standard; or b) A statement that a license for such implementation will be made available without compensation or under reasonable rates, with reasonable terms and conditions that are demonstrably free of any unfair discrimination. This assurance shall apply, at a minimum, from the date of the standard's approval to the date of the standard's withdrawal and is irrevocable during that period.

I disclosed an essential foundation patent for motor vehicle event data recorders to the IEEE-SA on April 25, 2003, in accordance with the patent policy stated above and furthermore, encouraged each and every working group member to do likewise. The first order of business at each of the thirteen IEEE-SA working group meetings was to review the patent policy and call for patents.

This is a critical time in land transport history. We are experiencing both a global road safety crisis of mass trauma and a technological revolution in automotive electronics. Over the course of the last century, any significant shifts in America's highway fatalities were "powered" by major societal forces—wars, recession, or periods of great economic growth. Despite serious attempts, no such shift can be attributed to an effective initiative employed by government or industry toward the end of increasing highway or vehicle safety. Simply said, there have not been any ways or means to break the stalemate in safety.

The public must not only be aware of the debate surrounding how to achieve true motor vehicle and highway safety but also must contribute to it. The public must have free access to the details of the debate in order to be educated participants in the process.

The information you need in order to offer your informed opinion about how to achieve vehicle and highway safety is in this book. However, you must understand that there is a strict adherence to a chronological exposition of events that often gets in the way of thematic discussion. Initially I consider it worthwhile to develop particular themes for each chapter, but at the end of the day I chose to retain the chronological model. What I'm actually doing is recording an important slice of transportation history—that time between 1974 and 2004 when black box technology was debated for the highway mode of transportation. So if you are trying to zero in on a certain stakeholder position the index will be your best bet. Next to the index, the bibliography is another example of how important chronology is to this story. The bibliography is presented from 1974 to 2004 in order to demonstrate the increased interest in this important topic. And of course there are extensive notes following each chapter. So now that I explained how and why I wrote *FATAL EXIT* it is your turn to read about the debate and express yourself. Until the voice of the public is presented with unmistakable clarity, the solutions will continue to be put on hold.

The time to make your informed voice heard is now.

PART ONE

SYMPTOMS OF A PROBLEM

CHAPTER 1

Question Everything

Modern Day Crash—Deer inside vehicle, an incident similar to what my family experienced.

My interest in automobile safety isn't driven by personal profit or political gain. It's driven by the simple desire to travel safely. Far too many times in my life, it's been made clear to me just how unsafe our nation's most common mode of transportation really is.

As a child, I recall the horrific scene that ensued when, while riding through the rolling hills of northeast Pennsylvania, a deer jumped from an embankment through the windshield of my father's 1956 Hudson automobile. Though the car itself was a tank, the glass shattered like thin ice around the doe, which suffered pitifully and spewed blood all over the car and us until a local game warden put her out of her misery.

A few years later, I remember screaming when my father, soaked in blood, returned to our car after pulling to the side of the road to help victims at the scene of a crash he kept hidden from our small eyes. As terrifying as that incident was for me, my father was in fact accustomed to helping victims of car crashes. Our house was situated on a mountain in rural Pennsylvania beside a dangerous road that was the scene of many automobile wrecks. People frequently knocked on our door to use the phone in the middle of the night, waking me and leaving me to wonder what had happened out there on the road.

How ironic it was, then, that after having helped so many victims of car crashes, my father himself would perish in one. I never was able to learn the details of his crash, but the gut-wrenching pain of losing him in such a sudden and senseless tragedy is something I'll never forget.

Those experiences might have been enough to move me from being merely curious about highway and vehicle safety toward being obsessed, but the incident that actually drove me to begin serious research came years later. It was a muggy June afternoon in 1992, and I was driving a Mazda van through Laurinburg, a small town in eastern North Carolina, heading to the world famous Spoletto Festival in Charleston, South Carolina. Suddenly, an anxious high school student ran a stop sign and crashed his car into the right side of my vehicle. The collision pushed my van into the opposing traffic lane where it was then sideswiped on the left side by a hit-and-run driver in a small truck. Nearly all the windows in my vehicle shattered, and the van was totaled in fewer than ten seconds. The shaved ice of their Sno-Cones chilled the four young children buckled in the back seats of my van, but thankfully, no one was injured in the crash. Any one of us could have been, though. It was, for me, a dramatic realization of how just one false move, one split second in time has the potential to put a family's life into an utter tailspin.

The fact that our society accepts the reality of frequent and devastating car crashes without much question is a reflection of just how desensitized and fatalistic we've become. The numbers alone are staggering:

- In the year 2003, 24 million people were involved in 6,267,000 crashes in the United States.
- In the last decade, 400,000 people died on America's highways, while 32 million people were injured.

- The overall economic cost of these crashes in the United States was one and a half trillion dollars.

- In the decade of the 1990s, we killed on our nation's highways more than 90,000 young people—from infant to 20 years old; 33 children under the age of ten died in car crashes every week over that time span; more than 110 teenagers died every week during those years.

- 42,643 died last year in the United States alone in motor vehicle crashes, or 118 people each day.

- During the past two decades, motor vehicles have accounted for over 94 percent of all transportation fatalities in the United States.

- World Health Organization (WHO) lists road crashes as the single leading cause of death due to injury in the world.

Unless we or someone we love happens to be one of those thousands of people involved in a car crash, we blithely go through our daily routines, driving from place to place without a thought toward what a gamble it is that we take with our lives each time we put the key in the ignition. We don't think about the truly tragic nature of car crashes, usually calling them "accidents" instead.

An "accident" is taken to mean an unpreventable tragedy, one where culpability is difficult if not impossible to discern or attribute. We don't ignore the grim reality of what happens when tragedy strikes other modes of transportation: planes crash, trains de-rail, and boats sink. But we insist that motor vehicles are involved in "accidents." Misnaming the event minimizes the true nature of what we're dealing with. An editorial in the June 2001 edition of the *British Medical Journal*, in which the editors make clear their reasoning for banning the use of the word "accident" from their articles, notes that, "For many years, safety officials and public health authorities have discouraged use of the word 'accident' when it refers to injuries or the events that produce them. An accident is often understood to be unpredictable—a chance occurrence or an 'act of God'—and therefore unavoidable. However, most injuries and their precipitating events are predictable and preventable."

Words are powerful, and if the word "accident" continues to convey the sense that we are powerless to do anything to stem the tide of increasing death and destruction on our roadways, then we remain passively resigned to the misconception that there is nothing we can do about it.

Nothing could be further from the truth.

The History of Road Safety

The simple fact is that there is no freedom to travel safely in America. And this has always been the case.

Like many pioneering inventions of the late 1800s, the automobile was introduced into an environment with no regulations to guide its design or use. Vehicles were sold to anyone who had the money to purchase one, and there was no formal training. There was no infrastructure in the beginning; the roads that did exist were mainly for horse-driven carriages. Eventually, driving laws such as speed limits were introduced, but by and large they were ignored. In fact, safety precautions in general were ignored, and both drivers and pedestrians were blamed for crashes. It took decades before any real movement was made toward initiating standards of safety on our growing roadway system.

Most of the improvements that were made were vehemently opposed by industries powerful enough to simply ignore the problems. It's important to note that those in power only considered solutions after it was made clear that the public demanded them.

A brief study of the history of vehicle safety indicates that there have been some short-lived periods of dramatic increases and decreases in traffic related fatalities in the United States. There have been three times when fatalities have increased by 5,000 or more deaths over a relatively short span of time; conversely, there are four periods of time in which deaths have fallen by 7,000 or more. B.J. Campbell,[4] writing about seat belt history, notes:

> It is interesting to speculate on the reasons for these massive changes. By the 1930's automobile accidents were already a considerable problem in the United States, and by 1937 deaths reached 39,643. The death rate per hundred million miles was more than five times greater than in 1986 (14.68 vs. 2.58). There was a large decrease (–7000) in 1937–38, seemingly not related changes in exposure, for the mileage exposure before and after the decrease appears to have stayed virtually the same. The change is puzzling since it happened more or less in the middle of the Great Depression. On the other hand, it appears more likely that the upswing in deaths just before WWII reflects the economic expansion on the eve of the war with a consequent increase in exposure. There was an increase of 10% in mileage during the same period, but fatalities went up even more (up 5000).

> The largest downswing in our nation's history, a decrease of 16,000 lives over a single year period, happened early in WWII when mileage exposure dropped by more than one third. Gas rationing, tire rationing, a 35-mph speed limit, and millions of young men in the armed forces and off the highways coincided with this period; all these factors presumably contributed. Actually, fatalities per unit exposure were about the same as earlier, thus the improvement appears to have been almost entirely exposure driven. After the war, there was an increase of 9,000 traffic fatalities within two years at the time of de-mobilization. As with the previous large decline, the increase in fatalities was largely exposure driven, but fatalities actually increased somewhat less than the exposure would have indicated, suggesting the simultaneous influence of other factors.

The next upswing was very large, though spread over a longer period—an increase of nearly 15,000 in fatalities from 1961 to 1966. In 1961, the actual number of deaths in the U.S. was 38,091, fewer than the 39,643 recorded 24 years earlier in 1937. Thus, despite the growth in population and cars from the 1930's to 1961, the death rate per hundred million miles had fallen so much that the raw number of fatalities remained relatively constant. Within the next five years, the rate soared such that in 1966 the raw number of deaths was 53,041. This rate is described by two phenomena: first, a great increase in cars and mileage exposure, and second, by a plateau in the improvement in mileage death rate. For approximately nine years the mileage rate did not fall. In 1961, the death rate per hundred million vehicles was 5.16. In 1969, it was 5.21. This long-term stagnation in the death rate was unique to that point in history—a time when car ownership was soaring, speed limits were high, and powerful cars were a central fact of car marketing and owner preference. It was probably no coincidence that during this period calls for an increased federal role in highway safety were growing more urgent, finally culminating in the activation of the National Highway Safety Board in 1967. The downswing in 1973–75 reflected a combination of the oil embargo, the related severe depression, and the 55-mph speed limit enacted in response. In that time, deaths dropped by about 9,000 despite the fact that exposure did not decrease proportionately. Likewise, during the recession of 1981–83, a drop in fatalities of 9,000 occurred though exposure remained much the same.[5]

Campbell's point is that there are many factors that have contributed to a dramatic increase or decrease in traffic related fatalities in our history. My point is that, with the exception of enacting the 35-mph speed limit in the 1940s and the 55-mph speed limit in the 1970's, not one of these factors has to do with a real safety driven initiative on the part of the automobile industry or the government

As human beings, we are prone to error. We can be diligent, careful, competent drivers, but it's unrealistic to assume that drivers will never make mistakes. In 2001, an article titled *Wrong Turn*, written by Malcolm Gladwell, appeared in the *New Yorker*. The article began by citing, "Every two miles, the average driver makes four hundred observations, forty decisions, and one mistake. Once every five hundred miles, one of those mistakes leads to a near collision, and once every sixty-one thousand miles one of those mistakes leads to crash. When people drive, in other words, mistakes are endemic and accidents inevitable."

Human error is inevitable.

A crash is faster than a blinking eye and is measured in milliseconds. Eyewitnesses can't even accurately relate what exactly happens in a crash. Of course, we have millions of crashes, but no two crashes are identical. In fact, no two versions of one crash are identical. Get two eyewitnesses to a motor vehicle crash together, and you're guaranteed to hear two different versions of events!

In the evaluation of motor vehicle crashes, too much emphasis has been placed on the driver and little if any on the vehicle and infrastructure. The reason for this imbalance in emphasis is the lack of scientific objective data about the vehicle or the infrastructure. NHTSA and others rely on the Indiana Tri-Level Study from the early 1970s and is in the process of commencing another study of crash causation. Could it possibly be true that drivers are primarily responsible for the majority of crashes on our roadways? The fact is, without objective data with which to measure crashes, we simply don't know the answer.

If we can't rely on the subjective accounts of eyewitnesses to provide reliable data, then we need more and better objective data. We also must make sure that this data is accessible to everyone. Motor Vehicle Event Data Recorders (MVEDRs) will provide the necessary information that will enhance vehicle and highway safety. Much of the information to be derived from MVEDRs is information that eyewitnesses could not observe even if they were accurate in all their observations. However, regardless of whether drivers make errors or have sufficient time at high speeds to make correct, safe decisions, vehicles and highways should be designed and built so that those errors do not necessarily result in fatalities.

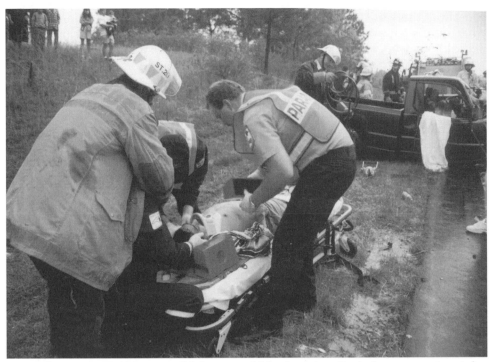

Every day, over 45,000 new vehicles are sold in America. Each day approximately 8,000 people are injured.

The Struggle for Safety

After the incident I endured in Laurinburg in 1992, my curiosity about crashes turned into serious research. I wanted to know how it could be that we suffer so much pain and loss on a daily basis in this country, and yet efforts toward achieving real safety seemed to have remained in neutral. I became obsessed and drove myself hard for a seven-year period between 1997 and 2004, learning about all the forces, both physical and political at work in the quest for true automobile safety. It was, to say the least, a sobering task. Never once could I forget that the ultimate stakes involve human life, human loss, and the millions of survivors whose lives are forever changed on America's roadways.

In my research, I discovered that the problem can, in a general sense, be attributed to two specific factors. First of all, the nation's apparent wholesale commitment to safety has not resulted in a significant reduction of deaths, injuries and crashes because of the increasing demand on our transportation infrastructure. Safety efforts by both government and private groups should be recognized, yes, there have been accomplishments—seat belts, air bags, and other performance standards are responsible for saving tens of thousands of lives, but regrettably, such efforts have been overwhelmed by an increasing tide of vehicles, miles drive, alcohol and speed. More people driving, more vehicles being driven, and few miles of new lanes or roads on which to conduct all that driving leads to the kind of congestion that makes crashes more likely.

The second factor, and the one with which this book is primarily concerned, is the fact that we have no standard, reliable means for determining what exactly happens in a motor vehicle crash. The technology exists in the form of motor vehicle event data recorders (MVEDRs), but it's not being widely used.

The bottom line is that the more we know about motor vehicle crashes— the better the opportunity to enhance vehicles.

A Motor Vehicle Event Data Recorder (MVEDR) is a device that is installed in a motor vehicle to record technical vehicle and occupant-based information for a brief period of time (*i.e.* seconds, not minutes) before, during, and after a crash. For instance, MVEDRs may record (1) pre-crash vehicle dynamics and system status, (2) driver inputs, (3) vehicle crash signature, (4) restraint usage/ deployment status, and (5) certain post-crash data such as activation of an Automatic Collision Notification (ACN) system.

In discussing automatic recording devices, notice that the terms EDRs, MVEDRs and black boxes are often intermingled. In some cases we are describing stand-alone devices—in other cases, embedded systems. It is possible to have an Original Equipment Manufacturer (OEM) system, that is one made by the vehicle maker and an after-market device, or both, onboard a vehicle.

Safety in a motor vehicle is not dependent on one system or feature. Safety is divided into three parts: 1) active safety, 2) passive safety and 3) personal safety. Terms applied to these are crash avoidance and crashworthiness. Each one has some features or equipment that can be described as either "active" or "passive."

When automakers design a vehicle that will help the driver avoid crashes, this sometimes advertise this as an example of "active safety." To them, it is better to avoid crashes than to have crashes, and many vehicles are equipped with innovative active safety features such as improved visibility, improved handling, an interior environment with accessible controls, dynamic stability and traction control, and anti-lock braking systems (ABS), just to name a few. Safety advocates, on the other hand, think Active refers to something that requires "active" participation of the driver or occupant—buckling a seat belt, installing a child safety seat, as example.

Then again, when automakers design a vehicle for the prevention or reduction of injury in a collision, they sometime advertise it as "passive safety." To them, passive refers to systems that protect without action of the driver or occupant. Vehicles incorporate elements designed for exceptional rigidity under enormous stress for a quiet ride, handling consistency, and occupant protection. The goal is to absorb, redirect, or dissipate the force of major impact and to protect the occupants. Safety advocates, on the other hand, view passive safety as features that include air bags, side impact protection systems, daytime running lights, and whiplash protection seating systems.

It should be noted that when seat belt use rates were low and "active" safety was not, as a result, working successfully, "passive" restraint systems such as automatic belts and air bags were developed as a means of achieving safety without changing behavior. Generally, active and passive refer to safety equipment rather than design features such as crumple zones that are built in to the vehicle and are necessarily passive. Thus, while some might associate "active" safety with crash avoidance and "passive" safety with crashworthiness, in truth, they are actually apples and oranges—but both important.

The idea of "personal safety" includes foolproof systems that help protect against vehicle theft, forcible entry, theft of personal property, and personal threats.

Necessary to the effective development of all the first two safety features is information about what exactly happens during a crash, both in order to keep occupants safe during one, and more importantly, to prevent them from occuring in the first place. Motor vehicles have markedly transitioned from machines with mechanical controls to highly technological vehicles with integrated electronic systems and sensors. Modern automobiles generate, analyze, and utilize electronic data to improve vehicle performance, safety, security, comfort, and emissions. It is completely possible to generate the kind of crash information that is critical to understanding what causes a crash, the physical motions of a vehicle's occupants, and vehicle performance both during a crash and during the post-crash events. Manufacturers, engineers, policy makers, researchers, and others rely on crash information to improve vehicle design, shape regulatory policy, develop injury criteria, detect vehicle defects, and resolve investigations and litigation. Capturing the data surrounding a crash on a motor vehicle event data recorder (MVEDR) makes important information readily available for medical responders, crash investigators and researchers.

The degree of social benefit from MVEDRs is directly related to the number of vehicles operating with an MVEDR and the ability to retrieve and utilize the data. It doesn't help to have only a few cars outfitted with the device, or to have different devices installed in different vehicles. Having standardized data definitions and formats allows for the useful capture of vehicle crash information.

It has been my privilege to participate in every major MVEDR initiative (60+ meetings) since 1997 and to meet and work with a wide range of professionals from government, industry, and academia. The tremendous amount of research and development achieved over the past seven years demonstrates that there is no technical impediment to making black box technology the standard in all automobiles.

The reason why this kind of technology must be standard on all vehicles is simply this: Motor Vehicle black boxes speak for the victims. They tell the truth in a way that nothing or no one else can.

Toward Real Change

The time has come to change the automotive industry's focus from luxury, speed, and ruggedness to real safety. Instead of praising the economic benefits that the automotive industry provides the U.S. economy, the emphasis should be on genuine industry involvement in reducing crashes, injuries and deaths. There are a few major initiatives underway now to address the problem of vehicular safety. Both the National Highway Traffic Safety Administration (NHTSA) and the National Academy of Sciences (NAS) are conducting research and development. Other federal agencies such as the National Transportation Safety Board (NTSB) have conducted extensive international symposiums to address the need for better crash data. Worldwide, research and development is underway focusing on systems that link highway infrastructure and telecommunications, using emerging technologies via computers, embedded electronics and advanced sensing technologies.

And yet, we still are woefully lacking in terms of making real progress on this issue. It's not because we don't have the technology. It's that we're not using it. Why? One reason is that money for research and implementation is lacking. The meager budget of the NHTSA is a good example of the lack of priority given to federal funding on highway safety. The current nationwide research effort is additionally compromised by a number of factors, including fragmentation of resources, minimal coordination and partnership activities, redundancies of effort, critical gaps and failure to implement research findings.

Our Current State of information About Crashes and How That Information Is Derived

Any safety system depends on good data. Globally, our road safety awareness of vehicle and highway safety issues has increased, nations and states have enacted new traffic, driver license and vehicle registration laws. Vehicle and highway design standards have incorporated significant safety improvements, and the general public has become less tolerating of drinking and driving. Although these changes, along with numerous other initiatives, have resulted in significant improvements in the safety of our highways, drivers and vehicles what they actually did was only to reduce the deaths and injuries that were easy to eliminate. They certainly haven't solved the overall problem and that problem continues to worsen. I forecast that we will soon have more pressure on the highway safety community (this book will help) to be more sophisticated in identifying problems as to why additional changes to laws, policies and system designs are justified in terms of potential saving of lives, injuries and property damage. All my research points to one key. The need for better information. Our current information is not timely, accurate or reliable. We can and must reverse this state of affairs. I can explain (simply) where aggregate statistical information comes from and why such information sources are deficient. Furthermore, I can demonstrate what information can be obtained from crash investigations/ reconstructions, and why the current state-of-the-art for individual crash events is deficient. Thus, I can make a rational argument about how MVEDR data will provide better information in both aggregate and individual instances that is currently available.

NHTSA is authorized by Congress (Volume 49 of the United States Code, Section 30166, 30168 and Volume 23, Section 403) to collect statistical data on motor vehicle traffic crashes to aid in the development, implementation and evaluation of motor vehicle and highway safety countermeasures. Data collected at the local level serve as the foundation for everything. Whenever a car is registered, a driver license is issued, a counter clicks to record traffic volume, or a car crashes, data are generated as part of an administrative function that also serves a safety purpose.

The Traffic Safety Information System (TSIS) encompasses the hardware, software, personnel and procedures that capture, store, transmit, analyze and interpret highway safety data. The data that are managed by this system include the crash, driver license, vehicle registration, roadway, injury control, Enforcement / Prosecution / Adjudication, motor carrier, exposure, and data analysis information that are described within these TSIS Guidelines.

A complete Traffic Safety Information System collects and stores data from a variety of sources. Much of the data involved are collected for purposes other than (or in addition to) safety. Thus, the TSIS overlaps other areas of data collection and analysis and is dependent also on the policies and practices used in collecting those data. Consequently, the TSIS is usually a collaborative effort spread across a number of agencies and organizations.

The primary purpose of the TSIS is to provide timely, accurate and complete information on highway safety issues. To accomplish this, the data that are collected and stored must also be timely, accurate and complete. The TSIS must be able to integrate information from a number of sources to identify problems and answer questions. The results of analyses must be presented in a manner that is useful to the intended user, which will often include maps, tables and written analyses.

To understand the current state-of-the-art process you need to know the nine components.

1. *Crash Data Component*

Crash data are the driving force for most traffic safety programs. They contain information from the law enforcement traffic crash report, operator crash reports and other sources. They also include information on person/vehicle/event/environment/ location as well as pre-, post-, and the actual crash event. Crash data can be derived from collected data or linked to other state data. Traffic crash data identify safety problems and traffic safety trends. They demonstrate which safety interventions work and which are less effective. They can be aggregated for reporting annual crash facts for a state, and can also be reported in response to inquiries from the public or legislature Crash data are a cornerstone of highway safety and injury prevention programs at all levels of government as well as the health community and private sector. Any changes to crash data that do not recognize the significant number and range of users could have a significant adverse impact upon many safety initiatives. At the same time, changes in crash data that improve their timeliness, completeness, accuracy, reliability and accessibility can facilitate more efficient and effective safety programs.

2. *Driver License Data Component*

The Driver License Data Component includes the license data that every state maintains for every driver license issued by that state. One component of the Driver License Data Component is the driver history file. Driver history data are a vital part of the safety picture when one is trying to identify problem drivers and to assess the impact of driver improvement programs or changes in licensing procedures such as Graduated Driver Licensing. The driver history contains such information as: convictions, administrative actions, driver demographics and residency, crash involvement, pending actions (citation file), state of record if not the current state, court dismissals/reductions, driver improvement and training/test results, and verification of impairment.

3. *Vehicle Registration Data Component*

The Vehicle Registration Data Component contains information such as: owner information (ID, address, type, home county), vehicle description (VIN, make, model), plate history, insurance carrier, brands on title (salvage, flood, etc.), vehicle history (crashes, safety inspection), odometer information (particularly with inspection renewals), and vehicle/owner sanctions.

4. *Roadway Data Component*

The Roadway Data Component contains information about resources and issues related to the right-of-way such as traffic volumes, traffic control device inventory, geometrics, bridges, road condition, and railroad grade crossings. Roadway inventory data are usually the basis for the statewide Geographic Information System (GIS) and the various location reference systems that are supported in the state. Roadway Inventories are typically built and maintained to provide inventory and history information to support construction and maintenance operations.

5. *Injury Control Data Component*

The Injury Control Data Component tracks the persons injured during a motor vehicle crash from the scene, enroute, in the emergency department, and after admission as a hospital in-patient. The tracking capability enables specific characteristics of the event, person, and vehicle to be linked to the medical outcome (type, area, severity of the injury) and the financial outcome (total charges). The linkage of crash and injury outcomes enables the health and highway safety communities to target interventions so they will have the most impact on reducing deaths, injuries, injury severity, and costs.

6. *Enforcement / Prosecution/ Adjudication Data Component*

The Enforcement / Prosecution / Adjudication Data Component tracks a traffic citation from the time it is issued to a law enforcement agency, through its issuance to an offender, through the court system and to the driver history file. In this respect it provides an accountability trail for traffic citations as well as information on the results of citations that are issued. Location information can provide the means to assess enforcement activities against crash experience. This component includes information on other citation activity such as electronic citations (red-light-running, etc.) and possibly even traffic warnings. Traffic citation and conviction data can be used to evaluate trends to determine the short-term and long-term impact of enforcement measures. This component provides the conviction data required in the driver history.

7. *Motor Carrier Data Component*

The Motor Carrier Data Component contains information generated by the licensing and administration of commercial carriers. These data, relative to the carrier and commercially licensed vehicles, are significant when linked to the other TSIS components. Elements within the Motor Carrier Data Component include: carrier, operator and vehicle identification, inspections, compliance reviews, routing, insurance, and so forth.

8. *Exposure Data Component*

The Exposure Data Component is actually a collection of data files and information support systems. Exposure data include information that is used to establish the exposure to risk for crashes. Popular exposure measures are:

- Vehicle-Miles of Travel (VMT) over a section of roadway
- Average Daily Traffic (ADT) at a location
- Census data for political subdivisions down to census tract level
- Distribution of licensed drivers within political subdivisions
- Distribution of registered vehicles by political subdivision

9. *Data Analysis Process Component*

The Data Analysis Process Component addresses the various analytical needs of the end-users and assures that they have access to the data and the analytical tools to support their information needs. The Data Analysis Process Component is composed of the system of data connections, data shares, networks and policies that allow the safety professional to access the data they need, in a form that they can use. It also includes the tools to assist in organizing, analyzing and displaying the data to allow for decision-making. These tools might include data integration, warehousing and linkage software, data analysis packages such as SAS or SPSS, and data analysis/display tools such as GIS systems. End users should have access to training and support in the use of the system, which may range from basic file documentation to sophisticated programming services. The end result of the nine components is data. Once there are data they can be processed. Listed below are the state-of-the-art processes.

The Problem Identification and Crash Analysis Process

All safety agencies, no matter what the primary business function (engineering, injury prevention, EMS, education, etc.) have a need to identify problems, select safety countermeasures, manage the implementation of those countermeasures and evaluate their effectiveness in reducing crashes and injuries. This safety management cycle starts with the identification of problems. A comprehensive problem identification process incorporates information from a variety of sources, and interprets these data against standards and norms to establish those areas that are most in need of attention. The Crash Analysis Process entails the support of those highway safety professionals who draw upon the crash data for decisions about prioritization of resources, design of highways and operations systems, design of vehicle and highway appliances, driver education, enforcement activities and many other areas. A key element of the Crash Analysis Process is the ability of the Traffic Safety Information System to access various data sources and link those data.

How MVDERs Will Improve the Collection of Data for Existing Crash Databases

The most important short-term use of MVEDR data is to enhance quality of the data. Some of the most widely used databases are:

- Fatality Analysis Reporting System (FARS)
- National Automotive Sampling System/Crashworthiness Data System (NASS/CDS/GES)
- Model Minimum Uniform Crash Criteria (MMUCC)

Fatality Analysis Reporting System (FARS) (NHTSA)

FARS: In order to improve traffic safety, the United States Department of Transportation (USDOT/NHTSA) created FARS in 1975. This data system was conceived, designed, and developed by the National Center for Statistics and Analysis (NCSA) to assist the traffic safety community in identifying traffic safety problems and evaluating both motor vehicle safety standards and highway safety initiatives. FARS is one of the two major sources of data used at the NCSA. Fatality information derived from FARS includes motor vehicle traffic crashes that result in the death of an occupant of a vehicle or a non-motorist within 30 days of the crash.

National Automotive Sampling System (NASS) (NHTSA)

NASS is composed of two systems—the Crashworthiness Data System (CDS) and the General Estimates System (GES). Both systems are based on cases selected from a sample of law enforcement crash reports. CDS data focus on passenger vehicle crashes, and are used to investigate injury mechanisms to identify potential improvements in vehicle design. GES data focus on the bigger overall crash picture, and are used for problem size assessments and tracking trends. Established in 1979, NASS was created as part of a nationwide effort to reduce motor vehicle crashes, injuries, and deaths on our nation's highways. NASS collects crash data to help government scientists and engineers analyze motor vehicle crashes and injuries. NASS collects detailed data on a representative random sample of hundreds of thousands of minor, serious and fatal crashes involving passenger cars, pickup trucks, vans, large trucks, motorcycles, and pedestrians. The two components which make up NASS, the Crashworthiness Data System (CDS) and the General Estimates System (GES), select cases from law enforcement crash reports (also known as PARS: Police Accident Reports) at law enforcement agencies within randomly selected areas of the country. These areas are counties and major cities that represent all areas of the United States. CDS data are collected by field researchers who carefully study and record aspects of selected motor vehicle crashes to include exterior and interior vehicle damage, occupant injury, crash scene investigation, environmental conditions at the time of the crash, etc. GES data come from a larger sample of crashes, but only basic information is recorded from the accident report and entered into the system.

Minimum Model Uniform Crash Criteria (MMUCC)

MMUCC are a voluntary set of guidelines that help states collect consistent, reliable crash data that are more effective for identifying traffic safety problems, establishing goals and performance measures, and monitoring the progress of programs. Some of America's leading traffic safety experts worked together to develop the MMUCC guidelines, including representatives from groups in safety, engineering, emergency medical services, law enforcement, public health, and motor carriers.

It is possible to identify data elements that could be collected more accurately and more effectively by an MVEDR. It is also feasible to capture some data elements in the pre-crash phase. The matrix used to classify crashes is called the Haddon Matrix. Basically, human, vehicle, and environment factors get grouped as either pre-crash, crash, or post-crash. Current crash investigation methods does not collect all data elements (provided by the MVEDR) in pre-crash. Current state-of-the-art methods are mostly concerned with collecting data elements in the post-crash phase. This leaves a giant void. With a MVEDR data can be obtained directly from the MVEDR and transferred to a database without any intermediate interference. Direct elements would include vehicle travel speed, time of the crash, and deployment of air bags during the event. These elements have great potential for increasing the accuracy and efficiency of the information present in crash databases. Indirectly, MVEDR data can be utilized to infer or derive a particular variable captured in a current crash database.

A national research agenda should be based on objective scientific knowledge. Such knowledge can be gained from MVEDRs. Data from MVEDRs can identify themes focused on areas with the greatest potential to reduce crashes. Their usage may reinvent the research process by encouraging sponsoring organizations to work collaboratively in the search for effective solutions to specific safety concerns.

The impact of improved crash data goes beyond just understanding the dynamics of a crash. It affects a myriad of important personal, societal, and business functions:

- The **Automotive Industry** can benefit from using safety data for data-driven design of vehicles by using a larger number of crashes across a continuum of severity in its early evaluation of system and vehicle design performance, and it can also work toward the international harmonization of safety standards.

- The **Insurance Industry** can benefit from using safety data to help identify fraudulent claims, which currently cost more than $20 billion annually, and to improve risk management. The data can also be used to expedite claims and decrease administrative costs. Insurers require accurate data for subrogation of claims and recovery of expenses, and event data recorders can supply the necessary information.

- **Government** can benefit from using safety data for promulgating and evaluating safety standards, identifying problem injuries and mechanisms, stipulating injury criteria, and investigating defects. State and local officials require crash information to identify problem intersections and road length to determine hazard countermeasures and to evaluate the effectiveness of safety interventions.

- **Researchers** can benefit from using safety data for human factors research such as man—machine interface, crash causation, age and medical conditions, fatigue, biomechanics research on human response to crashes, harmonized dummy development, and injury causation.

- **Medical Providers** can benefit from using crash data for on-scene field triage of motor vehicle crash victims, improved diagnostic and therapeutic decisions, automatic notification of emergency providers, and better organization of trauma and EMS system resources.

- **The Public** can benefit from using safety data to assure better policies, safer vehicle design, effective emergency response, improved roadway design, safer driving habits, lowered insurance costs, decreased possibility for fraud, fewer crashes, and more efficient systems.

Based on my research, there is every reason to believe that the MVEDRs will constitute one of the most powerful and cost-effective automotive safety devices ever conceived and implemented.

Every time a vehicle crashes, someone loses something, and someone else profits from the loss. There are many who struggle for safety, and several are mentioned throughout this book. Those who participate in the struggle know the technical capabilities, and some understand the practical, not to mention the moral, implications of doing nothing. Sadly, the vehicle and highway safety arena is full of silent observers and note-takers whose timidity and conformity to their affiliations blinds their vision. In the process, countless lives are lost forever, and many more unalterably changed.

The debate over whether to make black boxes standard in all vehicles has been raging for over thirty years. It's time to move from debate to action. Our lives all depend on it.

We are living in the second century of the motor vehicle. We can and must enhance vehicle and highway safety.

CHAPTER 2   Nothing Happens for the First Time: 1969–1998

The first attempt to introduce automotive black boxes was between 1974 and 1998. It was a dead end.

Mary Jo Kopechne died in an infamous late-night incident on Chappaquid-dick Island, Massachusetts. Kopechne was in the passenger seat on 18 July 1969, when a car driven by Senator Edward Kennedy flipped off the Dike Bridge and into a large pond. Kopechne was trapped in the car and died at the scene. Kennedy escaped, but his failure to report the crash until the next morning led to a public scandal that scuttled his plans to run for president in 1972.

I sometimes wonder if Senator Edward M. Kennedy had his radio blaring that hot July night in 1969, when he drove off Chappaquiddick Bridge. If the radio was on, then it's possible that he could have heard Bob Dylan sing *Blowin' in the Wind*.[1]

In that classic tune, Dylan asked, "How many years can a man turn his head, pretending he just doesn't see?"

Maybe Kennedy did have his radio on that night, and maybe he didn't. As with nearly everything else connected with that controversial crash, the evidence is lacking—still blowing in the wind.

One thing is for sure: Kennedy survived the crash and the political aftershocks. As he was chauffeured around Washington, the Honorable Edward M. Kennedy served as Chairman of the Technology Assessment Board in 1975. The Technology Assessment Board was then being asked to evaluate the automotive crash recorder program proposed by the National Highway Traffic Safety Administration (NHTSA). How ironic for Senator Kennedy to have been integrally involved in assessing a possible solution for one of the most urgent problems of our time: motor vehicle related injury and death.

The Office of Technology Assessment (OTA)[2] was created by the Technology Assessment Act of 1972 (2 U.S.C. 472) to serve the United States Congress by providing objective analyses of major public policy issues related to scientific and technological change. The Technology Assessment Advisory Council comprises 10 public members eminent in science and technology. The Council is appointed by the Board and advises the Board and OTA on assessments and other matters.

The Office's assessments explore complex issues involving science and technology, helping Congress resolve uncertainties and conflicting claims, identifying alternative policy options, and providing foresight or early alert to new developments that could have important implications for future Federal policy.

Each project is guided by an advisory panel of experts on a particular subject as a way of ensuring that reports are objective, fair, and authoritative. After approval for release by the Board, OTA assessment reports are distributed to the requesting committees, with summaries provided to all Members of Congress. The reports are available to the public through the Government Printing Office.

On November 19, 1974, Chairman Kennedy received a letter sent by George H. Mahon, Chairman of the U.S. House of Representatives Committee on Appropriations on behalf of Congressman John J. McFall, Chairman of the Transportation Subcommittee, and Congressman Silvio O. Conte, the Subcommittee's Ranking Minority Member transmitting the attached request for a technology assessment with regard to automobile crash recorders.[3] Chairman

Kennedy received his letter because an earlier Conference Report to H. R. 15405 (Department of Transportation and Related Agencies Appropriations Bill, 1975) stated that:

> The conference agreement contains no funds for the crash recorder program. The Committee intends to request an evaluation of this program by the Office of Technology Assessment.[4] The purpose of the program, as proposed by the National Highway Traffic Safety Administration (NHTSA), was to assemble detailed data on actual collisions so as to develop realistic automobile design standards.[5] NHTSA proposed the installation of 1,000,000 crash recorders in vehicles used in ordinary driving. Total cost of the 5 year program including installation of the recorders and monitoring and analysis of the data was estimated at $14.5 million in 1973. An alternate approach had also been proposed by NHTSA. This entailed the controlled crashing of unoccupied vehicles along with computer emulations of automobile crashes. The cost of this program had been estimated as approximately the same as the crash recorder program.

Although the committees of both Houses have heard extensive testimony on this program over the past three years, substantial questions and differences still existed on the necessity for gathering additional information through the installation and monitoring of the requested crash recorders. Since the issue remained unsolved, the Conference Committee on H.R. 15405 decided to call upon the OTA for assistance.

The objective was to undertake a study of the need for and means to assemble detailed data on actual automobile collisions so as to develop realistic automobile design standards.[6] The study examined the desirability, utility, design and cost of crash recorders and of the alternate approaches to collecting crash data, including computer crash simulation, controlled laboratory crashes and their correlation with observed vehicle deformations, and methods to improve the accuracy of crash investigation reporting and to increase the utility of national crash data files. Specific data collection programs previously proposed to Congress by the NHTSA were studied and evaluated.

The OTA report concluded that the current national crash data base was inadequate to resolve the uncertainties in NHTSA's current and proposed motor vehicle safety programs.[7] One of the major deficiencies was data relating collision forces and actual fatalities and injuries. The report included a quote from Professor B. J. Campbell (University of North Carolina):[8]

> ...when one is forced to use nonhuman subjects [in laboratory crashes] then one is left in the situation of knowing a great deal about the physics of the crash but knowing little of the actual injuries that might have occurred in such a crash. On the other hand, in real

world automobile crashes one can learn about the
actual outcome in terms of survival and injuries, but
the input variables mentioned before are unknown.
The need to link these two systems is apparent. Engineers who design protective systems need to know
about stopping distances, forces, decelerations, etc.
But knowing these things is of too little help unless
one has a way to relate them to real world injuries.[9]

The findings of the report indicated:

- That the existing national database was inadequate, and that only
  four of 40 existing standards have been shown to be beneficial based
  on statistical evidence.

- That the nationwide effectiveness of lap belts in mitigating fatalities
  was still unknown after five years since statistical evidence was
  available from only one state.

- That there was an immediate need for more and better data to support rulemaking and to estimate the benefits of proposed safety
  standards, to determine the effectiveness of existing safety standards, and to determine causes of crash, injury and fatality to aid
  crashworthy vehicle design.

- That data were needed to identify new safety problems as they
  develop and for predicting the impact of trends in motor vehicle
  design on crash incident and outcome.

- That larger crash data collection expenditures than the $5 million to
  $6 million then programmed annually appeared to be justified.

It was estimated in 1974 that motor vehicle crashes cost society $22 bil-
lion to $44 billion annually and that the present safety standards cost consumers
$2.5 billion annually. It was suggested that proposed and possible safety stan-
dards could cost an additional $4 to $12 billion annually.

For contrast, by 2000 motor vehicle crashes cost society over $230
billion.

Included in the OTA report was a study of twenty crash cases involving
vehicles equipped with crash recorders, summarized in Society of Automotive
Engineers (SAE) paper by S.S. Teel of the NHTSA.[10] In 1974, NHTSA Disc
Recorder Project equipped 1000 vehicles in several fleets that totaled 26 million
miles. Twenty-six crashes were analyzed, measuring delta-Vs up to 20 mph
Delta-V is a an important term to understand. Quite simply, it means changing
velocity. It is also important to understand that speed does not equal velocity.
Speed is merely how fast something is moving. Velocity, however, is speed in a
particular direction. Changing speed or direction, therefore changes velocity.
Finally, the definition of acceleration is changing velocity, and changing velocity,
once again is called delta V. In the 1974 project actual deceleration-time histories were collected. During the same year, General Motors (GM) introduced the
first regular production driver/passenger air bag systems in selected vehicles.

These units contained a data-recording feature for deploying air bags in severe crashes. This recorder was known as the Disc Recorder, and was installed in about 1,000 vehicles in several fleets.

During 1973 and early 1974, the fleets equipped with these recorders accumulated about 26 million miles. During that time, 23 crashes were analyzed, which included delta-Vs up to about 20 mph. Actual deceleration-time histories were collected. These devices were expensive to manufacture, and because installation of these recorders in a vehicle was a prerequisite to collection of crash data, data were limited to a few crashes. The results were an analysis comparing each case vehicle's velocity change, as reported by the police and/or an crash investigation team. The analysis were used to construct a sample of the population of differences between velocity changes estimated by an crash investigator and the velocity change experienced by the vehicle, as reflected in the crash recorder.

The findings of the OTA report indicate:

> We are 95% confident that ten percent of the reported impact speeds overestimate the true change in velocity by at least 35 mph while one-quarter of them overestimate the true change in velocity by at least 25 mph.

The significance of this report is the fact that important policy decisions were being made based on incorrect impact speeds estimated by police and crash investigators. The lack of a sound data base with which to evaluate the need for higher speed performance requirements further underscores the need for a large scale crash recorder program to evaluate actual crash dynamics.

The NHTSA is responsible, under the National Traffic and Motor Vehicle Safety Act of 1966, (Public Law 89-563) for the promulgation of Federal Motor Vehicle Safety Standards (FMVSS) to which vehicles manufactured for sale or in use in the United States must conform. The NHTSA has a legislative mandate under Title 49 of the United States Code, Chapter 301, Motor Vehicle Safety, to issue federal safety standards called the Federal Motor Vehicle Safety Standards (FMVSS). Generally, in administrative law, regulations refer to the entire body of agency "rules" that might vary in degree to which they are mandatory and enforceable. Guidelines, that are neither, are nevertheless regulations. Standards refer to a particular subset of regulations that are compulsory and must be compiled with. The initial FMVSS standards were issued by the Commerce Department within the predecessor to NHTSA—the National Highway Safety Agency, as it was housed there before the Department of Transportation was established. FMVSS 209 was the first standard to become effective on March 1, 1967. A number of FMVSS became effective for vehicles manufactured on and after January 1, 1968. Subsequently, other FMVSS have been issued. New standards and amendments to existing standards are published in the *Federal Register.*

These Federal safety standards are regulations written in terms of minimum safety performance requirements for motor vehicles or items of motor vehicle equipment. These requirements are specified in such a manner that the public is protected against unreasonable risk of crashes occurring as a result of the design, construction, or performance of motor vehicles and is also protected against unreasonable risk of death or injury in the event crashes do occur. The FMVSS's are put in place when they are believed to yield in aggregate a societal benefit greater than their consumer cost. The major problem has always been that the body of data to support or deny a standard is inadequate. The FMVSS does not govern most overall aspects of vehicle safety but only some features considered critical to driving and operating safety.

Thus an initial objective of crash data collection and analysis from the standpoint of the Government rulemaker is that of evaluating the efficacies of the existing standards to determine which should be kept on the books and which should be eliminated. The specialists in auto safety have, as their concerted objective, the reduction of this enormous waste. A body of collision data is needed that will provide a substantial part of the means to determine the causes of crashes, of injuries, and of damage. A second objective from the standpoint of rulemaking is that of providing the necessary statistical support to estimates of a projected safety or damage-limiting standard. A third objective is the early identification of problem areas in motor vehicle damage and injury so as to permit designing effective motor vehicle and highway safety programs.

The foregoing objectives from the standpoint of rulemaking have their parallel from the standpoint of motor vehicle manufacturers. C. Thomas Terry of General Motors summarized the objectives of gathering crash data in the field as 1) evaluation of production safety systems, 2) prediction of performance of proposed safety systems, 3) identification of problem areas and evaluation of proposed solutions on a cost/benefit basis, and 4) estimation of human tolerance to impact. It was further noted that automobile manufacturers are, of course, vitally concerned with the relative merits of specific alternative designs as well as with the validation of safety standards to which they are required by law to conform. A number of universities and institutes, both profit and non-profit, have been for years involved in research in crash causation, injury causation and designs of vehicles and roads that will reduce crashes and injuries. They need crash data to discover causes of crashes and injuries; armed with this information, they can accomplish and test design modifications in their laboratories and provide valuable advice to NHTSA and motor vehicle manufacturers. Finally, there was a need for national planners to predict the impact of new trends in motor vehicle designs. Fuel and resource conservation programs, encouraged if not mandated by the federal government, were predicted to lead to lighter vehicles-lower power-to-weight ratio automobiles. Data on collision frequencies and outcome are needed as a function of these parameters. It was determined thirty years ago, in 1974, that there were unsatisfied needs for crash data. The body of specialists concerned with motor vehicle collisions—the rulemakers, safety researchers, crash statisticians, car designers, insurers, and public interest people—overwhelmingly agreed that there was a grave and compelling need for more and better crash data. It was determined that the Federal Government, not the states, manufacturers or insurance companies, should support the central data collision activities since it was a national problem.

The report concluded that crash recorders provide data that may be admissible in a court of law.[11]

Once the report was complete, it was not acted upon. At that time, the technology was not economically feasible. Instead, the quest for safety traveled a number of different directions and by the next time around (1998) the infrastructure would be in place. The Internet, the Global Positioning System (GPS) satellites and cell phones are just a few examples of the technological infrastructure.

During the initial attempt to introduce automotive black boxes the technology was not economically feasible. Now it is—we have an infrastructure and there is no longer any technical impediment.

With the development of air bags in passenger vehicles, General Motors Corporation, Inc. (GMC) added DERM (Diagnostic and Energy Reserve Module) technologies to record closure times for both the arming and discriminating sensors, as well as any fault codes present at the time of deployment of the air bag.[12]

During the early 1990s, GM installed sophisticated crash data recorders on 70 Indy Formula One racecars.[13] By 1994, GM and other automakers updated to Sensing and Diagnostic Module (SDM) technology which incorporated connections with additional onboard vehicle computers allowing more information to be recorded.[14] By 1999, several more computer systems were incorporated allowing access to additional useful information.

In 1992, the European Union completed a Drive Project for Crash Data named "Samovar:"[15]

> In Great-Britain, the Netherlands and Belgium nine vehicle fleets with a total of 341 vehicles fitted with

data recording equipment participated in the research
program SAMOVAR (Safety Assessment Monitoring on
Vehicles with Automatic Recording) conducted by the
European Union in the framework of the Drive Project
V 2007. Together with a control panel involved in similar tests a total of 850 vehicles participated in the program. The data were collected over a period of 12
months. The result shows that the crash rate
decreased by 28.1% by the use of the vehicle data
recorder. The Samovar Report finally concluded that
the intelligent use of a vehicle data recorder is able to
make a considerable, distinctive, and independent
benefit to road traffic safety.

In 1994, The Johns Hopkins University Applied Physics Laboratory
issued a final report titled Technology Alternatives for an Automated Collision
Notification System (DOT HS 808 288) demonstrating the potential of Accident
Crash Notification (ACN) technologies:[16]

ACN is technology that will provide faster and smarter
Emergency Medical Services (EMS) response in an
attempt to save lives and reduce disabilities from injuries. Although ACN and EDR technologies are not
directly related, they share common aspects.

The goal of Automated Collision Notification (ACN) is
to use technology to provide faster and smarter emergency medical services (EMS) responses in an attempt
to save lives and reduce disabilities from injuries.[17]
This can be accomplished by both reducing the
response time for providing emergency medical assistance to victims of motor vehicle crashes and providing information, such as estimates of crash severity
and the probability of serious injury, to improve the
response. To attain this goal, an ACN system should
automatically determine that a motor vehicle has
been in a collision, notify emergency response personnel of the collision and the vehicle location, provide
information concerning the crash, and establish a
voice link between the vehicle and emergency
response personnel.

The in-vehicle system determines location using a
Global Positioning System (GPS) receiver, senses a
crash with accelerometers dedicated to the ACN function, and communicates with the PSAP via a cellular
phone. PSAP is short for Public Safety Answering
Point, a physical location where 911 emergency telephone calls are received and then routed to the proper

emergency services. The in-vehicle system applied the output of its accelerometers to an algorithm that computed a measure of the severity of a possible crash based on the vehicle acceleration history. This severity measure was compared to a threshold based on an estimate of injury risk being exceeded to determine the occurrence of a crash. The threshold varied depending on the change in velocity and principal direction of force for the crash. Once a crash was detected, a data message containing the vehicle location, information characterizing the crash (*i.e.*, change in velocity, principal direction of force, and rollover occurrence), and the vehicle cellular phone number was sent to the Erie County Sheriff's Office, the PSAP for the Field Operational Test (FOT). Once the data message was delivered, the system automatically switched to voice mode providing the vehicle occupants with a hands-free voice line with the PSAP.

ACN in-vehicle systems were installed in about 700 vehicles during the ACN test period (July 1997 through August 2000). About 500 systems were installed during the first year of the test period, and the high of slightly under 700 vehicles was reached in April 1999.

In summary, it can be stated: [18]

1. The ACN in-vehicle system worked as expected. It was able to sense that a crash had occurred, determine the vehicle's position, and deliver a crash notification message to the FOT 9-1-1 dispatch center via a cellular telephone call that was then switched to a voice line.

2. The crash detection algorithm detected all but one injury crash during the FOT and reduced the notification of property damage-only crashes by more than 85%.

In 1996, the NHTSA Special Crash Investigation (SCI) program began collecting crash data to support crash investigations activities. [19] The mission of SCI is to examine the safety impact of new, emerging, and rapidly changing technology (such as air bags and alternative fuel systems) and for exploring alleged or potential vehicle defects. Since 1972, NCSA's Special Crash Investigations (SCI) Program has provided NHTSA with the most in depth and detailed level of crash investigation data and insurance crash reports collected by the agency. The data collected ranges from basic data maintained in routine police and insurance reports to comprehensive data from special reports by professional crash investigation teams. Hundreds of data elements relevant to the vehicle, occupants, injury mechanisms, roadway, and safety systems involved are collected for each of the over 200 crashes designated for study annually. SCI cases are intended to be an ancedotal data set useful for examining special crash circumstances or outcomes from an engineering perspective. The benefit of this program lies in its ability to locate real-world crashes anywhere in the country, and perform in-depth clinical investigations in a timely manner, which can be utilized by the automotive safety community to improve the performance of its state-of-the-art safety systems. Individual and select groups of cases have triggered both individual companies and the industry as a whole to improve the safety performance of motor vehicles, including passenger cars, light trucks, or school buses. Most of these early cases were low speed air bag related fatalities that could not be accurately reconstructed by the current algorithm.

Another important player in vehicle and highway safety is the Transportation Research Board (TRB). The TRB is a division of the National Research Council, which serves as an independent adviser to the federal government and others on scientific and technical questions of national importance. The National Research Council is jointly administered by the National Academy of Sciences, the National Academy of Engineering, and the Institute of Medicine. The mission of the TRB—one of six major divisions of the National Research Council—is to promote innovation and progress in transportation through research. In an objective and interdisciplinary setting, the Board facilitates the sharing of information on transportation practice and policy by researchers and practitioners; stimulates research and offers research management services that promote technical excellence; provides expert advice on transportation policy and programs; and disseminates research results broadly and encourages their implementation.

TRB fulfills this mission through the work of its standing technical committees and task forces addressing all modes and aspects of transportation; publication and dissemination of reports and peer-reviewed technical papers on research findings; administration of two contract research programs; conduct of special studies on transportation policy issues at the request of the U.S. Congress and government agencies; operation of an on-line computerized file of transportation research information; and the hosting of an annual meeting that typically attracts 9,000 transportation professionals from throughout the United States and abroad.

Highlights of the 1996 TRB Special Report stressed the following.[20]

> Automobile crashes are complex events that result from the interaction of driver behavior, the driving environment (*e.g.* weather, time of day, type of road), and vehicle design. Experts agree that driver error or inappropriate driver behavior—drunk and reckless driving—is the dominant factor affecting the likelihood of being in a crash. Human characteristics, such as age and stage of health, also affect the likelihood of surviving crash injuries.
>
> The dominance of human factor in crash causation does not diminish the important effect of vehicle

design and safety features on crash likelihood or, in particular, on crash outcomes. Drivers cannot change their age or control the behavior of others.

Vehicle features affect safety in two ways: (a) they help the driver avoid a crash or recover from a driving error (crash avoidance) and (b) they provide protection from harm during a crash (crashworthiness). Characteristics such as vehicle stability and braking performance affect the probability of being in a crash, all else being equal. But the driver plays a more important role in determining the extent to which these crash avoidance features reduce crash likelihood. For example, the high fatality rates for drivers of sports cars—vehicles noted for their low center of gravity and stability as well as advanced handling and braking capabilities—attest to the importance of driver and use patterns, which largely determine crash involvement. Once in a crash, however, vehicle characteristics that contribute to crashworthiness, such as size and weight, how the vehicle absorbs energy, and restraint system attributes, play a large role in determining the likelihood and extent of occupant injury.

Because of the close coupling of vehicle characteristics and vehicle crashworthiness, the motor vehicle safety research program of NHTSA has given top priority to research on measures for improving vehicle crashworthiness. Many standards have been developed and injury mitigation measures introduced, such as air bags, which have been incorporated into most vehicles.

At present, development of a defensible summary measure of vehicle crashworthiness is feasible only if current knowledge is supplemented with expert judgment. The uncertainties of present knowledge preclude development of a measure constructed strictly on scientific grounds, but, the relation between vehicle characteristics and occupant protection is sufficiently strong that a useful measure of crashworthiness can be developed if expert judgment is used and the uncertainties are acknowledged. The most reliable estimates can probably be achieved if experts begin with information about the relationship between crashworthiness and vehicle weight and size, and then use analysis combined with the expert professional judgment to incorporate results from crash tests, highway crash statistics, and a variety of other factors, such as the presence or absence of specific design features. Over time the estimates can be improved by development of more field-relevant crash tests and test criteria, more reliable test dummies, and collection of more comparable and consistent field crash data.

The state of knowledge is not well enough advanced, even with expert judgment, to develop a corresponding summary measure for crash avoidance. A major problem is the limited role that vehicle characteristics (as opposed to driving behavior) currently play in predicting crash likelihood.[21]

The 1996 report suggested that research should be conducted on what people know and believe about automobile safety. Market surveys suggest the existence of a growing safety-conscious market segment of new car purchasers. Yet it is unclear what consumers understand about safety. It is difficult to explain or describe.

Vehicle safety is a multidimensional concept that is hard to capture in a single measure.

One way to conceptualize vehicle safety is to distinguish vehicle design characteristics and features related to the probability of being in a crash—referred to as crash avoidance—from those providing protection from harm during a crash—referred to as crashworthiness.

Crash Avoidance includes vehicle characteristics such as braking performance, vehicle stability, and visibility that can help drivers avoid a crash or recover from a driving error. For example, antilock brakes prevent the wheels from locking and the car skidding, thus helping the driver maintain control of the vehicle, particularly on wet and slippery surfaces. Vehicles that have a low center of gravity relative to their track width have less of a propensity to roll over if the car runs off the road or collides with a barrier or another vehicle.

In some cases the potential benefits of crash avoidance may be considerably muted by driver behavior. For example, sports cars, which are known for their stability and handling capabilities, have some of the highest fatality rates as a vehicle class.

Vehicle crashworthiness is important once a crash occurs. Safety characteristics such as weight and size play a critical role in determining the protection afforded vehicle occupants. The extent of injury is directly affected by the crash energy and the manner in which vehicle occupants experience the associated forces. Heavier vehicles typically have a larger interior space, thus providing a longer distance for the occupants to decelerate to stop and reducing the likelihood of injury. Larger vehicles, with more external energy-absorbing structures, do a better job of preventing intrusion into the occupant compartment and increasing the time the crash forces take to reach the occupants. All else being equal, occupants of a heavier, larger vehicle will fare better than the occupants of a smaller, lighter vehicle, especially if the two vehicles collide.

Vehicle size and weight thus help mitigate the effects of a collision with another vehicle or an object outside of the vehicle. Occupant protection features have been developed to reduce what are known as the "second "and "third" collisions, that is, the collision of the vehicle occupants inside the vehicle (against the dashboard or windshield) and the collision of internal organs within the human body, respectively. For example, collapsible steering columns and padded dashboards help deflect or cushion collision impacts. Safety belt systems, which are required in all vehicles, help avoid the second impact. The seat belt is intended to restrain the lower torso and help hold the occupant inside the vehicle, whereas the shoulder belt primarily keeps the upper body away from the steering wheel, dashboard, or windshield. Air bags further protect the occupant's upper body in a severe frontal crash by providing an energy-absorbing cushion between the driver or front seat passenger and the interior of the vehicle. Frontal air bags however provide no protection in rollovers or rear or side-impact crashes; only safety belt systems offer this protection. Seat and door mounted air bags are now being introduced by some manufacturers to provide better protection in side-impact crashes.

This brief summary of how vehicle characteristics and features affect highway safety illustrates the complexity of attempting to describe these factors in a simple yet meaningful manner as a basis for comparing the relative safety of individual vehicles. A major issue is the difficulty of isolating vehicle factors from driver behavior and environmental conditions, which all interact to affect crash likelihood and crash outcomes.

In studying crash causation and the role of vehicle-related factors, it becomes clear that vehicle crashes are complex events involving driving behavior, vehicle characteristics, and environmental conditions. Three in-depth studies of crashes dating from the 1970s (Perchonok 1972; Sabey 1973; Treat et al. 1979) attempted to assign causality to each of the major factors contributing to crash likelihood. The studies found that driver error or appropriate driving behavior was the major contributing factor in 60 to 90 percent of motor vehicle crashes. Environmental factors (*e.g.* weather, road conditions, signing, and lighting) played a major role in 12 to 35 percent of the crashes. Vehicle-related factors (*e.g.* brake failures) were dominant in only 5 to 20 percent of the crashes.

A General Accounting Office (GAO) study (GAO 1994) investigated the relative contributions of driver attributes and vehicle characteristics to crash likelihood. GAO found that driver characteristics such as age and traffic violation history far outweighed vehicle factors—including vehicle age, weight, and size—in predicting crash involvement. Thus, although vehicles differ in many of their characteristics and features, vehicle-related characteristics is only one factor and, to the extent prior studies are correct, a relatively small factor in crash likelihood.

Once in a crash, however, vehicle characteristics that contribute to crashworthiness, such as size and weight, how the vehicle absorbs energy, and restraint system attributes, play a large role in determining the likelihood and extent of occupant injury. In fact, because of the close coupling of vehicle characteristics and vehicle crashworthiness, federal regulations and research have placed a high priority on measures for improving vehicle crashworthiness. Numerous studies have documented that crashworthiness improvements have resulted in measurable reductions to fatalities and that the benefits of crashworthiness regulations on the average are greater than the costs.

Nearly 40 years of federal safety regulation and manufacturer design to comply with these standards has resulted in great improvements in vehicle design and performance. In addition, safety regulations have provided standards against which individual vehicles can be compared and their performance measured.

The National Traffic and Motor Vehicle Safety Act (PL 89-563) of 1966 authorized the then newly created NHTSA to set minimum vehicle safety performance standards. Within 2 years of its creation, NHTSA had issued 29 motor vehicle safety standards and had proposed 95 more.

The primary concern of NHTSA's early vehicle-related programs was to improve vehicle crashworthiness in frontal crashes, because of the large number of fatalities and injuries in this type of crash.

Early crashworthiness research focused on the methods for reducing injury in frontal collisions through demonstration of air bag technologies and development of anthropometric test devices (crash dummies). These efforts culminated in Federal Motor Vehicle Standard (FMVSS) 208 requiring that new automobiles not exceed certain injury thresholds measured in a 48-km/hr (30-mph) frontal crash test.

After a lengthy and contentious debate over the technical feasibility and reliability of automatic or passive occupant protection systems to improve vehicle crashworthiness, the U.S. department of Transportation (DOT) issued a final rule in 1984 requiring automatic protection on new vehicles. The automobile manufacturers could meet the amended FMVSS 208 with automatic safety belts systems or air bags. Phase-in requirements began in model year 1987.

NHTSA's initial emphasis on mitigation of frontal collisions, NHTSA's attention shifted to occupant protection in side-impact crashes during the mid-1970s. After 10 years of development, NHTSA promulgated its amended regulation on side impact in 1988. The standard was upgraded in 1993 so that all passenger vehicles must now meet a dynamic side-impact crash standard.

In recent years NHTSA has focused on rollover crashes, another major source of fatalities and injuries. An Advanced Notice of Proposed Rulemaking (ANPRM) was published in January of 1992, but no Notice of Proposed Rulemaking (NPRM) was issued, just a notice withdrawing the rulemaking. NHTSA concluded the that establishing a single minimum stability standard for passenger cars and light trucks could not be justified on cost-benefit grounds. Instead, NHTSA proposed a broad range of measures to address rollover crashes, including antilock brakes, increased roof strength, better window construction, improved door latches and consumer safety labels.

Federal vehicle safety regulations appear to have contributed to greater uniformity in safety performance, particularly in vehicle crashworthiness as measured by frontal crash test results for passenger vehicles of roughly equivalent weight. Moreover, safety standards now apply to all categories of vehicles. By 1998, all new passenger vehicles—light trucks, vans, and sport utility vehicles as well as cars—are required to have the same safety features and meet the same crash test standards.[22]

Perhaps the most important catalyst towards research and development of MVEDRs came from the National Transportation Safety Board (NTSB).

The National Transportation Safety Board (NTSB) describes its history and mission as:

> The NTSB is an independent Federal agency charged by Congress with investigating every civil aviation accident in the United States and significant accidents in other modes of transportation—railroad, highway, marine and pipeline—and issuing safety recommendations aimed at preventing future accidents.

> The Safety Board determines the probable cause of: all U.S. civil aviation accidents and certain public-use aircraft accidents; selected highway accidents; railroad accidents involving passenger trains or any train accident that results in at least one fatality or major property damage; major marine accidents and any marine accident involving a public and a non public vessel; pipeline accidents involving a fatality or substantial property damage; releases of hazardous materials in

all forms of transportation; and selected transportation accidents that involve problems of a recurring nature. The rules of the Board are located in Chapter VIII, Title 49 of the Code of Federal Regulations.

The NTSB is responsible for maintaining the government's database of civil aviation accidents and also conducts special studies of transportation safety issues of national significance. The NTSB provides investigators to serve as U.S. Accredited Representatives as specified in international treaties for aviation accidents overseas involving U.S.-registered aircraft, or involving aircraft or major components of U.S. manufacture. The NTSB also serves as the "court of appeals" for any airman, mechanic or mariner whenever certificate action is taken by the Federal Aviation Administration or the U.S. Coast Guard Commandant, or when civil penalties are assessed by the FAA. The NTSB opened its doors on April 1, 1967. Although independent, it relied on the U.S. Department of Transportation (DOT) for funding and administrative support.

In 1975, under the Independent Safety Board Act, all organizational ties to DOT were severed. The NTSB is not part of DOT, or affiliated with any of its modal agencies. Since its inception in 1967, the NTSB has investigated more than 114,000 aviation accidents and over 10,000 surface transportation accidents. In so doing, it has become one of the world's premier accident investigation agencies. On call 24 hours a day, 365 days a year, NTSB investigators travel throughout the country and to every corner of the world to investigate significant accidents and develop factual records and safety recommendations. The NTSB has issued more than 11,600 recommendations in all transportation modes to more than 2,200 recipients.

Since 1990, the NTSB has highlighted some issues on a Most Wanted list of safety improvements. Although it has no regulatory or enforcement powers, its reputation for impartiality and thoroughness has enabled the NTSB to achieve such success in shaping transportation safety improvements that more than 80 percent of its recommendations have been adopted by those in a position to effect change. Many safety features currently incorporated into airplanes, automobiles, trains, pipelines and marine vessels had their genesis in NTSB recommendations. At an annual cost of less

than 24 cents a citizen, the NTSB is one of the best bar-
gains in the government.

In 1997, NTSB issued Safety Recommendation H-97-18 to NHTSA to
"pursue crash information gathering using EDRs."[23]

> H-97-18 (NHTSA) Develop and implement, in con-
> junction with the domestic and international automo-
> bile manufacturers, a plan to gather better
> information on crash pulses and other crash parame-
> ters in actual crashes, utilizing current or augmented
> crash sensing and recording devices. (Source—Letter
> of recommendation issued as a result of a Safety Board
> 1997 public forum on air bags and child passenger
> safety.)

Then in April 1997, as a result of the concern for the growing number of
air-bag-induced injuries and fatalities, the administrators of the NHTSA and the
National Aeronautics and Space Administration (NASA) agreed to a cooperative
effort that "leverages NHTSA's expertise in motor vehicle safety restraint systems
and biomechanics with NASA's position as one of the leaders in advanced technology development...to enable the state of air bag safety technology to advance
at a faster pace..." They signed a memorandum of understanding for NASA to
"evaluate air bag performance, establish the technological potential for
improved (smart) air bag systems, and identify key expertise and technology
within the agency (NASA) that can potentially contribute significantly to the
improved effectiveness of air bags."[24] NASA Jet Propulsion Laboratory (JPL)
report investigating air bags recommended that NHTSA "study the feasibility of
installing and obtaining crash data for safety analyses from crash recorders on
vehicles."[25] It also stated that "crash recorders exist already on some vehicles
with electronic air bag sensors, but the data recorded are determined by the
OEMs. These recorders could be the basis for an evolving data-recording capability that could be expanded to serve other purposes, such as in emergency rescues, where the information could be combined with occupant smart keys to
provide critical crash and personal data to paramedics. The questions of data
ownership and data protection would have to be resolved, however. Where data
ownership concerns arise, consultation with experts in the aviation community
regarding the use of aircraft flight recorder is recommended."

Another finding was that "another source of real-world data may be
available. The crash sensors in most current production vehicles use some kind
of a single-point accelerometer that lends itself to crash data recording. OEMs
now use these systems in fleet test programs and in some production vehicles to
evaluate air bag systems...the feasibility of crash detection using recorders with
single-point sensing should be determined...the development of a low-cost
crash recording device is technologically feasible." The significance of this
report is that it established the scientific basis for the need for better real-world
crash data and set in motion a series of events toward that end.

Between 1975 and 1998 there were no significant initiatives towards
implementing automotive crash recorders by the federal government. However,

during that same time period there were significant research and developments in the automotive industry. 1998 would be a breakthrough year. Widespread use of air bags would make further progress possible.

Transportation injuries are a leading cause of workplace deaths.

CHAPTER 3

Shifting Gears: April 1998–September 2002

The National Highway Traffic Safety Administration (NHTSA) motto is "People Saving People."

In order to achieve a safe transport system, there must be a change in our views concerning responsibility, to the extent that the system designers are given clearly defined responsibility for designing the road system on the basis of actual human capabilities, thereby preventing the occurrence of those cases of death and serious injury that are possible to predict and prevent.

—The Swedish Committee of Inquiry Into Road Traffic Responsibility, 1997.

Although it would be possible to trace the research and development of motor vehicle event data recorders since the early 1970s, the current initiative was born on Wednesday, April 1, 1998, when NHTSA held a meeting before invited interested parties.

Early in 1998, NHTSA held several internal planning meetings, and it was decided to propose the creation of a working group (WG) within NHTSA Research & Development's Motor Vehicle Safety Research Advisory Committee (MVSRAC).

The purpose of these meetings was to "explore the possibility of establishing a committee to facilitate the collection and utilization of crash avoidance and crashworthiness data from on-board event data recorders." Seventeen people attended the meeting: ten from NHTSA, one from the National Transportation Safety Board (NTSB), one from the Transportation Research Board (TRB), one from the Federal Highway Administration (FHWA) and four from General Motors Corporation, Inc.

Presentations were made by NHTSA and General Motors. Position statements were made by the NTSB, FHWA and TRB personnel. NTSB provided a history of EDR technology and discussed the NTSB recommendation. The Highway community, represented by the TRB and the FHWA, expressed interest in collection of crash data for crashes into roadside safety devices such as guard rails. Along the same lines, the TRB representative announced they were considering funding an initiative to look into using EDRs to define vehicle crash characteristics for roadside hardware.

The outcome of the meeting was an agreement that a committee should be formed. Several possibilities were discussed, including forming a "Blue Ribbon" panel, setting up group through the Society of Automotive Engineers (SAE), and forming a working group within the Motor Vehicle Safety Research Advisory Committee (MVSRAC).

Twenty-seven days later, Raymond P. Owings, PhD, NHTSA's Associate Administrator for Research and Development and Acting Chairman of the MVSRAC convened the sixteenth meeting of the MVSRAC.[1] MVSRAC was formed to support the mission of NHTSA, which is to reduce injuries and fatalities associated with motor vehicle crashes.

The NHTSA MVSRAC Committee had sixteen members on its main committee. These members represented the Federal Government, the Automobile and Trucking Industries, medical and injury researchers, academia, and the public. There were two subcommittees: Crashworthiness Research and Crash Avoidance.

Dr. Owings began the meeting by explaining some of the general rules for the MVSRAC meeting. "The MVSRAC Committee is to meet at least once a year. It has to be a public meeting. We have to give advance notice in the *Federal Register*. We have to keep detailed minutes. The description in our case, we'll cover this whole meeting." He explained the scope of the meeting. "I want to go through what our scope is. MVSRAC provides information, advice and recommendations to the Administration on matters related to motor vehicle safety research. There are some exceptions. There have been some problems in the past, so I have legal counsel here advising me what things I can or can't do, so we had one or two lessons."

Late in the afternoon on April 28, 1998, the NHTSA staff presented a briefing to the MVSRAC full committee to seek their approval in forming a working group under the Crashworthiness Subcommittee. The presentation was made by John Hinch of the NHTSA Research and Development Office. The meeting was documented by Heritage Reporting Corporation, (202) 628-4888. A transcript of the event is available in the notes section:[2]

The meeting was successful in helping to launch further event data recorder initiatives.[3]

With a green light from the MVSRAC committee all of a sudden event data recorders were back on the research and development agenda of NHTSA. They were also in the capable hands of John Hinch, one of the agency's most qualified engineers. John, a ex-Marine, was a professional with a deep understanding of vehicle and highway safety. He had a reputation for allowing all views to be heard and of getting things done.

Soon afterwards, the 16th International Technical Conference on the Enhanced Safety of Vehicles (ESV) was held at Windsor, Ontario, Canada, from May 31 to June 4. The ESV conferences were held every four years. The ESV program originated more than thirty years ago under the North Atlantic Treaty Organization (NATO) Committee on the Challenges of Modern Society, and was implemented through bilateral agreements between the governments of the United States, France, the Federal Republic of Germany, Italy, the United Kingdom, Japan and Sweden. The participating nations agreed to develop experimental safety vehicles to advance the state-of-the-art technology in automotive engineering and to meet periodically to exchange information on their progress. Since its inception the number of international partners has grown to include the governments of Canada, Australia, The Netherlands, Hungary, Poland, and two international organizations—the European Enhanced Vehicle-Safety Committee, and the European Commission. Several papers were specially on emerging EDR technologies. These papers are listed in the bibliography.[4]

By October 1998, the initial meeting of the NHTSA Event Data Recorder (EDR) Working Group under John Hinch was conducted at Washington, DC.[5] The first meeting had several objectives:

1. Understand the status of EDR technology;
2. Understand the needs of crash data;
3. Review the privacy issues;
4. Develop the working group.

During this meeting members of the WG spoke about EDRs. NHTSA R&D presented operating rules for a MVSRAC working group, which included the public documentation process, a background presentation of EDRs, and a short discussion on privacy. A detailed data list was circulated for consideration.

It was pointed out that the NHTSA Office of Safety Performance Standards (NPS) received a petition for rulemaking in November which requested the government to require Event Data Recorder (EDR) technology on all new vehicles.[6] This petition was denied because of on-going initiatives.

By February 1999, the second NHTSA Event Data Recorder (EDR) WG meeting was held on February 17, at Washington, DC.[7] The second meeting's objectives were:

1. Refine working group objectives;
2. Review WG members' input for data elements;
3. Review of WG's privacy issue policy papers;
4. Other discussions regarding systems and data.

A set of objectives was developed by the WG. Then manufacturers, the government, and others presented short papers regarding their individual privacy policies. The WG also continued its effort to quantify data elements, including selecting a set of "Top-Ten" data elements which should be considered when developing a new EDR.

In May 1999, the National Transportation Safety Board (NTSB) held an "International Symposium on Transportation Recorders," May 3–5, at Arlington, Virginia.[8] The objective was to share knowledge and experience gained from the use of recorded information to improve transportation safety and efficiency.[9]

By June 1999, the NHTSA Office of Safety Performance Standards (NPS) received a second petition for rulemaking, which requested the government to require EDR technology on all new vehicles. This petition was again denied.[10]

The third NHTSA Event Data Recorder (EDR) WG meeting was held on June 9 at Washington, DC.[11] The third meeting's objectives were:

1. Review the objectives;
2. Review members' input for data elements;
3. Review privacy issue policy papers.

During this meeting, the WG continued to refine its position on data elements and privacy issues. Presentations included: Information regarding an upcoming NTSB symposium on data recorders, ACN, recent activities in ISO related to EDRs, and current and recent activities at Ford regarding EDRs.

Four months later, the fourth NHTSA Event Data Recorder (EDR) WG meeting was held on October 6, at Washington, DC.[12] The fourth meeting's objectives were:

1. Discuss insurance company issues;
2. Continue to learn about EDR systems;
3. Hold two group sessions—Data Elements, and Privacy and Legal Issues.

The session on data elements reworked the WG's top 10 data elements list from individual to categories of data elements. The privacy and legal issues session discussed WG members concerns and company and government practices related to EDRs. Presentations included: the I-Witness EDR system, VDO North America (now Siemens VDO, Automotive Trading & Aftermarket Division), and potential for EDR or EDR/ACN use in Massachusetts based on a study of fatal level crashes.

The NTSB issued Safety Recommendations H-99-45-54 to NHTSA by November, 1999.[13]

The NTSB Safety Recommendations for highway safety range from human service concerns such as driver fatigue, and alcohol and drug use, to engineering problems like school bus construction, seat belt usage, air bag concerns, and highway design.

> H-99-53 (NHTSA) Require that all school buses and motor coaches manufactured after January 1, 2003, be equipped with on-board recording systems that record vehicle parameters, including, at a minimum, lateral acceleration, longitudinal acceleration, vertical acceleration, heading, vehicle speed, engine speed, driver's seat belt status, braking input, steering input, gear selection, turn signal status (left/right), brake light status (on/off), Head/tail light status (on/off), passenger door status (open/closed), emergency door status (open/closed), hazard light status (on/off), brake system status (normal/warning), and flashing red light status (on/off) (school bus only). For those buses so equipped, the following should also be recorded: status of additional seat belts, air bag deployment criteria, air bag deployment time, and air bag deployment energy. The on-board recording system should record data at a sampling rate that is sufficient to define vehicle dynamics and should be capable of preserving data in the event of a vehicle crash or an electrical power loss. In addition, the on-board recording system should be mounted to the bus body, not the chassis, to ensure that the data necessary for defining bus body motion are recorded. (Source—Special Investigation Report—Bus Crashworthiness Issue, (NTSB/SIR-99/04).)[14]

> H-99-54 NHTSA. Develop and implement, in cooperation with other government agencies and industry, standards for on-board recording of bus crash data that address, at a minimum, parameters to be recorded, data sampling rates, duration of recording, interface configurations, data storage format, incorporation of fleet management tools, fluid immersion survivability, impact shock survivability, crush and penetration survivability, fire survivability,

independent power supply, and ability to accommodate future requirements and technological advances.
(Source—Special Investigation Report—Bus Crashworthiness Issue, (NTSB/SIR-99/04).)[15]

In February 2000, the fifth NHTSA Event Data Recorder (EDR) WG meeting was held on February 2 at Washington, DC.[16] Meeting objectives included:

1. Review OEM EDR systems.

2. Group sessions—"Status of EDR Technology," and "Who Are the
   Customers."

At this meeting, NHTSA announced that the MVSRAC had been termi-
nated because the charter under which it operated had expired and that all
activities within MVSRAC would need to be halted. Because the nature of the
WG was that of fact finding, NHTSA R&D agreed to continue the WG efforts
under an R&D-sponsored WG. Both group sessions discussed the two objectives
and their outcomes were shared with the WG. Presentations included: OEM discussions of EDR technologies and a NHTSA demonstration of the Vetronix crash
data retrieval device.

The Vetronix Corporation began selling its Crash Data Retrieval (CDR)
system in 2000[17] The CDR system was the first and only device available to the
public that allowed users to download data from the EDRs installed on passenger and light-duty vehicles.

The NTSB held a symposium titled Transportation Safety & the Law,
April 25–26, 2000 at Crystal City, Virginia.[18] The symposium focused on issues
related to improving transportation safety and the use of available information
in the 21st century. Some of the questions addressed included:

1. How can the generation of data and information enhance transportation safety?

2. What are the implications of government investigations and private
   litigation for information development?

3. What is the proper governmental approach to encourage the availability of data for legitimate uses?

The initial meeting of the NHTSA Truck & Bus Event Data Recorder
(EDR) WG was held on June 8 at Linthicum, Maryland.[19] The meeting included:

1. An explanation of the objectives;

2. Limitations of the NHTSA's role;

3. Emphasis of a fact finding mission;

4. Awareness that the WG cannot make recommendations to regulatory agencies;

5. Consent that the WG could compile information to provide input for
   future decisions.

Additional materials circulated included the agenda, a technical brief on SafeTRAC from Assistware Technology, a draft Recommended Practice from the TMC, a paper on Crash Survivable Modules from Smiths Industries, a system schematic from LMS, a paper on Accident Reconstruction from Eaton VORAD Technologies and VORAD Safety System, Inc., and a system schematic from Traxis System Components.

During October 2000, the United States presidential campaign was coming to an end. Vehicle and highway safety were not campaign issues but one candidate captured the attention of the *Detroit Free Press* when he visited the motor city. On October 11, 2000, in an article titled, *Nader Says Automakers Sabotaging Safety,* Free Press staff writer Alexa Capeloto noted:

> Thirty-five years ago, consumer crusader Ralph Nader took on the American automobile industry with his expose "Unsafe at Any Speed." He revived the battle Tuesday at a campaign stop in Detroit, blasting the industry during the few hours he spent in the birthplace of the Big Three.
>
> Nader, the Green Party's presidential candidate, stood before the Economic Club of Detroit in Cobo Center and accused automakers of deliberately blocking advances in auto safety and fuel efficiency.
>
> Denouncing the "nice statements" auto executives have recently made on the industry's responsibility to the public, he also said Ford Motor Co. and Bridgestone/Firestone Inc. initiated a cover-up when they failed to report tire problems to the government. Tread separation on Firestone tires has been linked to 101 deaths in the United States.
>
> "Both companies knew of the problems they had for several years and got feedback from abroad and in this country that people were being killed," Nader told his audience of 300. "They thought they could get away with covering it up because they had a weak auto safety agency in Washington."
>
> That was a reference to the National Highway Traffic Safety Administration, which he called a "consulting firm" for the auto industry. Nader said the NHTSA had not upgraded safety standards for decades.
>
> "In the area of safety, the agency has been nothing short of disgraceful," he said.
>
> Nader was no less disdainful of the Clinton-Gore administration, which he blamed for promising increased fuel efficiency in 1992 and then failing to

follow through. Lax regulation, coupled with a lack of fuel efficiency research, has led to increased importing of oil, higher cost to consumers and serious damage to the environment, he said.

Although he touched on other issues such as urban renewal, the Middle East conflict and school vouchers, which he opposes, Nader devoted most of his appearance to the automobile industry.

He called on Congress to pass a series of amendments to the 1966 Motor Vehicle Safety Act now before the Senate and the House. The amendments include adding criminal penalties for violations of safety laws, increasing civil penalties the government can collect from automakers, stretching the time automakers are liable for defective products, and strengthening testing and research requirements.

"All of the major auto companies are armlocked to weaken or defeat on Capitol Hill the kinds of strengthened amendments that would make motor vehicles safer," Nader said.

Auto industry representatives have said they are in favor of increased safety legislation, but the amendments have so far been blocked on the Senate floor.

Nader said automakers have worked to block legislation for years. He read from notes taken at a 1986 Ford policy meeting, during which executives said safety legislation was successfully stopped by a "broad-based industry coalition led by Ford."

"Doesn't it look even more crazy and short-sided and myopic for the leaders of one of the great industries in our country to behave in this way, in contrast to what they say, in terms of flowery speeches about corporate responsibility?" Nader asked.

Although auto company executives sit on the Economic Club's board of directors, none attended Tuesday's event.

No high-ranking officers from the United Auto Workers, which endorsed Democrat Al Gore after considering backing Nader, attended, either. But two men approached Nader after the speech and one photographed the other next to Nader, a UAW jacket displayed between them.

At a press briefing before his speech, Nader said he doesn't understand why he is so often considered a foe of the automobile industry, because he simply wants it to protect people and the environment. Auto workers identify with his cause because he wants better work standards and improved products, he said.

Keith Crain, a member of the club's board and publisher of *Automotive News* and Auto Week magazines, said he was "sad, but not surprised" that no auto executives attended. "Who can argue with safer cars, cleaner cars?" Crain asked. "The only argument is how to get there."

In October 2000, the second NHTSA Truck & Bus Event Data Recorder (EDR) WG meeting was held on October 25 at Washington, DC.[20] This second meeting gave participants an opportunity to see an example of a NTSB accident analysis using EDR data, an overview of an available event-recording product from VDO, and a presentation of crash statistics and types of crashes where EDRs may show benefit. There was also a discussion of numerous issues surrounding EDRs.

On December 1, 2000 *USA Today* published an article titled *GM Sued Over Automobile "Black Boxes."*

Excerpts of the article note:

General Motors is battling a lawsuit alleging that it violates customer privacy by installing a "black box" in its vehicles to record critical data about speed, braking, and seat belt use, in the moments before a crash. The suit, which seeks class-action status, alleges that the world's largest automakers never told motorists about the devices, known as sensing diagnostic modules (SDM). GM says SDMs, which are part of the air-bag systems, help it design safer cars and aid investigators in reconstructing accidents. But the complaint claims GM failed to inform motorists about the existence of the devices or what they do.

Motorists never learned that "it would be possible for GM, or anyone else to whom GM provided the data surreptitiously recorded by the SDM, to invade the driver's privacy and monitor the particular driver's driving characteristics and seatbelt use habits at a particular point in time," the lawsuit alleges. The lawsuit targets the 1999 model year or later for eight cars: the Chevrolet Corvette and Camaro; the Pontiac Firebird; the Cadillac DeVille, El Dorado, and Seville; and the Buick Century and Park Avenue.

This week, GM filed a motion to transfer the lawsuit from state court in Middlesex County, NJ to federal court in Newark. The complaint alleges GM violated New Jersey's Consumer Fraud Act, invaded the privacy of drivers, and failed to get their consent for the devices. A GM spokeswoman said it is the only case of its type in the country.

GM began installing the devices, which are similar to black boxes on airplanes, in the early 1990s. Since then, their capabilities have grown, and, since 1999, the amount of data recorded has greatly increased.

GM disclosure

Sensing diagnostic modules gained widespread attention after GM disclosed their existence at a technical conference in May 1999.

GM spokeswoman Kelly Cusinato denied that the company failed to inform customers, saying that owner's manuals indicate that some vehicles are equipped with devices that record data about the integrity of the air bag system.

The manual for the 1999 Cadillac DeVille said that some modules also record speed, engine RPM, brake, and throttle data. "We think that our collection and use of the data is legal and appropriate," Cusinato said. "When accidents occur, the device is providing a level of precision that you may not have otherwise in a crash situation."

She said GM engineers or other people can buy software from Vetronix of Santa Barbara, Calif., enabling them to download and interpret the data. GM only seeks to gain access to the data after the company learns of an accident and gains permission from the car's owner or leasee, she said.

Owner permission

"Our policy is that we have to get the vehicle owner or leasee's permission," Cusinato said. "If the owner requests it, they can get a copy of downloaded data. But we're not in the business of providing this information to third parties."

However, attorney Roy A. Katriel, who represents plaintiff Sherry Valan, said that despite the sentence in the owner's manual, GM never adequately disclosed what data was collected and how it may be used.

"In most instances, the recording has been taking place without the owners ever knowing about it," said Katriel. "It raises very troubling questions about informed consent."

The lawsuit alleges that GM also has used the data against at least one car owner to defend a product liability suit.

Katriel said that GM employees also may have access to the data without a driver's consent if a car is totaled in an accident and an insurance company takes ownership of the car.

But Cusinato said that GM would only gather the data from an insurance company if the driver gave his consent. She also said that the company may seek to subpoena data in some instances to defend against a lawsuit.

One New Jersey attorney not involved in the lawsuit said that police and prosecutors, who could benefit from the data in accident reconstructions, are unaware of its existence.

Thomas J. Vesper, a past president of the New Jersey chapter of the Association of Trial Lawyers of America, said he organized a seminar for prosecutors and police early this year after learning of the device.

"Every one of the police officers and prosecutors that I spoke to didn't have a clue about this," said Vesper.

In a December 11, 2000 article in *Traffic World* by John D. Schultz titled *Class Warfare: OOIDA to Sue if Government Mandates On-Board Recorders*, Schultz quoted Jim Johnson, president of the 60,000-member Owner-Operator Independent Drivers Association (OOIDA): "Totally (expletive deleted) unconstitutional, Can I help you spell that?"

By February of 2001, the third NHTSA Truck & Bus Event Data Recorder (EDR) WG meeting was held on February 15–16 at Florida Atlantic University, Boca Rattan, Florida.[21] The agenda items for this meeting included:

1. Event Data Recorder (EDR) Issues and Recommendations paper presented by the Smiths Group;

2. An update on current EDR technologies;

3. Status of EDR technology;

4. The VDO crash recorder;

5.   Emerging technologies and application;

6.   Safety Intelligent Systems (SIS);

7.   Solutions for a Dynamic Marketplace presentation by Insurance Services Office (ISO).

On May 3–6, 2001, The Institute of Electrical and Electronics Engineers (IEEE) Vehicular Technology Society (VTS) conducted its 53rd Conference in Rodos, Greece.[22] I presented a paper titled, "Pros and Cons of Event Data Recorders in the Highway Mode of Transportation." I also had the opportunity to meet the IEEE President Joel Snyder. After informing him of the initiatives under way, I asked if the IEEE might sponsor a standards initiative. He reply was that "I was pushing on a open door." So, once I returned from the trip I began to research the standards process.

At this time there was a need for a standards initiative. The NHTSA working groups had completed their work. A few months later, I conferred with the IEEE Standards Association and received information from staff who explained the standards process.

In June 2001, the 17th International Technical Conference on the Enhanced Safety of Vehicles (ESV), took place at Amsterdam, The Netherlands, June 3–7.[23] I traveled to Amsterdam to present a poster/paper titled, "Real World Perceptions of Emerging Event Data Recorder (EDR) Technologies." This research focused on what college-age motorists perceive to be the positive and negative aspects of implementing Event Data Recorders (EDRs) in the highway mode of transport. The research was later accepted by NHTSA as part of the docket on EDRs.

By August 2001, the Event Data Recorder (EDR) Working Group Final Report (100 pages) was issued emphasizing that "Event Data Recorders (EDRs) offer great potential of improving vehicle and highway safety." [24]

EDRs—Summary of Findings—NHTSA EDR Working Group (8/01):

> ABSTRACT: This report documents the findings of the Event Data Recorder (EDR) working group established by the National Highway Traffic Safety Administration's (NHTSA) Motor Vehicle Safety Research Advisory Committee. In 1997, the National Transportation Safety Board issued recommendations to pursue vehicle crash information gathering using Event Data Recorders. In early 1998, NHTSA's Office of Research and Development launched a new effort to form a working group comprised of industry, academia, and other governmental organizations. The members of the working group participated in the forum to study the state-of-the-art of EDRs. Meetings were held on a regular basis, culminating in this EDR findings report. The following selected findings present the highlights of the report: [25]

1. EDRs have the potential to greatly improve highway safety, for example, by improving occupant protection systems and improving the accuracy of crash reconstructionists.

2. EDR technology has potential safety applications for all classes of motor vehicles.

3. A wide range of crash related and other data elements have been identified which might usefully be captured by future EDR systems.

4. NHTSA has incorporated EDR data collection in its motor vehicle research databases.

5. Open access to EDR data (minus personal identifiers) will benefit researchers, crash investigators, and manufacturers in improving safety on the highways.

6. Studies of EDRs in Europe and the U.S. have shown that driver and employee awareness of an onboard EDR reduces the number and severity of driver crashes.

7. Given the differing nature of cars, vans, SUVs, and other light-vehicles, compared to heavy trucks, school buses, and motor-coaches, different EDR systems may be required to meet the needs of each vehicle class.

8. The degree and benefit from EDRs is directly related to the number of vehicles operating with an EDR and the current infrastructure's ability to use and assimilate these data.

9. Automatic crash notification (ACN) systems integrate the on-board crash sensing and EDR technology with other electronic systems, such as global positioning systems and cellular telephones, to provide early notification of the occurrence, nature, and location of a serious collision.

10. Most systems utilize proprietary protocols and require the manufacturer to download and analyze the data.

The significance of this report is that it would influence further research and development. The pace would intensify. The next steps towards implementing automotive black boxes would involve the process of standardization.

There would also be an important call for comments from NHTSA.

By December 2001, the IEEE-Standards Association Standards Board approved a standards project until December 2005.[26]

The Institute of Electrical and Electronics Engineers is a non-profit, technical professional association of more than 380,000 individual members in 150 countries. The IEEE ("eye-triple-E") helps advance global prosperity by promoting the engineering process of creating, developing, integrating, sharing, and applying knowledge about electrical and information technologies and sciences for the benefit of humanity and the profession.

The IEEE Vehicular Technology Society (VTS) sponsored Project 1616: Draft Standard for Motor Vehicle Event Data Recorder (MVEDR): [27]

> **Project scope:** Motor Vehicle Event Data Recorders (MVEDRs) collect, record, store and export data related to motor vehicle pre-defined events. This standard defines a protocol for MVEDR output data compatibility and export protocols of MVEDR data elements. This standard does not prescribe which specific data elements shall be recorded, or how the data will be collected, recorded and stored. It is applicable to event data recorders for all types of motor vehicles licensed to operate on public roadways, whether offered as original or aftermarket equipment, whether stand-alone or integrated into the vehicle. [28]

> **Project purpose:** Many light-duty motor vehicles, and increasing numbers of heavy commercial vehicles, are equipped with some form of MVEDR. These systems, which are designed and produced by individual motor vehicle manufacturers and component suppliers, are diverse in function, and proprietary in nature. The continuing implementation of MVEDR systems provides an opportunity to voluntarily standardize data output and retrieval protocols to facilitate analysis and promote compatibility of MVEDR data. Adoption of the standard will therefore make MVEDR data more accessible and useful to end users. [29]

Like all IEEE standards projects, P1616 would be developed by a process governed by five imperatives principles:

1. Due process
2. Openness
3. Consensus
4. Balance
5. Right of Appeal. [30]

The IEEE has a Code of Ethics: "We, the members of the IEEE in recognition of the importance of our technologies in affecting the quality of life throughout the world, and in accepting a personal obligation to our profession, its members and the communities we serve, do hereby commit ourselves to the highest ethical and professional conduct and agree:

1. to accept responsibility in making engineering decisions consistent with the safety, health and welfare of the public, and to disclose promptly factors that might endanger the public or the environment;

2. to avoid real or perceived conflicts of interest whenever possible, and to disclose them to affected parties when they do exist;

3. to be honest and realistic in stating claims or estimates based on available data;

4. to reject bribery in all its forms;

5. to improve the understanding of technology, its appropriate application, and potential consequences;

6. to maintain and improve our technical competence and to undertake technological tasks for others only if qualified by training or experience, or after full disclosure of pertinent limitations;

7. to seek, accept, and offer honest criticism of technical work, to acknowledge and correct errors, and to credit properly the contributions of others;

8. to treat fairly all persons regardless of such factors as race, religion, gender, disability, age, or national origin;

9. to avoid injuring others, their property, reputation, or employment by false or malicious action;

10. to assist colleagues and co-workers in their professional development and to support them in following the code of ethics."

An IEEE press release [31] titled, "WORLD'S FIRST MOTOR VEHICLE 'BLACK BOX' STANDARD UNDERWAY AT IEEE: Industry and Government Experts Join to Standardize Event Data Recorders for Highway Crash Data" noted:

> Driven by the lack of uniform scientific crash data needed to make vehicle and highway transportation safer and reduce fatalities, the Institute of Electrical and Electronics Engineers Standards Association (IEEE-SA) has begun working to create the first universal standard for motor vehicle event data recorders (MVEDR) much like those that monitor crashes on aircraft.
>
> According to National Safety Council statistics, motor vehicle crashes are the leading cause of death between the ages of one and 33 in the U.S. There is a death caused by a motor vehicle crash every 12 minutes and a disabling injury every 14 seconds. These crashes,

and the injuries and fatalities they cause, are the nation's largest public health problem.

Since the first road crash fatality in 1896, motor vehicles have claimed an estimated 30 million lives globally. On average, someone dies in a motor vehicle crash each minute worldwide. The IEEE standards project, IEEE P1616 "Motor Vehicle Event Data Recorders," brings together industry and government experts to formulate a minimum performance protocol for the use of onboard tamper and crash-proof memory devices for all types and classes of highway and roadway vehicles.

This international standard will help manufacturers develop devices the public commonly refers to as "black boxes" for autos, trucks, buses, ambulances, fire trucks and other vehicles. The MVEDR standard will define what data should be captured, including date, time, location, velocity, heading, number of occupants and seat belt usage. It will also define how that information should be obtained, recorded and transmitted. "The more accurate the data we gather on highway crashes, the better chance we have to reduce the devastating effects of crashes," says Jim Hall, co-chair of the IEEE P1616 Working Group and former head of the National Transportation Safety Board (NTSB). "That's why it's so important to have recorders that objectively track what goes on in vehicles before and during a crash to complement the subjective input we now get from victims, eye witnesses and police reports. The NTSB considers this so important that it features 'automatic crash sensing and recording devices' high on its current list of the 'Most Wanted' transportation safety improvements."

The IEEE P1616 project builds on more than a decade of ongoing MVEDR research and development. Major studies in this field have been or are being done by the Department of Transportation (USDOT), the National Highway Transportation Safety Administration (NHTSA), the Federal Motor Carrier Safety Administration (FMCA), the Federal Highway Works Administration (FHWA), the Transportation Research Board (TRB), the National Academy of Sciences (NAS), and many of the world's automotive, truck and bus manufacturers.

"This research has taught us to appreciate the significance of MVEDRs," says Tom Kowalick, co-chair of the IEEE P1616 Working Group and professor at Sandhills Community College in Pinehurst, N.C. "In providing essential crash information, these devices can help accelerate the deployment of emerging safety technologies, such as collision avoidance systems, driver assisted technologies, onboard vehicle diagnostic systems and advanced medical response capabilities. The next step is to build what we've learned so far into a series of global standards that fills a gap in our overall transportation system." Kowalick notes that highway vehicles are the only major mode of transportation in the U.S., which also includes air, rail, marine and pipeline transport, without an adequate event data recorder standard. IEEE P1616 aims to rectify this. "The IEEE is the logical group to lead this effort," says Kowalick. "Since the use of electronic components in motor vehicles has grown dramatically in the last decade, the challenge lies in integrating communication and information technology to improve transportation safety.

The IEEE is well positioned to take the lead in bringing these areas together, especially through its 37 technical societies. For instance, the IEEE Vehicular Technology Society is sponsoring the MVEDR standard and recently completed a similar effort along with other IEEE Societies for a rail event data recorder standard (IEEE Std. 1482.1). The IEEE is also playing a major role in developing standards for Intelligent Transportation Systems." IEEE P1616 has attracted interest and participation from a diverse range of public and private sector organizations and individual IEEE volunteers. These include NHTSA, TRB, FHWA, NTSB, the American Public Transportation Association, the American Automobile Association, the Alliance of Automobile Manufacturers (which represents 13 automakers), Transport Canada, Booz Allen Hamilton, General Motors, Honda and Visteon. IEEE-SA Working Groups often contain volunteers from industry, government, academia and trade, scientific and IEEE organizations. Anyone with expertise in automotive electronics, embedded systems, telematics, global positioning systems, solid state recorder technology and automotive software is invited to help develop the IEEE P1616 series of standards.

The IEEE Standards Association (IEEE-SA), a globally recognized standards-setting body, develops consensus standards through an open process that brings diverse parts of an industry together. These standards set specifications and procedures to ensure that products and services are fit for their purpose and perform as intended. The IEEE-SA has a portfolio of more than 870 completed standards and more than 400 in development. Over 15,000 IEEE members worldwide belong to IEEE-SA and voluntarily participate in standards activities. About the IEEE: The IEEE has more than 375,000 members in approximately 170 countries. Through its members, the organization is a leading authority on areas ranging from aerospace, computers and telecommunications to biomedicine, electric power and consumer electronics. The IEEE produces nearly 30 percent of the world's literature in the electrical and electronics engineering, computing and control technology fields. This nonprofit organization also sponsors or cosponsors more than 300 technical conferences each year.

The initial meeting of IEEE Project 1616 was on January 22, 2002, at IEEE/USA in Washington, DC.[32] There were thirteen in attendance. IEEE Standards developing staff members provided a presentation on the importance of practices, policies and procedures and the significance of IEEE standards. Afterwards there were discussions about the involvement of other standards developing organizations (SDOs) including the Society of Automotive Engineers (SAE). It was agreed to increase the notification of the project and to keep the doors wide open to government and industry. The sense amongst the group was that the IEEE standards initiative would be an excellent way to involve everyone. Everyone would be permitted to contribute to the effort.

The second meeting was on February 26, 2002 at The National Academy of Sciences in Washington, DC.[33] There were eighteen in attendance. Jim Hall, former Chairman of the National Transportation Safety Board (NTSB) offered some informal remarks.[34] He noted that he was also involved in another activity at the Transportation Research Board (TRB) that addressed terrorism and security. Hall said that both activities represented the value we place on human life—by preventing terrorism and supporting EDRs. He cited that if we did not learn from our mistakes we will repeat them, and that most advances over the past 40 years occurred because we learned from our mistakes.

He talked of his experience at NTSB and the tangible benefits he enjoyed when he met a survivor. He stressed the need to structure the responsible use of data in recorders. He stated that we are all sharing the highway—and if we could more defensively design our vehicles, and use data in emergency response once we have a crash, then society would benefit. He noted that the government has determined that we all need to share the same road. He said the hard work would have to come from the participants in the room and he hoped that to make what we are doing a reality. He noted that we continue to lose more life, that there are more vehicles, more congestion, and that we should use technology and information to save lives. He remarked it would be a shame not to do so.

Finally, he concluded by stating that almost all the safety in airplanes is the direct result of information collected from the black box.

Chairman Hall praised the IEEE for sponsoring the project. He stated that this is a rare opportunity and offered his continual support. His message stressed the need for 1) technical standards development and, 2) informing the American public of the inherent safety value of utilizing automatic recorder technologies.

From March to June 2002, the IEEE P1616 Working Group held four more meetings, with attendance growing to nearly forty people.[34] Throughout this period, the Working Group grappled with difficult procedural matters. One group of issues concerned understanding and conforming to the IEEE Patent Policy. Other issues revolved around how to best achieve balance in both the Working Group and the Sponsoring Ballot Group, which would be formed later as the group moved forward in the formal consensus process. At the end of this difficult series of meetings, the group decided to appoint one of its members as a Parliamentarian in hope that this would deal with procedural issues more effectively.

The seventh meeting was on July 29–30, 2002 at IEEE headquarters in Piscataway, New Jersey. Forty individuals were in attendance.[35] During this meeting the amended "Scope" and "Purpose" of the project were agreed upon. An item on the agenda was to review a draft work plan for the group. All the members were requested to provide input. The Alliance of Automobile Manufacturers representatives on the committee decided to present their own version titled "MVEDR Standardization Work Plan." After reviewing both plans, a motion was made to establish three sub-committees: 1) EDR definitions, 2) Data Definitions and 3) EDR output compatibility. Another highly debatable issue was real time medical download of crash data. Finally, despite the best efforts of all involved, the issues surrounding the policy and procedures of the working group were still not settled and were causing divisions among the group.

The eighth meeting was on September 23–24, 2002 at NTSB in Washington, DC. There were forty-five individuals in attendance.[36] During this meeting the sub working groups reported on their progress. An interesting sidebar to the meeting was the attendance of a *New York Times* reporter who monitored the meeting and interviewed several of the participants.

At this point in the IEEE project, Dr. Ricardo Martinez, former Administrator of the NHTSA (1994–2000) became actively involved. Dr. Martinez suggested that the words "improved vehicle design, performance, and safety" be reflected in the purpose, since he believes that they speak to the problems that this activity is supposed to solve.[37]

Approximately 500,000 crashes each year occur because of vehicle malfunctions.

PART TWO

SAFETY DEBATE

CHAPTER 4 # NHTSA Call for Comments: October 2002

T*raditionally, road safety has been assumed to be the responsibility of the transport sector, with the main focus within the sector limited to building infrastructure and managing traffic growth. With the sharp increases in motorization in the 1960s in many developed countries, traffic safety agencies were often set up, usually located within a government's transport department.*
—World Report on Road Traffic Injury Prevention, 2004.

The NHTSA is responsible for reducing deaths, injuries and economic losses resulting from motor vehicle crashes. This is accomplished by setting and enforcing safety performance standards for motor vehicles and motor vehicle equipment, and through grants to state and local governments to enable them to conduct effective local highway safety programs.

NHTSA investigates safety defects in motor vehicles, sets and enforces fuel economy standards, helps states and local communities reduce the threat of drunk drivers, promotes the use of safety belts, child safety seats and air bags, investigates odometer fraud, establishes and enforces vehicle anti-theft regulations and provides consumer information on motor vehicle safety topics.

NHTSA also conducts research on driver behavior and traffic safety to develop the most efficient and effective means of bringing about safety improvements.

On October 11, 2002, the NHTSA issued a call for Comments on Event Data Recorders in the *Federal Register.*[1] The call was in response to the three petitions received. NHTSA's response to two earlier petitons noted:

> In this document, we deny a petition for rulemaking submitted by Marie E. Birnbaum, a private individual. The petitioner asked us to initiate rulemaking to require passenger cars and light trucks to be equipped with "black boxes" (data recorders) analogous to those found on commercial airliners. We agree with the petitioner that the recording of crash data can provide information that is very valuable in understanding crashes, and which can be used in a variety of ways to improve motor vehicle safety.

59

However, we are denying the petition because the motor vehicle industry is already voluntarily moving in the direction recommended by the petitioner. Further, we believe this area presents some issues that are, at least for the present time, best addressed in a non-regulatory context.

We received a petition for rulemaking from Marie E. Birnbaum, a private individual, asking us to initiate rulemaking to require passenger cars and light trucks to be equipped with "black boxes" (data recorders) analogous to those found on commercial airliners. The petitioner stated that the purpose of the devices would be to record speed and possibly other data in order to (1) improve public safety by encouraging responsible driving, and (2) provide records of pre-crash speed and possibly other information. Ms. Birnbaum stated that this pre-crash information would work to improve driver accountability through better crash investigations, enforcement and adjudication.

We note that we received Ms. Birnbaum's petition just after we had denied another petition making essentially the same request. Price T. Bingham, a private individual, had asked us to initiate rulemaking to require air bag sensors to be designed so that similar information is recorded during a crash and can be read by crash investigators.

In responding to Mr. Bingham's petition, we noted that the safety community in recent years has shown considerable interest in the concept of crash event recorders. Such recorders can, in conjunctions with air bag and other sensors already provided on many vehicles, collect and record a variety of relevant crash data. These data include such things as vehicle speed, belt use, and crash pulse.

While we agreed with Mr. Bingham that the recording of crash data can provide information that is very valuable in understanding crashes, and which can be used in a variety of ways to improve vehicle safety, we nonetheless denied the petition. One reason for denying the petition was the fact that the motor vehicle industry is already voluntarily moving in the direction recommended by the petitioner. Another was our belief that this are presents some issues that are, at least for the present time, best addressed in a non-regulatory context.

We issued our denial of Mr. Bingham's petition on November 3, 1998, and published it in the November 9, 1998 edition of the *Federal Register* (63 FR 60270). Ms. Birnbaum's petition was dated November 7, 1998.

After reviewing Ms. Birnbaum's petition, we conclude that our reasons for denying Mr. Bingham's petition are also applicable to her petition. A full explanation of those reasons is provided in our November 9, 1998 *Federal Register* notice, which we incorporate by reference.

The November 1998 notice included a discussion of ongoing work in the area by NHTSA's Motor Vehicle Safety Research Advisory Committee (MVSRAC). The agency noted that MVSRAC had set up a working group on event data recorders under the Crashworthiness Subcommittee and that the first meeting of the working group had taken place in October 1998. Since publication of the November 1998 notice, another working group meeting has been held, and a third is planned for this summer. The Event Data Recorder Working Group is considering a wide variety of subjects related to crash event recording devices and anticipates producing a report by the end of calendar year 2000.

Minutes of the Event Data Recorder Working Group meetings are being placed in the public docket. For the reasons discussed above, we are denying Ms. Birnbaum's petition for rulemaking.

Many safety advocates noted a sense of irony in the fact that the first two petitions were denied by a NHTSA administrator who filed his own petition with the agency after he left office. The third petition to NHTSA was submitted from Safety Intelligence Systems Corporation. The petition was submitted on October 29, 2001, by Dr. Ricardo Martinez, former Administrator of NHTSA from 1992 to 2000. Dr. Martinez wrote:

With this letter, I am requesting that the National Highway Traffic Safety Administration initiate rulemaking to mandate the collection and storage of onboard vehicle crash event data, in a standardized data and content format and in a way that is retrievable from the vehicle after the crash.

Motor vehicle injuries continue to be the leading cause of death for persons age 4 to 33 years old and account for than 90% of all transportation-related fatalities. In 2000, 41,821 people were killed and an estimated 3.2 million people were injured in over 6 million police-reported vehicle crashes. A motor vehicle crash occurs every 5–6 seconds.

Understanding what happens in a crash is essential to preventing these injuries and deaths. This information is the cornerstone of safety decision-making, whether it is designing the vehicle, making policy, identifying a potential problem or evaluating the effectiveness of safety systems. There is no substitute for objective, accurate data from real-world crashes. Emerging technologies have provided crash reconstructionists and investigations and reconstructions. Validation of the complex safety and technology systems placed in vehicles today requires evaluation of the electronic information generated and utilized by these vehicle systems.

Despite the high-tech nature of motor vehicles today, current methods of crash investigation rely on analyzing the "archeology of the crash," subjective witness statements, and expert opinion to determine the "facts." Increasingly, the movement from mechanical to electrical systems and sensors means that physical evidence of the crash is diminishing. For example, anti-lock brakes, which measure the rotation speed of each wheel, also decrease the skid marks used as indirect evidence of wheel and vehicle behavior. Advanced air bags use multi-level deployments based upon various measured inputs, yet crash investigators may not be able to directly evaluate the performance of those air bags after the crash.

Field investigations of motor vehicle crashes are costly, time consuming, laborious, and notoriously inaccurate. Indirect measures of vehicle crashes, especially at the crash site, erode over time.

Because of costs and limitation of current crash investigations and reconstructions, the total number of cases available for analysis are limited and skewed toward the more serious crashes. As a result, current data bases are recognized to have major deficiencies because of the small number of crashes they contain and the bias of the information. For example, many crash analyses are based on police reports, which often rely on of subjective witness statements and self-reported information.

Today's vehicles generate, analyze and utilize tremendous amounts of vehicle-based information for operations such as engine and speed control, braking, and deployment of safety systems. Increasingly sophisticated air bags make "decisions" based on vehicle speed, crash direction and severity, occupants size and position, and restraint use. Additional parameters such as brake and throttle position, engine information and vehicle systems status can be captured to help better understand crashes and their causes. Capture and storage of this information is not in all vehicles, nor are the data elements or formats for this information standardized. As such, this information loses its value and is relatively unused or unavailable. The degree of societal benefit from EDRs is directly related to the number of vehicles operating with an EDR and the ability to retrieve and utilize these data.

This lack of knowledge of what happens in real-world crashes severely limits the ability of policy-makers and vehicle designers to save lives. The relative lack of credible real-world information—leading to delays in understanding, evaluating, and improving safety issues—surfaced during the problems with first generation air bags. As children were being killed by passenger front air bags, the Agency required almost a year or more to gather enough cases to better analyze the effectiveness of these air bags. Though the agency chose sled testing as the fastest way for manufacturers to re-design air bags, it soon became clear that there was relatively little knowledge of what a representative real-world crash impulse looked like.

NHTSA's own crash investigations have shown that the difference between derived crash severity calculations and those directly measured by a vehicle may differ by more than 100%. Yet, the Agency, manufacturers, researchers, and others rely on crash severity information in order to better design vehicles, understand crash performance, make policy, develop injury criteria, and understand the biomechanics of injury.

The increasing sophistication and decreasing costs of information technology has created the opportunity to now mandate the capture, storage and retrieval of onboard crash data. Rulemaking would standardize the collection of existing information as a minimal data set in a standardized format for storage and retrieval. In the simplest form, flash memory would simply collect information from the onboard diagnostic module, the air bag sensing and diagnostics module (SDM), and the engine control module. The NHTSA Working Group on EDRs, IEEE and the Society of Automotive Engineers already have suggested or begun work on minimum data sets. NHTSA would need to propose standards that ensure the crash survivability of this collected data.

The NHTSA has previously denied similar petitions based upon the belief that the automotive industry is already voluntarily moving in the direction recommended by the petitioners and that some issues associated with this mandate are best addressed in a nonregulatory context.

While some action has been taken by the various motor vehicle manufacturers since the mid-1990s, overall the industry response has been sluggish and disjointed. Much of the information is proprietary to each individual manufacturer and there is no standardization of the data elements or format of information, either by OEMs or suppliers. While some manufacturers have deployed units in their vehicles, others have stated that they will only do so if mandated by the government.

Views of the value of crash data recorder information within the automotive industry vary widely. Some manufacturers admit that having access to real-world crash information will help them better design cars, while others prefer that this information not be available for legal and liability reasons. In fact, some manufacturers have even stated that more accurate crash information would not be usable to them or others. With such a diversity of opinion, it is obvious that federal leadership on this important public safety issue is warranted.

There are other reasons why a minimum data set of standardized information is a critical keystone in continuing to save lives and improve motor vehicle safety. The FCC is implementing rules to require automatic location information for emergency calls made from wireless phones. The nexus between vehicles and communications provide the basis for Automatic Collision Notification (ACN). OnIy a small amount of vehicle information such as crash severity, restraint use, direction of force, and location (if available) will be of use to emergency providers. The advent of advanced automatic collision notification systems is dependent upon the standardized collection of crash information in the vehicle. Creating this uniformity will greatly accelerate the deployment of ACN, helping medical providers respond quicker and make better diagnostic and therapeutic decisions.

There are other important opportunities that come from greater amounts of more accurate, objective information including the ability to compare safety standards internationally, enhance roadway design, better understand crash causation and biomechanics, and to more quickly evaluate the effectiveness of policy decisions and engineering designs.

NHTSA has raised the fact that there are a variety of social issues to be addressed. While this is true, they are not insurmountable. Millions of vehicles in the fleet already have some form of an event data recording, so many of these social issues are already at work. Increasing awareness will create a need for society to address some of these issues, though many are beyond the purview of the Agency. The EDR Working Group report has created a substantive basis for addressing these issues.

Much of the privacy issues can be addressed by ensuring that the vehicle owner also owns the vehicle information and can provide permission for its use, including transmission for Automatic Collision Notification. Unlike telematic service providers that send vehicle tracking information and personal information back to a third party, vehicle crash information does not have personal identifiers and is only stored should a crash occur In addition, this information is collected without distracting or requiring the driver to interact with the data collection system. Current crash information in the form of police reports and insurance claims have much more personal identifying information than vehicle crash recorders collect.

In 1997, the National Transportation Safety Board issued recommendations to pursue vehicle crash information gathering using Event Data Recorders. In 1997, the National Aeronautics and Space Administration's (NASA) Jet Propulsion Laboratory (JPL) assisted the NHTSA in evaluating the state of advanced air bag technology. The JPL recommended NHTSA study the "feasibility of installing and obtaining crash data for safety analyses from crash recorders on vehicles." In 1998, the NHTSA's Office of Research and Development, with the support of the Motor Vehicle Safety Research Advisory Committee (MVSRAC), formed a Working Group to gather information to better understand and facilitate the collection and utilization of crash data from onboard recorders. This Working Group, comprised of members of academia, the industry, and other government organizations, finalized its report in September 2001. These events, coupled with NHTSA's, academia's, and the industry's recent experience, provide a solid foundation for rulemaking.

I do hope that the NHTSA will look favorably upon this petition and grant rulemaking on this important matter.

The rulemaking process will allow opportunity for adequate input from various constituents and allow NHTSA to craft a final rule that will provide tremendous societal benefits while observing the rights of individuals.

Please let me know how I may be of further assistance.

Dr. Ricardo Martinez

As follow-up, the DOT/NHTSA [Docket No. NHTSA-02-13546; Notice 1] noted:

Over the past several years, there has been considerable interest in the Safety Community regarding possible safety benefits from the use of Event Data Recorders (EDRs) in motor vehicles.

Types and uses of EDRs: EDRs collect vehicle and occupant-based crash information. They can be simple or complex in design, scope, and reach. Some systems collect only vehicle acceleration/deceleration data, while others collect these data plus a host of complementary data, such as driver inputs (*e.g.*, braking and steering) and vehicle systems status.

The information collected by EDRs aids investigations of the causes of crashes and injury mechanisms, and makes it possible to better define safety problems. The information can ultimately be used to improve motor vehicle safety.

EDRs have been installed as standard equipment in an increasingly large number of light motor vehicles in recent years. Moreover, these devices have become more advanced with respect to the amount and type of data recorded. We estimate that essentially all model year 2002 passenger cars and other light vehicles have some recording capability, and that more than half record such things as crash pulse data.

Since the term "EDR" can be used to cover many different types of devices, we believe it is important to define the term for purposes of this document. When we use the term "EDR" in this document, we are referring to a device that is installed in a motor vehicle to record technical vehicle and occupant-based information for a brief period of time (*i.e.*, seconds, not minutes) before, during and after a crash. For instance, EDRs may record (1) pre-crash vehicle dynamics and system status, (2) driver inputs, (3) vehicle crash signature, (4) restraint usage/deployment status, and (5) certain post-crash data such as the activation of an automatic collision notification (ACN) system. We are not using the term to include any type of device that either makes an audio or video record, or logs data such as hours of service for truck operators.

Research and Development: In 1997, the National Transportation Safety Board (NTSB) issued Safety Recommendation II-97-18 to NHTSA, recommending that we "pursue crash information gathering using EDRs." Also, in that year, the National Aeronautics and Space Administration (NASA) Jet Propulsion Laboratory (JPL) recommended that NHTSA "study the feasibility of installing and obtaining crash data for safety analyses from crash recorders on vehicles." In 1999, NTSB issued a second set of recommendations to NHTSA related to EDRs, H-99-53 and 54, recommending that we require EDRs to be installed on school buses and motor coaches.

In early 1998, NHTSA's Office of Research and Development (R&D) formed a Working Group comprised of industry, academia, and other government organizations. The group's objective was to facilitate the collection and utilization of collision avoidance and crashworthiness data from on-board EDRs.

The NHTSA EDR Working Group held six meetings between October 1998 and December 2000. The

Working Group explored both original equipment manufacturers (OEM) and aftermarket systems, and also looked into data collection and storage.

In August 2001, the NHTSA EDR Working Group published a final report on the results of its deliberations. Highlights of the Working Group findings were the following:

1. EDRs have the potential to greatly improve highway safety, for example, by improving occupant protection systems and improving the accuracy of crash reconstructions.

2. EDR technology has potential safety applications for all classes of motor vehicles.

3. A wide range of crash related and other data elements have been identified which might usefully be captured by future EDR systems.

4. NHTSA has incorporated EDR data collection in its motor vehicle research databases.

5. Open access to EDR data (minus personal identifiers) will benefit researchers, crash investigators, and manufacturers in improving safety on the highways.

6. Studies of EDRs in Europe and the U.S. have shown that driver and employee awareness of an on-board EDR reduces the number and severity of driver crashes.

7. Given the differing nature of cars, vans, SUVs, and other lightweight vehicles, compared to heavy trucks, school buses, and motor coaches, different EDR systems may be required to meet the needs of each vehicle class.

8. The degree of benefit from EDRs is directly related to the number of vehicles operating with an EDR and the current infrastructure's ability to use and assimilate these data.

9. Automatic crash notification (ACN) systems integrate the on-board crash sensing and EDR technology with other electronic systems, such as global positioning systems and cellular telephones, to provide early notification of the occurrence, nature, and location of a serious collision.

10. Most systems utilize proprietary technology and require the manufacturer to download and analyze the data.

Meanwhile, in 2000, NHTSA sponsored a second working group related to EDRs, the NHTSA Truck & Bus EDR Working Group. This Working Group collected facts related to use of EDRs in trucks, school buses, and motor coaches. The record of this second Working Group is in Docket NHTSA-2000-7699. Its final report was published in May 2002.

Event Data Recorders, Summary of Findings by the NHTSA EDR Working Group, May 2002, Final Report, Volume II, Supplemental Findings for Trucks, Motorcoaches, and School Buses. (Docket No. NHTSA-2000-7699-6).

In 2001, NHTSA developed a website for highway-based EDRs located at the following address: http://www-nrd.nhtsa.dot.gov/edr-site/index.html.

Federal Register notices: On two previous occasions, the agency has published documents in the *Federal Register* addressing particular questions about its role with respect to EDRs. Both occasions involved the denial of a petition for rulemaking asking us to require the installation of EDRs in new motor vehicles. (63 FR 60270; November 9, 1998 and 64 FR 29616; June 2, 1999.) The first petitioner, Mr. Price T. Bingham, a private individual, asked the agency to initiate rulemaking to require air bag sensors to be designed so that data would be recorded during a crash, allowing it to be read later by crash investigators. The petitioner cited a concern about air bag deployments that might be "spontaneous," but did not limit the petition to that issue. The second petitioner, Ms. Marie E. Birnbaum, also a private individual, asked us to initiate rulemaking to require passenger cars and light trucks to be equipped with "black boxes" (*i.e.*, EDRs) analogous to those found on commercial aircraft.

In responding to these petitions, NHTSA stated that it believed EDRs could provide information that is very valuable in understanding crashes, and that can be used in a variety of ways to improve motor vehicle safety. The agency denied the petitions because the motor vehicle industry was already voluntarily moving in the direction recommended by the petitioners, and because the agency believed "this area presents some issues that are, at least for the present time, best addressed in a non-regulatory context."

The agency has also received a third petition, from Dr. Ricardo Martinez, President of Safety Intelligence Systems Corporation, asking us to require the installation of EDRs in new motor vehicles. We have not yet responded to that petition. Copies of our responses to the two earlier petitions, and a copy of the petition submitted by Dr. Martinez, are being placed in the docket for this document.

Future actions: In light of the foregoing, the agency believes that it is appropriate to consider what role the agency should now be taking regarding the continued development of EDRs and their installation in motor vehicles.

II. Discussion of Issues:
This section discusses a range of issues and presents a series of questions for public comment to aid the agency in evaluating what role it should take at this time relating to EDRs. The issues and questions are grouped as follows: (a) safety benefits, (b) technical issues, and (c) privacy issues. Finally, in section (d), we ask a general question about NHTSA's role in this area.

a. Safety benefits:
As we noted earlier, the information collected by EDRs aids investigations of the causes of crashes and injury mechanisms, and makes it possible to better define safety problems. This information can ultimately be used to improve motor vehicle safety. By way of illustration, the more that is known about such things as the change in velocity in real crashes and the more that is known about how key safety countermeasures work in real crashes (*e.g.*, which stage of a multi-stage air bag fired), the better the chances are of developing improved safety countermeasures and test procedures.

We invite comments on the following questions related to safety benefits:

(1) Safety potential. The NHTSA EDR Working Group concluded in its August 2001 final report (section 11.1) that EDRs have the potential to improve highway safety greatly. Do you agree with this finding? What do you see as the most significant safety potential of EDRs?

(2) Application. EDR technology has potential safety applications for all classes of motor vehicles. Do you believe different types of EDRs should be used for different vehicle types, such as light duty vehicles, heavy trucks, intercity motor coaches, city transit buses and school buses? If so, why? If not, why not? Do you believe different types of EDRs should be used for different applications, such as private vehicles and commercial vehicles? If so, why? If not, why not?

(3) Use of EDR data. NHTSA has used EDR data primarily to improve its investigations and analyses of crashes. In some cases, EDR data includes information that the agency could not otherwise obtain; *e.g.*, which stage(s) of a multi-stage air bag deployed in a crash and when. In other cases, EDR data provide a more accurate indication of matters, *e.g.*, level of crash severity, that have previously been estimated based on crash reconstruction programs. NHTSA includes the new or improved information from EDRs in its crash databases as appropriate. We request comments concerning how other parties, including government agencies, vehicle manufacturers, insurance companies, and researchers, are using these data. We also request comments concerning other potential uses of these data, by NHTSA and/or other parties, which are related to improving vehicle safety, either in the short term or long term.

(4) Future safety benefits. What additional safety benefits are likely from continued development, installation, collection, storage, and use of EDRs?

(5) Research databases. NHTSA acquires EDR data in its Special Crash Investigations (SCI), National Automotive Sampling System Crashworthiness Data System (NASS-CDS), and Crash Injury Research and Engineering Network (CIREN) and incorporates them in its motor vehicle research databases. Have you ever used the EDR data stored in these databases? How could the presentation and/or use of EDR data be improved?

(6) Prevention of crashes. Several researchers have documented that the use of EDRs could have the potential to prevent crashes. Some studies of European fleets found that driver and employee awareness of an on-board EDR reduced the number of crashes by 20 to 30 percent, lowered the severity of such crashes,

and decreased the associated costs. (See section
2.5.1.1 of the August 2001 NHTSA EDR Working Group
final report.) These studies have generally been based
on small samples and concentrated on commercial
application of EDRs. We request comments on other
studies of this type and on this potential benefit from
EDRs, particularly for the U.S. driving population.

(7) Possible new databases. As more and more vehicles
are equipped with EDRs, more EDR crash data will be
generated. Collection of these data is likely to increase
as state and local officials collect these data as part of
their investigations. Do you have any recommendations for storing and maintaining a national or other
database? Do you believe maintaining a database
would be beneficial to motor vehicle safety? Please
provide specific examples.

(8) Standards. What standards exist for collecting EDR
data? The Society of Automotive Engineers (SAE) has a
recommended practice (SAE J211) that provides guidance for collecting crash test data. Would it be possible
to use this or similar standards for collecting EDR data
regarding real-world crashes? The Institute of Electrical and Electronics Engineers, Inc. (IEEE) has recently
initiated a new program to develop a standard for
motor vehicle EDRs. We request comments on the current activities of SAE, IEEE, and other standards organizations (U.S. and international) in developing
standards for EDRs, and on what types of standards
should be developed.

(9) Standardization. We request comments on whether
there would be any safety benefits from standardizing
certain aspects of EDRs, *e.g.*, defining specific data elements such as vehicle speed, brake application, air
bag deployment time, etc. Would such standardization
promote further development and implementation of
automatic crash notification systems or other safety
devices?

b. Technical issues
(10) Data elements. The NHTSA EDR Working Group
identified many data elements that could be collected
by an EDR. See section 4 of the August 2001 final
report. More recently, the Truck & Bus EDR Working
Group generated a list of 28 data elements. See section
4 of the May 2002 final report.6 What data elements
should be considered for inclusion in an EDR? Should

they vary by vehicle type and/or application? Please provide a rationale for each element, with particular emphasis on how it would lead to improvements in safety. What costs are related to each of your proposed data elements?

(11) Amount of data. Many late-model vehicles are equipped with OEM installed EDRs, but even among the vehicles of a given manufacturer, the type and amount of data collected vary. Do you have any recommendations for the amount of data to collect; *e.g.*, how long before the crash occurs should the data be collected? How should the data integrity be maintained?

IEEE's program is titled IEEE Project 1616: Draft Standard, Motor Vehicle Event Data Recorders (MVEDRs). The web address for this program is http://grouper.ieee.org/groups/1616/home.htm.

Docket No. NHTSA-99-5218-9.

Docket No. NHTSA-2000-7699-6.

(12) Storage and collection. Currently, data are accessed by a physical connection to the EDR unit. Manufacturers are developing wireless connections, *e.g.*, using a wireless probe near the crashed vehicle, or by having the on-board device upload the stored data to a central location using a telecommunications link, but such devices are not in widespread production. How should data be collected and stored in a motor vehicle? What measures should be in place to control traceability of EDR data to an actual vehicle or crash, such as EDR IDs or location and date stamping?

(13) Training. What training is needed for EDR data collection officials?

(14) Survivability. Recording and power systems need to withstand temperature and environmental effects, power failures, and the forces of different types and modes of crashes. They also need to be tamper proof. How can all these be accomplished? What needs to be done to ensure survivability of an EDR? What level of crash severity should an EDR be able to survive? What are the costs associated with producing an EDR with this level of crash survivability?

(15) Effect of EDR technologies on your responses. Indicate how the nature of the EDRs currently being installed in motor vehicles affects your answers to the questions in this notice. To the extent that future EDR technologies are foreseeable, how would the implementation of those technologies affect your answers?

c. Privacy Issues

The recording of information by EDRs raises a number of privacy issues. These include the question of who owns the information that has been recorded, the circumstances under which other persons may obtain that information, and the purposes for which those other persons may use that information.

We recognize the importance of these privacy and related legal issues. The EDR Working Group, too, recognized their importance and devoted a considerable amount of time to discussing them. It also included a chapter on them in its August 2001 final report. Among other things, the chapter summarizes the positions that various participants in the EDR Working Group took on privacy issues.

We also recognize the importance of public acceptance of this device, whether voluntarily provided by vehicle manufacturers or required by the government. We note that General Motors informed the EDR Working Group (Docket No. NHTSA-99-5218-9; section 8.3.5) that it believes the risk of private citizens reacting negatively to the "monitoring" function of the EDR can be addressed through honest and open communications to customers by means of statements in owners' manuals informing them that such data are recorded. That company indicated that the recording of these data is more likely to be accepted if the data are used to improve the product or improve the general cause of public safety.

While we believe that continued attention to privacy issues is important, we observe that, from the standpoint of statutory authority, our role in protecting privacy is a limited one. For example, we do not have authority over such areas as who owns the information that has been recorded, or the circumstances under which other persons may obtain and use that information. These areas are covered by a variety of Federal and State laws not administered by NHTSA.

We note that, in some press articles and op-ed pieces, persons have cited privacy issues as a reason for opposing the basic concept of EDRs.

In our own use of information from EDRs, we are careful to protect privacy. As part of our crash investigations, including those with EDRs, we often obtain personal information. In handling this information, we are careful to comply with applicable provisions of the Privacy Act of 1974 and other statutory requirements that limit the disclosure of personal information by Federal agencies. In order to gain access to EDR data to aid our crash investigations, we obtain a release for the data from the owner of the vehicle. We assure the owner that all personally identifiable information will be held confidential.

We invite comments on the general topic of privacy as it relates to EDRs.

(16) Privacy. What organizations are analyzing privacy issues in the context of roadways, vehicles, and vehicle owners? Are any additional types of analyses needed?

Are privacy concerns adequately met by the current Federal and State law and practices relating to the collection and use of the information recorded by EDRs? Are there significant differences in privacy and/or liability law among states, in the circumstances under which persons or institutions other than vehicle owners may obtain that information, and the purposes for which those other persons or institutions may use that information? In what circumstances are police officers and crash investigators (from government agencies or the private sector) allowed to access EDR data? What damages may result from inappropriate access to EDR data? What roles do technical solutions, such as data partitioning, encryption, and secure databases/vaults, play in addressing privacy concerns?

d. Role of NHTSA.
(17) Role of NHTSA. Over the past several years, NHTSA has been actively involved with EDRs, through the two working groups discussed above, as part of its crash investigations, and in research and development. Particularly since one working group has completed its work and the other is nearing completion, we request comments on what future role the agency

should take related to the continued development and implementation of EDRs in motor vehicles.

III. Rulemaking Analyses and Notices

NHTSA has considered the potential impacts of this request for comments under Executive Order 12866 and the Department of Transportation's regulatory policies and procedures. This document was reviewed by the Office of Management and Budget under E.O. 12866, "Regulatory Planning and Review." This document has been determined to be significant under the Department's regulatory policies and procedures. This document seeks comment on what future role the agency should take related to the continued development and implementation of EDRs in motor vehicles. The agency could take a variety of nonregulatory and/or regulatory actions.

This document does not contain any regulatory actions. Further, this agency has not identified any regulatory actions sufficiently likely to warrant calculation of possible benefits and costs. The EDRs currently installed in motor vehicles cost very little as they take advantage of the existing sensors, processors and memory that the vehicles have. We estimate that an EDR that records basic air bag related data such as air bag deployment, deployment timing, and seat belt status, with moderate survivability, typically costs five dollars or less. We believe that a substantial percentage of light vehicles are already being equipped with such an EDR. However, EDRs with additional sensors, processing capability and memory, and greater survivability capabilities, could cost more.

Given the costs associated with various EDRs, and the fact that 17 million light vehicles are produced each year, a rulemaking proposal for EDRs could, but would not necessarily, have cost impacts that exceed $100 million annually. If NHTSA were to initiate rulemaking and develop a rulemaking proposal, the agency would calculate the costs and benefits associated with the specific proposal and place its analysis in the docket for that proposal. The agency would also conduct the various other rulemaking analyses required by applicable statutes and Executive Orders.

This IEEE standards project and the NHTSA call for comment served as the catalyst for future research and development of EDRs. The safety vs. privacy debate intensified.

CHAPTER 5

Things Are Further Away in the Dark: December 2002

This vehicle split into hundreds of pieces.

One dark and rainy night I was traveling on a slick road toward the college in Pinehurst, North Carolina where I teach Holocaust Studies. Riding along with me was Dr. Susan Chernyk-Spatz, an Auschwitcz survivor who was scheduled to speak on her experiences. On the way to the college we passed a horrific crash. I lamented to Susan about how often people are injured and killed in motor vehicle crashes. Her response was that "things are further away in the dark." It took me years to understand what she meant.

On December 3–4, 2002, the ninth IEEE Motor Vehicle Event Data Recorder (MVEDR) meeting was conducted at Booz-Allen-Hamilton headquarters in Tyson's Corner, Virginia.[1] There were fifty-eight individuals in attendance and several guests including Joan Claybrook of *Public Citizen*, Professor Michael Edmund O'Neill of George Mason University, and Matthew L. Wald, national correspondent for the *New York Times*.

At the beginning of the meeting there was an announcement that the Peoples Republic of China was interested in joining our project. Mr. Bao Yong Qiang from the Traffic Management Research Institute of the Ministry of Public Security of the People's Republic of China sent a notice that China was working on a vehicle traveling data recorder (VDR). The main function of the Chinese VDR is speed and measurement, distance measurement, monitoring driving activities, displaying, printing, data storage, data downloading and uploading and downloading and uploading protocol. The copyright of this standard belongs to the Standardization Administration of China (SAC).

A recent study (2003) conducted by the National Academy of Engineering titled "Personal Cars and China" notes that:

> In 1991 the Chinese government published its eighth five-year plan (1991–1995), which designated the automotive industry as a "pillar industry" that would drive the economy in the twenty-first century. Accordingly, in its most recent five-year plan for the Chinese automotive industry (2001–2005), the government stated that its immediate goal is to produce over 1 million cars a year. [2]

At present, China has relatively few motor vehicles per capita and, of the cars on its highways, few are privately owned. In 2001 China had 18 million vehicles, of which 5 million were cars. If China's number of motor vehicles per capita were comparable to the world average, its fleet would have to total 160 million, with 10 million new and replacement vehicles acquired each year.[3]

109,363 people died on China's roads in 2002. This fatality figure is more than double the number of people killed on U.S. highways annually. In 2002, traffic crashes accounted for 79% of China's accidental deaths. In the first ten months of 2003, traffic crashes killed 86,000 people and seriously injured an additional 418,000.

China's love affair with automobiles is turning dangerous. The explosion in the number of new cars and new drivers is creating havoc on the roads. Besides resorting to tougher rules, new emission standards, and new national traffic laws, the Chinese would greatly benefit from incorporating event data recorder technology in vehicles.

At the end of the December 4th meeting, the host from Booz-Allen-Hamilton suggested that two of the three sub working groups be grandfathered. His view was that we were over-reaching. He lost the argument but gained the attention of those that were interested in seeing it happen.

NHTSA began receiving comments to Docket NHTSA-2002-13546 on November 19, 2002.[4] By October 1, 2003 approximately 82 responses were posted on the USDOT website titled "Docket Management System."[5]

This chapter provides an overview of the comments received. It does not include all the comments, for to do so would require a much larger book. However, if you want to read the entire docket, it is available through the footnotes and the Docket Management System (DMS) website; you may also refer to the footnotes for more information.

It is important to note that I was "selective" in choosing what docket submissions to include, and even "more selective" in the amount of text that I included from each docket. In some cases I chose to submit everything and in others I simply tried to capture the essence of the submission and only included partial text. My objective was to be objective and provide you (the reader) with an overview of the issues. Once again, the footnotes will direct you to the entire submission if you wish for a fuller understanding.

DOCKET # NHTSA-2002-13546-10 12/26/02 Turner & Associates, PA[6]

Tab Turner has represented clients in three different cases with verdicts in excess of $20 million.[7] He has litigated a wide variety of automotive product cases including sudden acceleration, ABS malfunctions, park to reverse cases, and cases involving child and occupant protection issues.[8]

Turner commented: [9]

> I believe that all passenger-carrying vehicles should be equipped with Event Data Recorders (hereinafter EDR). EDRs provide valuable information about the performance of the on-based safety systems of the vehicle. Analysis of the data will help improve these safety features thus further enhancing the safety of the traveling public. It is a fact that the industry analyzes the data on a routine basis. When analysis demonstrates that one or more safety-related components did not operate properly or as designed, the Agency should require that the failure be reported to NHTSA as part of the early warning system requirements. Given that failures are oftentimes hidden defects within the operating system; this should be mandated by the Agency. As noted in the Agency's notice, many of the currently available EDRs collect information regarding accident avoidance. I believe the Agency should require that the industry provide the accident

avoidance information to the Agency for further analysis and for the purpose of maintaining an active database on steering maneuvers, steering inputs, the role
of braking, and maneuver-oriented vehicle responses.
The information will surely be beneficial in the ongoing research being carried out by the Agency on accident avoidance. Finally, but most importantly, I
believe that all information collected from EDRs and
provided to the Agency should be publicly available to
consumers who have an interest in improving automotive safety. The Agency, safety advocates, and attorneys for victims who work daily to improve
automotive safety are oftentimes criticized for advocating positions without adequate objective information about crash performance. The type of
information that EDRs record can be extremely important to those interested in identifying defective systems, improving existing systems, and helping
alleviate much of the unnecessary suffering resulting
from automobile crashes.

DOCKET # NHTSA-2002-13546-11 12/27/02 University of Alabama (UAB)[10]

The University of Alabama at Birmingham has served as the state's premier medical and dental school for the past 25 years. UAB was established in 1969 as an
autonomous university within the University of Alabama system. It now serves
as one of the nation's top ranked universities in research support, higher education and provider of world-class medical care. UAB strives to be an internationally renowned research university—a first choice for education and health care.[11]

    UAB commented:

> This letter is in response to your request for com-
> ments. I am writing in support of Dr. Ricardo Mar-
> tinez's petition to NHTSA for the mandated collection
> and storage of onboard vehicle crash event data in a
> standardized and easily retrievable format. Dr. Mar-
> tinez's request would enhance safety decisions making
> nationally and would be of particular benefit to the
> State of Alabama. Specifically, standardized EDR for-
> mat will facilitate ongoing efforts in Alabama to inte-
> grate automatic crash notification (ACN) technology
> into an organized statewide trauma system. We
> strongly agree with the NHTSA EDR Working Group's
> conclusion that EDRs have the potential to improve
> highway safety greatly. Alabama has the fourth highest
> injury mortality rate in the nation, twice the national
> average. The rate of fatal motor vehicle collisions, the
> largest component of all injury events in Alabama, is

particularly high relative to the rest of the U.S., as is the rate of morbidity associated with injury. The predominately rural nature of state roads is a contributor to this ranking with prolonged crash notification and response times resulting from motor vehicle collisions in more remote areas. Research has shown that outcomes are dependent upon the time it takes to transport a patient to the most appropriately staffed and equipped facility. By uncovering the injury mechanisms involved in certain crashes it is true that improved safety countermeasures may be developed within vehicles. However, standardized EDR data may also allow investigators to develop algorithms that can predict the likelihood of certain injuries occurring in certain types of crashes.

The automatic transmission of this information to a central communications center immediately following a collision will allow response personnel to make an advance assessment of the crash scene, dispatch optimal transport (helicopter vs. ground unit) to the exact location, alert the most appropriate hospital to begin coordinating its resource capabilities and assume medical control. For the purposes of current efforts in Alabama, this is the most significant safety potential of EDRs.

A far-more sophisticated "black box" technology will be needed for the ultimate goal of automatic notification of computer-generated injury prediction that is available immediately upon crash to emergency care providers. In addition to the change in velocity at impact (delta-V), a more-detailed description of the crash pulse, the direction of impact, and forces acquired from occupant compartment contacts (such as seatbelt load, steering wheel load, knee bolster forces, etc) are needed to accurately and automatically predict injury type and severity.

Alabama has much to gain from a formal statewide trauma system In order to fully capitalize on the potential safety benefits of ACN technology, EDR data must be standardized, easily retrievable, and widely available. For researchers to collect the information needed to perfect an ACN system and to fully realize the positive impact on mortality rates and injury severity, each vehicle on the road should have an onboard EDR capable of recording a consistent set of data elements that can be automatically transmitted to a central communication center in the event of a collision. The only way to assure this uniformity and industry standardization is through a NHTSA mandate and industry oversight.[12]

On December 29, 2002, the *New York Times* ran a front page article: *Automakers Block Crash Data Standard*.[13] The following entire article is reprinted from permission of the *New York Times*:

Highway safety could be vastly improved if black boxes that record information about car crashes were standardized, experts say, but they contend that vehement objections from the automobile industry are thwarting efforts to set a standard.

About 25 million late-model cars and trucks, most built by General Motors and Ford, carry the boxes, which record crash information including how fast a vehicle was moving, whether the seat belts were buckled and how big a jolt the occupants suffered at impact.

Other manufacturers say they will install the boxes, small, inexpensive recording devices connected to the system that deploys the air bags. The companies use the data to determine how well the car safety systems work.

But safety and medical experts say benefits would be broader if the data were easier to collect. An immediate benefit, they say, would be fewer deaths.

Accessible data would enable ambulance crews to determine quickly whether a crash was likely to have caused serious internal injuries and help paramedics make more accurate lifesaving decisions, like whether to call for a medivac helicopter.

First, though, the industry needs a data standard, so ambulance crews will not have to carry a different cable and computer for each make of car. Without a standard, some data might be indecipherable except by the manufacturer.

Advocates of the standard say automakers are dragging their feet. The companies say they are defending the privacy of drivers.

"The privacy issues will have to be addressed," said Vann H. Wilber, director for safety and harmonization of the Alliance of Automobile Manufacturers, a trade group. "That's something we think needs to be debated and resolved."

Legal experts, however, say that many of the privacy issues have been settled and that courts have concluded that data recorded in a crash are subject to the rules governing other evidence.

In a lawsuit, for example, the data are subject to pretrial discovery just as other physical evidence is. If the car is totaled, ownership of the data goes with the wrecked car to the insurance company.

Concerns about the unauthorized use of the data can be met, safety experts say, and some have suggested that automobile industry executives are hiding their distaste for regulating a standard behind a feigned concern for drivers' privacy.

Beyond helping ambulance crews make better decisions, the safety researchers say, information from scores of data recorders could reveal design flaws and strengths.

Dr. Jeffrey W. Runge, administrator of the National Highway Traffic Safety Administration, called the prospect of having such precise information on big crashes "very tantalizing."

A committee representing the automakers and others has been meeting for more than a year, under the auspices of the Institute of Electrical and Electronics Engineers, a global association that offers help in setting standards for electronic devices. But progress toward standardizing the data recorders has been slow. In its ninth meeting this year, on Dec. 3 and 4, the companies would not allow the committee's co-chairmen to submit a progress report to the Department of Transportation, which is considering whether to impose a standard. It has requested public comment of the issue. Even if the agency were to decide to impose a standard, drafting and adopting it would take months.

The report incorporated an April 13 press release that said the standard would "define what data should be captured, including date, time, location, velocity, heading, number of occupants and seat belt usage."

The press release added, "It will also define how that information should be obtained, recorded and transmitted."

The DaimlerChrysler Corporation representative on the committee, Barbara E. Wendling, said, "That's really outrageous."

Ms. Wendling is also the chairwoman of an industry committee on crash data recorders. Of the progress report, she said, "It completely misrepresents what's going on in this committee."

A Toyota representative also objected.

The automakers had already made sure that the committee would not specify a "minimum data set." That means none of that data has to be preserved and available for downloading.

The progress report had already been signed by the committee co-chairmen, James E. Hall, a transportation lawyer and former chairman of the National Transportation Safety Board, and Thomas M. Kowalick, a professor of history at Sandhills Community College in North Carolina.

A cover letter to the Transportation Department described the report as their opinion. But facing opposition from the automakers, Mr. Kowalick promised that the report would not be sent.

Ms. Wendling also found fault with a presentation from a trauma medicine researcher, Elizabeth Garthe, who said that ambulance crews often underestimated the extent of internal injuries in car crashes and that a quick way to retrieve data about crash severity would save lives. The boxes, Ms. Garthe said, were "inexpensive and reliable."

But Ms. Wendling countered: "That's a value judgment. It's not been established that it's inexpensive and reliable."

The costs will depend on the box's capabilities and whether it has to be protected against fire. Research shows that more than 90 percent of existing boxes survive wrecks, and proponents say that may be enough to show patterns that will help safety studies.

Component suppliers, who are also represented on the committee, suggest that the cost would be a few dollars per car, although the automakers say they are not certain.

Joan Claybrook, of Public Citizen, a consumer group, said the automakers had also opposed vehicle identification numbers. Ms. Claybrook issued the rule requiring standardization of those numbers 20 years ago, when she was the administrator of the National Highway Traffic Safety Administration.

She said the agency must "come to some conclusion," because without a regulation, nothing ensured that the essential elements of the data would be available.

But John Hinch, a safety specialist with the agency who has attended most of the meetings, said: "The agency is a long way away from doing anything. We're just trying to figure out what the government role is."

The technical questions are complex. For example, should the box record data only for collisions that trigger air bags or for multiple collisions, which are common in serious crashes? General Motors' box records two collisions; Ford's records one. How many seconds of data should be preserved on, say, use of brakes and turn signals, steering wheel position, throttle position and skid? How many times each second should each be monitored? Should the system have independent power in case the first collision knocks out the car's electricity? Should the system permit wireless retrieval of data, so ambulance crews would not have to pry open a door or the hood to look for a data jack? Would that make the data vulnerable to interception by outsiders? The committee is to continue working for another year, but there is no assurance that it will develop a standard that will lead to widespread use of crash boxes to record useful data. Meanwhile, advocates of standardized boxes say, millions of new vehicles will be built without them. If a standard existed, vehicles could be equipped with uniform devices and greatly enhance highway safety, they say.

Ms. Garthe said a study she helped conduct recently in Massachusetts found that 15 percent of the people seriously injured in wrecks were transported to hospitals by helicopter, while perhaps four times as many should have been and would have had greater chances of survival with faster trips. With information from crash boxes available promptly to paramedics, she said, "lives can be saved."

The *New York Times* article changed the tone and direction of future activities. Some believed that having a *New York Times* reporter sitting in the room did not create an environment conducive to the difficult work of achieving consensus, while others welcomed the attention that motor vehicle black box technologies were receiving.

In every crash there is a very fine dividing line between death and injury.

Back in 1965, Ralph Nader wrote the following passage in *Unsafe at Any Speed:*

> The American automobile is produced exclusively to the standards which the manufacturer decides to establish. It comes into the marketplace unchecked. When a car becomes involved in an accident, the entire investigatory, enforcement and claims apparatus that makes up the post-accident response looks invariably to the driver failure as the cause. The need to clear the highway rapidly after collisions contributes to burying the vehicle's role. Should vehicle failure be obvious in some accidents, responsibility is seen in

terms of inadequate maintenance by the motorist. Accommodated by superficial standards of accident investigation, the car manufacturers exclude presumptions of engineering excellence and reliability, and this reputation is accepted by many unknowing motorists.[14]

Commenting on standards development, Nader noted:

The principal institution for the industry coordination of decisions concerning the technical issues in vehicle safety is the Society of Automotive Engineers (SAE)...The control by the automobile industry of SAE's motor vehicle standards work is so complete that the engineering community does not consider the society as anything more than a ratifier of industry policies and decisions. The Automobile Manufacturers Association is SAE's traffic light.[15]

Driver Safety Progresses on Several Fronts

Incident data recorders, commonly referred to as "black boxes," are required on all cars in NASCAR's three national series. They are used to tabulate, among countless other things, the G-Force load drivers withstand upon impact. In 2002, data recorders produced a wealth of information that NASCAR collected and built into an "incident database."

This database, written specifically for stock car racing, provides an in-depth history of what drivers and cars experience during impacts. They also serve as a guide for further safety enhancements.

By the year's end it was uncertain what direction the standards initiatives would take. The lingering issue was the appeal to re-write the IEEE P1616 Working Group policies and procedures, and at the same time the federal government was receiving comments on how to proceed. The year ahead would be decisive, with plenty of horn blowing.

CHAPTER 6

Blowin' the Horn: January 2003

Highway safety affects us all.

R*oad traffic injury prevention is a highly politicized issue. Most people have their opinions on what could make the roads safer. Ancedotal information and its reporting by the media all too often allow issues to be understood as major traffic safety problems requiring priority action, which in turn puts pressure on policy-makers to respond. Policy decisions for effective road injury prevention need to be based on data and objective information, not on anecdotal evidence. First, data on the incidence and types of crashes are needed. After that, a detailed under-standing of the circumstances that lead to crashes is required to guide public safety policy. Furthermore, knowledge of how injuries are caused and what type they are is a valuable instrument for identifying interventions and monitoring the effectiveness of interventions.*

—World Report on Road Traffic Injury Prevention, 2004

On January 7, 2003, the School Bus Manufacturer's Technical Council (SBMTC) commented to NHTSA:

DOCKET # NHTSA-2002-13546-16 01/07/03 School Bus Manufacturers Technical Council (SCMTC)[1]

The School Bus Manufacturers Technical Council (SBMTC) represents the major school bus body and chassis manufacturers. SBMTC is a council within the National Association of State Directors of Pupil Transportation Services and provides technical expertise to association members and the entire school bus transportation industry. SBMTC also serves as technical advisor to the National Conference on School Transportation, which publishes the National School Transportation Specifications and Procedures. It is the intent of these specifications to accommodate new technologies and equipment that will better facilitate the transportation of students. In order to be acceptable, the new technology or equipment shall generally increase efficiency and/or safety of the bus, generally provide for a safer or more pleasant experience for the occupants or pedestrians in the vicinity of the bus, or generally assist the driver and make his/her many tasks easier to perform.

SBMTC believes those comments are applicable to EDRs. Additionally, SBMTC believes that the National Highway Traffic Safety Administration (NHTSA) should continue to encourage and facilitate development of EDRs, but should not consider mandating them until they have been proven cost beneficial. As NHTSA knows, school bus accidents are already closely monitored and studied by both NHTSA and the

National Transportation Safety Board. The data from those crash investigations have provided extensive information upon which NHTSA developed the Federal Motor Vehicle Safety Standards (FMVSSs) for school buses. The same data has provided the school bus industry with the foundation for many safety improvements that are not covered by the FMVSSs.

Given the outstanding safety record of school buses (fewer than 10 passenger fatalities per year), it is clear that further improvements in the safety of school buses will result in small incremental benefits. As noted in a recent report from the National Academy of Sciences' National Research Council, more than 800 school-age children are killed each year during normal school transport hours. Of these, only 20 are passengers in or pedestrians outside of school buses. The others are either drivers or passengers in other types of motor vehicles, pedestrians, or bicyclists. SBMTC believes the 800 fatalities per year to children not in school buses are a national tragedy that must be addressed at all levels of government.

While SBMTC certainly supports technologies that can make school bus transportation safer, it also recognizes that the limited school transportation resources that are available make it imperative to assess the costs and benefits of each potential action. From an overall safety perspective, unless funding is made available to improve the safety of the 24 million students that ride school buses each day, as well as the safety of the 24 million students that do not ride school buses each day, it is important to only expend resources when proven safety benefits can be achieved.

SBMTC members were active participants in the two working groups that NHTSA organized to study EDRs. Of the two working groups, the second was focused specifically on heavy trucks and buses. The work of that group did not indicate that there are bases at this time to mandate EDRs in such vehicles, including school buses.

However, the EDR technology does have the potential to allow for a better understanding of motor vehicle crashes, which could lead to new or improved safety requirements for motor vehicles. As such, SBMTC supports the views and comments made by the Truck

Manufacturers Association (TMA) to the referenced docket.

DOCKET # NHTSA-2002-13546-18 01/07/03 Truck Manufacturers Association[2]

TMA represents all of the major North American manufacturers of medium and heavy duty trucks (greater than 8,845 kilograms, or 19,500 pounds, gross vehicle weight rating). Its members include: Ford Motor Company; Freightliner LLC; General Motors Corporation; International Truck and Engine Corporation; Isuzu Motors America, Inc.; Mack Trucks, PACCAR Inc.; and Volvo Trucks North America, Inc. The stated purpose of TMA is:

- To establish confidence between manufacturers of truck trailers, cargo tanks, intermodal containers and their suppliers to bring about a mutual understanding of the problems confronting all manufacturers.
- To conduct programs and activities which will further the interests of TTMA member companies and the truck trailer industry.
- To provide a means of cooperating with the various agencies of the U.S. Federal, U.S. state, and international governments in any appropriate manner which will best serve both the public and the members' interests.

TMA stated:

> EDRs offer the potential to improve highway safety through better crash data and subsequent vehicle and infrastructure development. However, the NHTSA Working Group needs to provide quantitative data to support this statement. What remains to be proven is that the data from EDRs are in fact useful in better understanding the causes of crashes. Should EDRs be proven to provide useful, accurate data, and depending on what data are collected and who has access to the data, such data could: assist vehicle manufacturers in improving the designs of their vehicles; provide hard, objective data to a crash database which could be useful in developing accident mitigation programs; and provide information to roadway designers should road design be identified as a causal mechanism. Potential future safety benefits include: increasing the accuracy of accident reconstruction, improving injury mechanism detection, providing data to vehicle manufacturers for improved vehicle design, providing a means of measuring improvements in vehicle design, and focusing resources where they are most needed.

EDRs could provide a better understanding of real world crash conditions, thus providing new information for vehicle and highway infrastructure design. The study cited utilized EDRs in police cars. The drivers were aware of the fact that their driving behavior was being monitored and there was the possibility of punishment and/or firing if improper behavior was observed. Therefore, the conclusions may apply to fleet applications, but probably can not be applied to private drivers. Long-term effectiveness was not studied.

The amount of data recorded should be sufficient to enable reconstruction of a large percentage of all crashes. One would have to understand the crash pulses typical for the various vehicle types and select pre-trigger and post-trigger such that the entire crash event is recorded. The technical basis for determining pre-triggers and/or post-triggers, data collection frequency or duration does not currently exist. These pre- and post-triggers would need to be vehicle specific and based on solid research.

NHTSA is the rulemaking body of DOT with jurisdiction over new vehicle motor vehicle safety. NHTSA has the authority to mandate EDRs. However, prior to exercising this authority, the agency should: perform research on the need for the multiple EDR configurations for various vehicle types; perform research to prove that EDRs will successfully increase accident reconstruction efforts through staged crash tests of varying complexity; conduct field operational tests, publish recommendations for EDR configurations; obtain feedback from OEMs Tier 1 suppliers, and the public; and develop an implementation plan for EDRs.

DOCKET # NHTSA-2002-13546-22 01/09/03 Chalmers University of Technology[3]

Chalmers University of Technology carries on broad-based education and research within engineering science, natural science and architecture.
They stated:

Researchers from Chalmers University of Technology Crash Safety Division have reviewed the request for comments in NHTSA Docket 02-13546. The numbered headings below correspond to specific information items listed in the NHTSA Docket text. The information provided represents the opinions of the researchers and not necessarily the official position of Chalmers University of Technology.

There are no doubts that road safety will benefit from the application of EDR data in vehicle safety research. The specific aspects of highway safety that will gain from the data collected by EDRs are:

1. Safety equipment development: Documentation of detailed crash dynamics will allow better design of air bag, seatbelt, active safety systems etc. based on real world collision data. The crash pulses can be used to simulate real occupant kinematics during a collision that will allow better understanding of injury mechanisms and injury thresholds.

2. Vehicle structural requirements: Crash pulses captures by EDR systems will allow real world car structure performance to be compared to laboratory crash dynamics.

3. Rule making / safety evaluation: Detailed EDR data will enhance the knowledge of injury thresholds and risk curves for injuries. These risk curves are needed to predict the influence of new safety standards.

4. Road design and redesign strategies: The collection of pre- and post-crash data will allow existing road and roadside designs to be evaluated. This knowledge can then be applied to improve or optimize existing road design strategies.

The potential of EDR data is not limited to the improvement of passive safety systems of vehicle occupants. As one of the most influential rule making bodies in the world, NHTSA should consider playing a role in the standardization of EDRs. A global standardization for EDRs could be a preferred approach to a national approach. While local road and traffic conditions may require differing safety equipment among nations, EDR recording systems and minimum data requirements should have no regional dependencies, nor be subject to national policies. Thus a world standard would impose an equal financial burden on consumer and manufacturers. By maximizing the number of affected vehicles the unit price should be minimized resulting in little or no change to vehicle price.

DOCKET # NHTSA-2002-13546-23 01/09/03 Bendix Commercial Vehicle Systems LLC[4]

Bendix Commercial Vehicle Systems LLC of Elytria, Ohio, commented:

> NHTSA needs to drive standardization on the type of data that should be recorded to permit quick and accurate accident analysis. As long as all EDRs are recording the same data at the same sampling rate, the analysis can be less complex, time consuming and ambiguous. This is especially true where more than one vehicle (and more than one set of data) is involved. BCVS supports the use of EDRs as it had been involved in accident/vehicle investigations. Some of which were performed with and in support of government agencies such as the National Transportation Safety Board (NTSB). In most cases, additional vehicle/circumstance information would have been extremely helpful.

DOCKET # NHTSA-2002-13546-28 01/13/03 Insurance Institute for Highway Safety (IIHS)[5]

The Insurance Institute of Highway Safety (IIHS) of Arlington, Virginia, mission and goals are:

> The Insurance Institute for Highway Safety is a non-profit research and communications organization funded by auto insurers. For over 30 years the Insurance Institute for Highway Safety has been a leader in finding out what works and doesn't work to prevent motor vehicle crashes in the first place and reduce injuries in the crashes that still occur. The Institute's research focuses on countermeasures aimed at all three factors in motor vehicle crashes (human, vehicular, and environmental) and on interventions that can occur before, during, and after crashes to reduce losses. In 1992 the Vehicle Research Center (VRC) was opened. This center, which includes a state-of-the-art crash test facility, is the focus of most of the Institute's vehicle-related research. The Institute's affiliate organization, the Highway Loss Data Institute, gathers, processes, and publishes data on the ways in which insurance losses vary among different kinds of vehicles.
>
> Before the 1960s, highway safety advocates focused nearly all their efforts on preventing crashes, primarily by trying to change driver behavior. Engineering

attracted some attention, but it was engineering to prevent crashes. Reducing the consequences of crashes didn't get much notice.

Because of the focus on crash prevention, many life-saving vehicle designs were overlooked. For example, a few physicians advocated safety belts in the 1930s, but U.S. automakers didn't begin installing lap belts as standard equipment until the 1960s—and then in response to state mandates. Shoulder belts didn't become standard until the 1968 model year when they were mandated by federal law.

Why did the safety belt, which today is universally recognized as a lifesaving device, take so long to be adopted? In large part because of the nonscientific approach to highway safety that prevailed until the 1960s—an approach that focused on accident prevention. The modern scientific approach really began with William Haddon, Jr., M.D., who developed the first systematic methods of identifying a complete range of options for reducing crash losses.

Through the mid-1960s, many advocates continued to emphasize accident prevention. Automakers, for example, said vehicle characteristics were irrelevant because people caused crashes, so people, not vehicles, needed to change. Since the late 1960s, however, the scientific approach has dominated, in part because of two 1966 laws that authorized the federal government to set vehicle safety standards and provide for a national highway safety program. Dr. Haddon was appointed the first federal highway safety chief and based the program on the scientific approach.

Soon after, U.S. auto insurers began joining the effort to transform the highway safety field into one based on science. In 1969, Dr. Haddon became president of the Insurance Institute for Highway Safety with a mandate to convert it into a research-oriented organization. Supported by auto insurers, the Institute is uniquely positioned to influence highway safety issues because the interest of insurers in reducing highway losses coincides with the public interest. Institute research covers three distinct areas:

1. Human factors research addresses problems associated with teenage drivers, alcohol-impaired driving, truck driver fatigue and safety belt use, to name a few.

2. Vehicle factors research focuses on both crash avoidance and crashworthiness. Crash tests are central to crashworthiness research, and the Institute has been conducting such tests for decades to illustrate, for example, the importance of safety belts and air bags. This work expanded with the opening of the Institute's Vehicle Research Center and an ongoing program of frontal offset crash tests.

3. Research aimed at the physical environment includes, for example, assessment of roadway designs to reduce run-off-the-road crashes and eliminate roadside hazards. These and other programs help reduce deaths, injuries, and property damage from motor vehicle crashes. Reducing such losses is why the Institute exists.[6]

IIHS commented to NHTSA:

The Insurance Institute for Highway Safety welcomes the opportunity to connect on the role of the National Highway Traffic Safety Administration (NHTSA) regarding the continued development and installation of event data recorders (EDRs) in motor vehicles.[7] Manufacturers increasingly are installing EDRs in passenger vehicles. NHTSA estimates that all 2002 and newer model will have some recording capability. These devices are becoming more sophisticated in the amount and type of data they record. EDRs have enormous potential to aid researchers in understanding the circumstances and precursors of crashes as well as in providing more reliable information on crash severities. A better understanding of these issues ultimately could lead to improve vehicle safety.

Although automobile manufacturers now have incorporated EDRs widely into their vehicle, there is considerable variation in the amount and type of data these devices capture. Unless there is standard set of key data elements, EDRs will have limited usefulness. NHTSA should take a central role in establishing minimum data requirements. Manufacturers may choose to record a common set of data elements. The Institute recommends the following data elements for routine recordation by all EDRs for all impact directions:

5 seconds prior to crash, belt use, throttle position, whether driver was braking, whether antilock was activated, and vehicle speed. During the crash: longitudinal and lateral vehicle acceleration (1,000 data points/ second), delta V by time (100 data points/seconds), delta V and delta T for the crash event, if delta V time is not feasible time of air bag deployment (including time of different stages of deployment).

Objectives data for these variables could lead to improvements in vehicle design, in both crash avoidance and crashworthiness technologies.

When investigating real-world crashes, researchers and crash investigators estimate parameters related to crash severity (such as velocity change) using information about vehicle damage, mass, stiffness, and principal direction of force. Crash severity estimates derived from these data are subject to considerable error. The current methodology by which delta Vs are estimated (for example, the methodology used in the National Automotive Sampling System/Crashworthiness Data System (NASS/CDS) relies on physical models that are very simplified representations of vehicle crashes; in reality crashes are very complex events. More problematic is that the principal parameter currently available to characterize crash severity (or velocity change, estimated delta V) is only a partial measure of severity; there is no information available on the time (delta T) over which the velocity change occurred. For example, in a crash into a rigid barrier at 30 mph, the delta T will be much shorter than a 30-mph crash into a roadside impact attenuator (the latter involves a greater deceleration distance equal to the crush of the attenuator plus the crush of the vehicle). Thus, these two crashes with the same delta V differ quite a lot in terms of severity; the decelerations experienced by occupants in the crash into a rigid barrier will be much greater because of the much shorter delta T.

Another important factor in understanding crash outcomes for occupants is whether a person was using a seat belt. Currently, we have to rely on the opinions of crash investigators or on self-reports of vehicle occupants to determine whether belts were used. Crash investigators can examine a seat for signs of routine wear or look for evidence such as D-ring scuffing or striations on the belt webbing for indications of use during the crash. Other indicators include evidence of lower severity, such evidence of occupant contacts throughout the vehicle, which may indicate an occupant was unbelted. But in many crashes, particularly those of lower severity, such evidence is not reliable. And because of reduced forces on seat belts since the introduction of air bags, evidence of the belt use often is not as available in even severe crashes. One result of the unreliability of seat belt use information in crash database is that, despite the fact that seat belts have been standard equipment in vehicles for more than 30 years, there still is an ongoing debate in the research literature about their effectiveness (Robertson, L.S., 2002, Bias in estimates of seat belt effectiveness, Injury Prevention 8:263). Early research on the effectiveness of seat belts estimated they reduce injury and the death by about 40–45 percent. This was prior to laws in the United States that mandate belt use. More recent research has estimated belts to be 60–65 percent effective. These likely are the overestimates because of over-reporting of belt use by vehicle occupants (as a result of laws requiring use). For the spectrum of crash types and severities, the only way to know reliably about belt use is through EDRs.

The availability of information from EDRs raises a number of potential privacy issues. These include data ownership, for example how and by whom the data may be used. The critical issue for highway safety is how the government can encourage the availability of such information for legitimate use by highway safety researchers without compromising individuals' rights to privacy. The Privacy Act of 1974 recognizes the government's need to maintain information about individuals; at the same time the Act requires that individuals be protected against unwarranted invasion of their privacy arising from use of information. To this end, NHTSA maintains national crash databases the provide valuable and detailed information about vehicle crashes without disclosing personal information to indicate the identities of the occupants involved. This policy should be continued with data obtained from EDRs to keep the identity of the crash victims confidential.

It is important for NHTSA to continue to collect and expand the information available from EDRs in the agency's crash databases. NHTSA already is collecting EDR information for some crash investigations on a limited set of vehicles (Special Crash Investigation program and NASS/CDS). These efforts should be expanded as soon as vehicle and data availability will allow. Furthermore, NHTSA should begin collecting EDR-related information as a routine part of the Fatality Analysis Reporting System (FARS). FARS provides detailed information, derived from police crash reports, on almost every crash that occurs on public roads in the United States. FARS data are invaluable in estimating the safety benefits of many vehicle, driver, and environmental countermeasures, but some critical information such as belt use and crash severity is either unavailable or unreliable. NHTSA should encourage police departments, through the use of grant monies, to explore the possibility of downloading EDR data as a routine part of fatal crash investigations.

One stumbling block to widespread use of EDR information for research purposes is that accessing such information from a crashed vehicle can require different approaches, depending on the vehicle manufacturer.

In some cases data can be collected directly from a vehicle using either a device provided by the manufacturer or a commercially available device; in other cases the EDR must be sent to the manufacturer in order to access the information. NHTSA should encourage manufacturers to develop and establish standard practices to download and interpret information from EDRs so it is readily and easily available.

In summary, EDR technology holds tremendous potential to improve our understanding of how vehicle crashes happen, how injuries are sustained, and how well occupant protection features are working. In turn, this should lead to significant improvements in occupant protection in future vehicles. But there is a lack of standardization across manufacturers in the type and amount of information stored and the way data are accessed and retrieved. NHTSA needs to develop standards for the types of data that are being recorded and require standard means of accessing these data. In the short term, NHTSA should work with manufacturers to increase the availability of data that currently are recorded and include this information in NASS/CDS and FARS databases.[8]

Vehicle aggressivity and compatibility in multi-vehicle crashes is a growing problem.

DOCKET # NHTSA-2002-13546-30 01/09/03 Owner-Operator Independent Drivers Association, Inc. (OOIDA)[9]

Owner-Operator Independent Drivers Association, Inc. of Washington, D.C., is a not-for-profit corporation inorporated in 1973 under the laws of the State of Missouri, with its principal place of business in Grain Valley, Missouri. The more than 88,000 members of OOIDA are small business men and women in all 50 states and Canada who collectively own and operate more than 150,000 individual heavy-duty trucks. Owner-operators represent nearly half of the total number of Class 7 and 8 trucks operated in the United States.

OOIDA commented: [10]

> OOIDA is an international trade association representing the interests of independent owner-operators and professional drivers on all issues that affect small business truckers. The Association actively promotes the views of small business truckers through its interaction with state and federal government agencies, legislatures, the courts, other trade associations, and private businesses to advance an equitable environment for commercial drivers. OOIDA is active in all aspects of highway safety and transportation policy, and represents the positions of small business

truckers in numerous committees and various forums on the local, state, national, and international levels. NHTSA's request for comments mentions the possibility event data recorders could directly affect owner-operators, motor carriers and professional drivers, including members of OOIDA.

NHTSA asks many questions related to the practicality and privacy concerns surrounding the use of event data recorders on motor vehicles. NHTSA has spent considerable time exploring the use of EDRs to fulfill its own regulatory agenda. OOIDA does not believe, however, that NHTSA has sufficiently explored the more generalized privacy implications of EDRs and the many ways that data collected by EDRs could be used, and misused, outside of accident investigations and vehicle safety activities. In this regard, OOIDA's comments highlight three general areas of concern.

First, OOIDA is concerned that while the definition of EDRs for this request is narrowed to device that records information in the "seconds" prior to an impact event, the choice of this increment will artificially limit the comments that are submitted to NHTSA. An increment of "seconds," while seemingly assuring, is arbitrary and not tied to limitations inherent in the equipment anticipated. Data storage technology has improved dramatically such that the capability for collecting and storing data for seconds will not differ substantially from a device that can collect and store such data for minutes, hours, and even longer periods of time. Thus, it is OOIDA's view that any discussion of EDR privacy concerns should presume EDRs that can collect and store data for long periods of time not associated with a vehicle accident unless the law puts limits on EDRs.

Second, OOIDA is concerned that NHTSA may well take regulatory action with regard to EDRs without a requisite accountability for its action. Specifically, while NHTSA portends to recognize the impact of regulating in this field, it instructs the citizenry to look elsewhere for privacy protections. It has long been expected that the Federal Government would enter and dominate the privacy protection field. The HIPAA statute anticipated comprehensive privacy legislation that would, in part, protect the information collected and disseminated in the health care sector. Such legislation never materialized. Where the Federal

Government has declined to legislate in a way to protect privacy in a manner uniform to 50 State to State.

Third, OOIDA calls attention to a dangerous paradigm shift that has changed the environment in which the current comments are being solicited. NHTSA seeks comments regarding the interplay between privacy laws, criminal conduct, and access to information by government agents. However, the current state of criminal permits the stop, arrest, and seizure of a mother—in the presence of her young children—and the impoundment and full inventory search of her vehicle. For what heinous crime was she arrested? She failed to adequately secure her children with seat belts. Thus, while it is important to discuss privacy issues associated with EDRs, such discussion is taking place in an environment that appears to have exhausted its patience with the time-honored view that there should be shelters from the powerful reach of government. This is especially true in a society sprinting toward greater security in the wake of 9/11. Relative to this sprint toward safety, any entity, including OOIDA, that advocates maintaining our current level of privacy and personal liberty will appear to be running away from security and safety concerns. This is an unfortunate state of affairs.

For these reasons, OOIDA encourages NHTSA to propose only the use of EDRs on a limited a case by case basis to investigate representative samples of vehicle models already under investigation and owned by volunteers or compensated members of the public. Through this limited use, NHTSA could develop all the data it needs to analyze the safety of vehicles, without requiring the creation of massive amounts of data that it will never use, but would create a serious opportunity for privacy violations.

OOIDA appreciates the intention of NHTSA's involvement in the development of EDR technology and applications. To the extent that NHTSA would like to use EDRs to advance its safety mission, however, OOIDA urges the agency to tailor the development and deployment of EDRs for purposes falling solely within its own jurisdiction. Without such a focus, few protections exist to guard against the abuse of the vast quantities of data that would be created by EDRs but never used by NHTSA.

OOIDA supports the mission of NHTSA to improve the safety of vehicles on our roadways. Truck drivers experience the highest annual number of fatalities of any profession in the United States. While vehicle quality contributes to the safety or danger in a trucker's work place, vehicle defects are a minute percentage of reasons attributed to accidents involving trucks each year. OOIDA does not think that professional truck drivers and members of the public are willing to give up more privacy than necessary to have safe vehicles. The association urges NHTSA to make a concerted effort to focus the ability and use of EDRs to its mission of vehicle defect investigation. We also would welcome the opportunity to participate in future forums and panels to discuss the issues implicated by EDRs.[11]

DOCKET # NHTSA-2002-13546-31 American Trucking Associations, Distribution & LTL Carriers Conference and Truckload Carriers Association[12]

The American Trucking Associations (ATA), Truckload Carriers Association (TCA) and the Distribution & LTL Carriers Association (DLTLCA) hereby submit the following comments in response to the National Highway Traffic Safety Administration's (NHTSA) Request for Comments published in the October 11, 2002, *Federal Register* Volume 67, Number 198 entitled "Event Data Recorders".

The ATA, DLTLCA, and the TCA are trade associations of the American trucking industry and are vitality interested in matters affecting the nation's trucking fleet, highway safety and productivity including vehicle and component safety performance standards. In addition, we strongly support reliability performance standards for safety related equipment.

The American Trucking Associations, Truckload Carriers Association and the Distribution & LTL Carriers Association support the voluntary use of technology and devices to enhance highway safety and productivity. We do support mandated performance standards including reliability performance standards of safety related equipment regardless of whether the equipment is required or optional.

The ATA, DLTLCA, and TCA appreciate the care that NHTSA has placed upon using the definition of electronic Event Data Recorders (EDR) and generally concurs with the use of the term to describe devices that are installed in motor vehicles to record technical vehicle and occupant- based information for a brief period of time (*i.e.* seconds, not minutes) before, during, and after a crash. We suggest that with the further development and proliferation of smarter vehicle systems, the definition should be expanded to include road/ environmental based information as well. Times of day, icy conditions, precipitation, temperature, et cetera are well within the realm of available technology. In addition, the definition should be expanded to include events that may not constitute a "crash" but some other event that if re-occurring, could result in a crash or harmful event. Such events or incidents could include near roll over events, hard braking events, active braking events, and trailer jackknifes.

The American Trucking Associations believe that some information collected by EDRs can aid in the investigation of causes of incidents and crashes and could therefore be used constructively, to develop preventive countermeasures, that would increase highway safety and trucking productivity. Information recorded and retrieved would be helpful, if and only if, the data fully complied with a data reliability and accuracy standard which still must be developed. Erroneous and inaccurate data would be misleading and very harmful to the development of safety enhancement and incident prevention programs.

In the context of crash reconstruction, data from all vehicles (either involved or in proximity) to the incident or crash is essential for accurate causation and prevention analysis. Data from only commercial or for hire vehicles would result in false conclusions and inaccurate liability assignment. Because of the differing architecture and performance of vehicle types, appropriate recorded data would indeed be type specific. For instance, a small passenger car EDR would not record trailer air brake pressures. In addition, because of the open communication data busses in heavy vehicles, the use of J1939 and other appropriate protocol, not based upon the existing and successful SAE standards would result in added complexity, costs, operational and reliability problems.

The ATA is concerned with how the EDR data will be analyzed and managed. In Aviation crashes, comprehensive EDR data is available from a multitude of aircraft systems. Important points to remember are that extremely specialized/highly trained scientists and engineers analyze a very limited number of events annually. Sadly, there are over 900 fatal crashes per week in the USA involving vehicles. This would result in an enormous analysis burden. Typically, local or state law enforcement officers investigate crashes and are currently not trained or equipped to analyze the level of data that is possible to collect from a highway vehicle EDR, nor are they trained to form conclusions from much of the data. Some of the data recorded in currently available EDRs are used by the heavy vehicle operator/motor carrier, but often with the technical assistance from Original Equipment Manufacturer's (OEMs) and suppliers, and on a limited basis. Use of such data at times has been and may be helpful to operators to define vehicle specifications, assist with equipment purchase decisions, and to develop safety programs and incident countermeasures. Inappropriate use of data can include misunderstanding or incorrect interpretation of data, use of erroneous data, and obtaining and using data for purposes other than to improve vehicle, driver, and highway safety.

There may be certain safety benefits to motor carriers, other motorists and highway users, and the ATA suggests that the NHTSA engage in a dialog with those entities currently attempting to use EDR data for safety enhancement. ATA would welcome the opportunity to help facilitate such dialogue.

EDR data would be much improved if a reliability/data quality standard was in place and enforced. Erroneous data, false or inaccurate data points, performance levels, these indications impede and harm the analysis process of incidents and crashes and could result in inaccurate conclusions concerning liability.

We suggest that NHTSA approach the measuring of effectiveness of driver behavior on EDR presence alone, in a similar manner as discussed in Question 4—by engaging in dialogue with those that may have direct and applicable experience with the devices used in the appropriate application.

The ATA firmly believes that prior to any national data collection/ database expansion, the NHTSA, in conjunction with motor carriers that currently use EDRs. A must evaluate the accuracy of the data supplied to and recorded by EDRs. A performance standard for the recorded data must be established.

The ATA has been monitoring the Institute of Electrical and Electronic Engineers (IEEE) process but has not directly participated in their activities. Actual participant comments on this question would be more appropriate and useful to the NHTSA.

Standardization of certain elements of the data sets would be extremely helpful. This would facilitate motor carriers interested in using EDR for safety management/development activities the added capability of complete comparability of their data should they operate in a mixed fleet. In fact, without standardized parameters, EDR data can vary by make, model, year, if the motor carrier does not require or achieve continuity in their purchase specification.

We support the Society of Engineers Truck and Bus EDR working group's recommendations of a data set and as such suggest that the NHTSA continue to participate in that group to further refine the 28 parameters as technologies emerge. In addition we suggest that NHTSA work with the Technology & Maintenance Council (TMC) of the ATA, Study Group S.12 and use the recommended practice RP 1214 "Guidelines for Event Data Collection, Storage and Retrieval" as the foundation of further development.

The ATA encourages NHTSA to consider the duration of events and how they differ among types of vehicles. NHTSA has extensively evaluated durations of events including first harmful events and subsequent events. EDRs may be developed with decision capabilities that will enable them to capture or filter appropriate data sets based upon the type of event. Also, beyond the duration is the sampling rate, which will differ by parameter. The duration of certain parameters may appropriately vary. This is not a trivial analysis. Please refer to TMC RP 1214.

The present EDRs require physical connection to the vehicle for a variety of technological and operational reasons. One important point to remember is that the data belongs to the vehicle owner, and access to the vehicle device should not occur without the owner's knowledge and consent. Unlike aircraft data, which, when and incident occurs, becomes the property of the Federal Aviation Administration (FAA), and upon a crash, the property of the National Transportation Safety Board (NTSB), for motor vehicles, private or commercial, no transfer exists. Also, no such protection that is afforded the aviation industry regarding the analysis and use of the data exists for the highway vehicle fleet. There are many technologies that can support the access of data from an EDR. These technologies will evolve as improvements are made in telemetry, networking and communications. Please refer to TMC RP 1212 for the current industry recommended practice of "PC-to-User Interface Recommendations for Electronic Engines. We look for the industry to continue to self develop recommended practices and standards that are appropriate for the application. Methods to access contents of EDRs may be a part of a standard and if so, protection of the contents against unwarranted access must also be assured. We recommend that the NHTSA work with the appropriate Federal agencies, including the FAA and NTSB, to develop the necessary privacy protection statues afforded other industries like commercial and general aviation.

The level of training required in order to access, collect, and protect data must be at an appropriate level, in consideration of the types and numbers of events that might warrant event data collection. The NHTSA will likely not have the protection for the survivability of EDRs is an issue that requires a balance between cost effective and manufacturability solutions and the desire to maintain data integrity for a variety of physically harmful events.

The ATA suggests that the NHTSA look at the existing data of physically harmful events that vehicle components are subject to, both force, chemical, temperature and other parameters, along with durability and reliability performance parameters and work directly with vehicle operators and vehicle manufacturers to determine appropriate recommended practices or standards for such devices. The ATA will support the NHTSA effort to identify and determine reliability, durability, and data quality performance parameters.

If the NHTSA works solely at the component manufacturer level, the motor carrier and total vehicle manufacturer's expertise of the vehicle system performance requirements will be excluded in the identification process.

In conjunction with existing motor carrier in-house safety, maintenance, and operations programs, many motor carriers use EDR data for both vehicle and driver performance data collection to further improve their overall fleet or individual driver safety performance. As technology evolves, we envision improved access to pertinent data, through reduced cost, simplified motor carrier analysis, and increased data accuracy. The ATA believes that the most important improvement needed as EDRs evolve is the improvement of the quality and reliability of the data provided to, and stored by, the EDR.

The ATA is concerned that privacy concerns are not adequately met, and that further deterioration of privacy will occur as EDRs proliferate. The NHTSA complies with the provisions of the Privacy Act of 1974, and other appropriate statutory requirements that limit the disclosure of personal information by Federal Agencies and uses such information for the purpose of improving highway safety, either from a strategic, problem/definition/causation analysis, or detailed analysis of a specific event with specific countermeasure definition as a result. We support the continued practice of obtaining a release from the owner of the vehicle for any data NHTSA wishes to collect. We are concerned that access to data by individuals and entities will occur for purposes other than improving highway safety. However, there may be certain data elements defined in the ultimate architecture of an EDR that would enhance the capability of first responders to a crash or incident. (such as fire and rescue professionals), for instance, whether an air bag deployed, vehicle power or fuel supply/pressure is "off." We would recommend that data of this type be as accessible as possible to the appropriate individuals and the ATA looks forward to working with the NHTSA to develop criteria for this category, should NHTSA desire.

We recommend that the NHTSA remain technically engaged in the continued development of EDRs. We suggest that NHTSA request direction from the

stakeholders after the completion and dissemination
of results from the IEEE activity.[13]

On January 13, 2003, Jim Hall, former Chairman of the National Transportation Safety Board (NTSB) made a presentation titled, "Use of Event Data Recorders in Crash Reconstruction: Past, Present and Future" at the Transportation Research Board (TRB) Annual Meeting EDR session in Washington, DC. The Transportation Research Board (TRB) is a unit of the National Research Council, a private, nonprofit institution that is the principal operating agency of the National Academy of Sciences and the National Academy of Engineering. The Board's mission is to promote innovation and progress in transportation by stimulating and conducting research, facilitating the dissemination of information, and encouraging the implementation of research results. TRB fulfills this mission through:

- The work of its standing technical committees and task forces addressing all modes and aspects of transportation;
- Publication and dissemination of reports and peer-reviewed technical papers on research findings;
- Administration of two contract research programs;
- Conduct of special studies on transportation policy issues at the request of the U.S. Congress and government agencies;
- Operation of an on-line computerized file of transportation research information;
- And the hosting of an annual meeting that typically attracts 8,000 transportation professionals from throughout the United States and abroad.

TRB's varied activities annually draw on more than 4,000 engineers, scientists, and other transportation researchers and practitioners from the public and private sectors and academia, all of whom contribute their expertise in the public interest. The Board is supported by state transportation departments, the various administrations of the U.S. Department of Transportation and other federal agencies, industry associations, and other organizations and individuals interested in the development of transportation.

DOCKET # NHTSA-2002-13546-32 01/15/03 Automotive Occupant Restraints Council (AORC)[14]

> The Automotive Occupant Restraints Council (AORC) is pleased to submit the attached comments to the National Highway Traffic Safety Administration (NHTSA) on the Request for Comments on the Event Data Recorders (EDR) and the future role NHTSA should take related to the continued development and implementation of EDRs in motor vehicles.

The Automotive Occupant Restraints Council is an industry association of 47 supplies of occupant restraints, components/materials and services to the automobile industry. The mission of the Council is to reduce highway casualties and injuries by providing the motoring public with reliable and effective occupant restraint systems, components and services, and to promote public acceptance and proper use of their restraint systems.

The Automotive Occupant Restraints Council believes that the installation of event data recorders and the capture of data related to vehicle accidents had the potential to greatly improve highway safety by providing crash data that can be utilized in designing improved occupant restraint systems. The AORC Electronics and Sensing Technical Committee have reviewed the Request for Comments by the national Highway Traffic Safety Administration on Event Data Recorders and we provide the following comments (see Attachment) to the specific questions requested in Docket No. NHTSA-02013546; Notice 1.

The Automotive Occupant Restraints Council appreciates the opportunity to comment on the Request for Comments concerning Event Data Recorders and hopes that our submission will assist NHTSA in future decisions in this area.

EDRs have the potential to improve highway safety by aiding us to understand the real world crash events/scenarios. EDR data gives us a better picture of the real world crashes as compared to the laboratory-based crashes, thus improving occupant systems. In addition, it also benefits accident/crash reconstruction. Some of the vehicle manufacturers (OEMs) are already using this data to improve the safety of their vehicles.

AORC is limited in addressing the applicability of EDRs to different vehicle types. Our knowledge base is predominantly in light motor vehicles and hence all our comments should be treated as such.

EDRs help us (AORC members) to understand the real world crash events/scenarios. EDR data gives us a better picture of crash severity, occupant characteristics/behavior prior to the crash, which helps in the development of better safety systems. Data from EDRs could also be used in the calibration studies to improve the systems performance. In addition, EDR data would help NHTSA in comparing the regulatory tests to the real world conditions, benefiting the development of future regulations. In addition EDRs generate data for pre-crash and crash avoidance technologies.

The summaries generated from the databases mentioned have been extensively used by the industry. But we have not used the specific EDR data *per se*. We believe that maintaining a standard database would be beneficial to motor vehicle safety as seen in our comments to question

Most AORC members are part of the SAE and IEEE organizations and our interests are being represented by their activities. We recommend that both the SAE and IEEE committees converge to a common agreed upon EDR standard for the industry.

Standardization. We believe there would be safety benefits from standardizing EDRs. It ensures that the data is taken in the right way for maximum utilization. Standardization of the rate and resolution of EDR data is critical for our analysis. We believe standardization at the present time will reap benefits later.

The list generated by the Truck & Bus EDR Working Group and NHTSA's EDR Workings Group's may not be technically (complexity) or economically feasible at the present time. We recommend that the SAE and IEEE committees should address the specifics of the data elements. In addition, we recommend that both the committees converge to a common EDR standard.

EDR data should be collected at a sample rate efficient to simulate the crash algorithm's reaction to the event. In addition, the duration of the event and inclusion of satellite sensory data is necessary. Driver reaction data must also be at a sample rate sufficient to judge his/her behavior.

Wireless connections make more sense for emergency rescue operations, which conveys information like severity of crash, number of people *etc*. The EDR main data may not be practical for use using wireless communications since there could be loss of power to the module due to the crash. Traceability could be achieved by vehicle's VIN number and date stamp, which are already implemented in vehicles using EDRs.

Standardization will result in common procedure to retrieve the data from EDRs. We believe once the EDR data is standardized not much training may be required. Pamphlets and brochures easily available to crash investigators, rescue personnel, and other related people might be sufficient.

Typically these environmental conditions are addressed in a vehicle's validation program. Some of our customers have enough backup power for a single vehicle crash. Typically, EDRs are located such that they will survive almost all the crashes that the vehicle might encounter. It may not be realistic to protect the EDR units from fire situations, mainly because fire incidents that destroy vehicle's passenger compartment (especially where restraint systems could be of any benefit) are low probable incidents.

EDRs are already in a number of vehicles in the USA from various manufacturers. The vehicle manufacturers who have this data are already using it in their vehicle development process to develop safer vehicles, in addition to understanding the real world crash scenarios. We believe that the industry (through SAE, IEEE, Truck and Bus Working Group) is addressing various ways the EDR data can be put to maximum use. We recommend that both these committees issue a single (light motor vehicles) common EDR standard. AORC recommends that NHTSA should wait and see how the industry implements the standard. We would prefer that NHTSA act more like a catalyst in this process.

DOCKET # NHTSA-2002-13546-36 01/28/03 State of New Jersey, Department of Transportation[15]

It is the mission of the New Jersey Department of Transportation to provide reliable, environmentally and socially responsible transportation and motor vehicle networks and services to support and improve the safety and mobility of people and goods in New Jersey.

NJDOT stated that, "Thank you for the opportunity to comment regarding the future role that NHTSA should

take related to the continued development and installation of Event Data Recorders (EDRs) in motor vehicles. The following are our comments to the seventeen
questions that NHTSA presented in the above-refer-
enced docket.

Event data recorders can be incorporated into a
broader class of devices, accident notification systems,
which can have direct and immediate benefit to accident victims in their survivability during a serious
accident. Victims' survival is based on receiving aid in
the first hour following an accident. This is referred to
as the "Golden Hour." Victims who receive assistance
in this first hour have a much better chance of survival. This is critical in rural and remote areas in
instances when the victim is rendered unconscious by
the crash and there are no witnesses to call for
assistance.

NHTSA should consider encouraging new vehicle
manufacturers to include this technology as standard
equipment and other suppliers to develop inexpensive
after-market devices that can be available for older
vehicles.

In all likelihood, a variety of EDR devices will be necessary depending on the type of vehicle. For instance,
school buses and long-distance trucks may require
very different data collection for post-crash analysis.
Differing equipment, operating conditions, vehicle
uses, and accident data information needs would justify a variety of EDR devices. Further, benefit-cost
analysis would likely show that for certain types of
vehicles that are in continuous use (revenue or commercial service) additional benefits would inure to
those operators, justifying costs for more enhanced
technology and measurements, compared to that
which might be useful for individual passenger vehicles. Vehicles in public transportation (buses, taxis,
etc.) should be considered for a greater application of
EDR technologies for the safety benefits that will inure
to the public.

The data should clearly be used for preventative safety
efforts. The purpose for using EDR and attendant
information should be properly set forth and faithfully
followed; otherwise the information goes beyond
safety to areas of potential controversy, such as, accident liability, increased insurance costs, greater

litigation, etc. The data should be used for overall safety and engineering purposes; it should not be considered dispositive of causation of individual accident occurrences.

Through the tracking of trends in the information, emphasis can be given to increased focus on safety effects to reduce trends in accidents.

New ITS technologies are evolving in crash prevention, which would include proximity sensors to warn drivers in advance of crashes. Safe following distances between vehicles at high speed is a major concern in the effort to prevent crashes. Such ITS technology employed with an EDR device could conceivably assist in tracking reckless driving at high speeds and could be very helpful in reducing serious accidents. Such would be a matter for study.

A national database makes sense, recognizing the significant interstate commerce and mobility on our nation's road systems. Safety agencies of the government, including commercial vehicle safety oversight, should be the main entities maintaining the databases. Highway safety experts should structure the databases to discern trends in repetitive crashes, noting injury severity, for best use.

It is suggested that the American Association of State Highway and Transportation Officials (AASHTO) and the American Association of Motor Vehicle Administrators (AAMVA) be included in the development of standards, particularly regarding the elements of data that are to be collected, how the data are to be collected, and how the associated expenses are to be covered.

Standardization is desired. In addition, we should consider consultation with our NAFTA partners, Canada and Mexico, in the standardization of EDR devices for vehicles in operation in North America.

Data elements deemed advisable by highway and vehicle safety experts, with proper safeguards regarding personal identifiers, would be appropriate.

The period of time for retention of data in the device should be dictated by the shortest period reasonably needed for useful safety analysis of the most complex or time-consuming element.

Storage and collection decisions should be dictated by the safety benefits derived. Identification to specific vehicles of the location and time of occurrence should be essentially limited for purposes of timely emergency response to accident victims. Location information could also be useful in identifying places for road safety improvement, if EDRs are widely implemented.

Training is dependent upon the complexity of the EDR system. Training is deemed important to assure consistency and reliability in data collection activities. EDR systems should be crash-hardened, tamper resistant and weatherproof. Seals or other indicators may be included to indicate whether the device has been disturbed. New technology should always be explored, developed, and implemented to enhance public safety.

Care must be taken to assure that only information needed for public safety is collected. Too much information not only raises greater privacy concerns; it also can be overwhelming and reduce the benefits of the data, for safety efforts. Consideration must be given to the laws that will apply regarding these new technologies. Clear distinction between transportation safety and other uses must be set forth. Clear distinction between civil safety endeavors and criminal law enforcement must be made, with protections accorded for constitutional rights against warrant less searches and invasions of privacy. Federal and state laws should be reviewed, perhaps through a national conference, and a uniform system developed regarding the legal system to surround such new technologies. The data collection may be stymied for safety purposes if there is a concern over litigation or use of data for liability purposes. Uses other than safety may have to be precluded by law in order to assure the integrity and usefulness of the data collected for public safety. The public's input is critical on these issues.

NHTSA should continue to meet its mandate for vehicle safety. Technical consultation, such as the outreach of this docket, should continue. Standardization and harmonization should be encouraged. The very significant policy matters on privacy should be addressed by the public through its representatives in the legislative branch.

NHTSA continued to receive comments.

PART THREE

SOCIETAL ISSUES

CHAPTER 7

Cruise Control: February 2003

NHTSA estimates that 11,889 lives were saved in the year 2000 by the use of safety belts.

E*very parent's worst nightmare is when a highway trooper comes pulling up in the front yard.*

It doesn't matter what time, day, week, month or year—for the moment becomes frozen. It is never forgotten. When he comes to tell you, a part of you dies. Although not the first to know, you will always be the last to understand. Nothing about the crash will ever make sense, except maybe the reason your earlier attempts to cell phone failed.

The tenth IEEE Motor Vehicle Event Data Recorder meeting was conducted at Ramtron Corporation in Colorado Springs, Colorado, on February 10–11, 2003.[1] There were fifty-two individual in attendance. There were two special presentations. Dearborn International provided an overview of in-vehicle networks and embedded systems. Lt. Colonel Scott Henderson, Commanding Officer of the United States Air Force (USAF) 2ND Space Operations Squadron presented an overview of the Global Positioning System (GPS).

A letter written on February 10, 2003 by the DaimlerChrysler Corporation (DCX) representative addressing the IEEE/SAE relationship was received on February 11, 2003.[2] She asked to post it to the public record and in the final minutes of the meeting. The letter explained her position toward the voluntary standards initiative for EDRs. It is provided verbatim as follows:

Dear P1616 Members,

> EDRs raise a number of sensitive issues related to our sense of guilt and innocence, and truth and justice, which—as we have amply proven over the past year—are almost impossible to ignore. For this reason, virtually every person exposed to the issue has an opinion (sometimes quite strong) on the matter, whether or not they understand it in all the particulars. Discussion of EDR standardization therefore tends to stir up questions of appropriate policy over which rational minds may—and inevitably do—disagree. The only way to avoid this pitfall when writing an otherwise-technical standard is to construct a strict barrier between those aspects of standardization that do not involve setting policy, and those that do. I believe the current scope of P1616 accomplishes this goal, and I'll go out on a limb and say that those who voted in favor of it agree, and approved it for this reason.

> What the current scope and purpose of P1616 says, in effect, is: We will not in this standard prescribe what data should be recorded, or how to record it, because we recognize that choosing a minimum data set and associated "required" recording parameters would set a policy for a certain type of EDR. This would not only exclude other types of possible EDRs, but it would impose hardware and development costs and burdens

on vehicle manufacturers, for which consumers might not be willing to pay. On the contrary, some consumers are quite vocal in their opposition to having *any* type of EDR installed on their vehicle, let alone one designed by business interests perceived to be contrary to those of consumers. Therefore, the only fair way to set EDR policy is for the issue to be taken up in a political forum, rather than a technical forum. In a political forum, consumer qua voter interests are (in some fashion, anyway) represented in the decision-making process, and policy concerns can be raised, vetted and decided according to a process we all more or less endorse as "fair."

However, for all of the many reasons we have discussed, debated, and otherwise squabbled over, the reigning political authorities (the legislatures and the regulatory agencies) have been reticent to impose an EDR policy under the current circumstances. I personally believe that their reticence to act so far is a wise reaction to the fact that there is insufficient public consensus on this issue at this time. I believe that policy makers in general recognize that an effort to develop a public policy on EDRs in the current climate will quickly degenerate into a firestorm of controversy, from which no good policy can be expected to result. In short, I believe (and I'm not alone) that as a society we are better off waiting some indefinite period of time while EDRs are implemented voluntarily, thus allowing more of a public consensus about the value of EDR data to evolve on the basis of real-world experience, including the development of legal remedies and judicial precedence to deal with the thornier aspects of blame and liability.

Therefore, contrary to what some of you may have unfortunately come to believe on the basis of some ill-worded exchanges during these meetings (for which I accept partial blame—sorry!), please understand that I am *not* personally or professionally opposed to EDRs or to the standardization of EDRs. On the contrary, I am a strong proponent of carrying out the current scope and purpose (at least for light vehicles), which is why I have also been a proponent of taking up this aspect of EDR standardization under SAE. I think a strong business case can be made that this standard belongs in SAE. SAE is the technical liaison to ISO for automotive standards, and the official technical organization for issuing U.S. automotive standards. While EDRs do encompass electronic aspects and protocols, they are nevertheless inextricably linked to motor vehicles, components and systems, and have no purpose or function independent of a motor vehicle. Moreover, SAE already has multiple standards and protocols established and under development that can be readily incorporated or adapted to EDR use. The expertise behind these existing and developing standards resides in SAE, rather than in IEEE, to which this expertise is, by virtue of subject matter, less accessible. (For similar reasons, SAE would be ill-advised to undertake the standardization of a hardware protocol for computer chips.) Finally, IEEE as an institution appears to have a much higher tolerance for committee dysfunction and instability than I have seen at SAE, which I believe would have dealt with the underlying problems of P1616 before things degenerated to this point.

Having said all that, however, I will not stand in the way of an intelligent and workable solution for establishing a professional relationship between P1616 and the newly-constituted SAE activity. If a joint standard, or mutually-compatible standards, can be developed without undue delay and further frustration, I will work as hard as anyone (with the possible exception of Hideki Hada, who is a hard act to follow) to make things work. However, I will not endorse the setting of an EDR policy under the auspices of this—or any other—technical standards setting body, and I am not confused about where that line is drawn.

Finally, to the person who asked why I (or anyone else) should care what IEEE does by way of setting a voluntary standard—particularly if I plan to turn my attention to SAE—please understand that technical standards in the automotive industry generally establish the state-of-the-art for the purposes of determining liability under tort. Therefore, vehicle manufacturers who deviate from them without technical justification for doing so, do so at their peril. Since there is no technical justification for deviating from a policy proclamation, it is important that policies not be written into technical standards, which otherwise impose unwarranted risks for manufacturers who choose not to follow them. (Note that this is not an opinion I hold just because I work for

> DaimlerChrysler, either, but because I understand the
> implications of misapplied policies and judge them to
> be unreasonable. Therefore, it is consistent with a pro-
> fessional ethic, and appropriate for a member of a
> professional society to act on this opinion in the
> course of developing technical standards.)
>
> I hope this letter clarifies apparent misunderstanding.
> Sorry for the 'ruffled feathers.' Thanks for listening.
>
> Best regards, Barbara Wendling

This DCX representative previously informed the working group that she served as the chair of the Alliance of Automakers Event Data Recorder Committee. The Alliance of Automobile Manufacturers is a trade association of 10 car and light truck manufacturers who account for more than 90 percent of U.S. vehicle sales. Members include BMW Group, DaimlerChrysler, Ford Motor Company, General Motors, Mazda, Mitsubishi Motors, Nissan, Porsche, Toyota and Volkswagen.

Another issue that surfaced at the February meeting was media representation of the work of the committee. While some disagreed, it was the consensus of those in attendance that this issue was best settled offline and that the group would not make a statement about media coverage.

This would be the last IEEE meeting that Barbara E. Wendling would participate in. She and others pursued the procedural appeal they had filed. When the appeal was lost, Wendling resigned from the Working Group, on 22 April 2003. [3] As acclaimed leader of the Alliance of Automobile Manufacturers EDR committee, other automakers would soon follow her lead. General Motors Corporation, Inc. dropped out two days later, and Toyota followed on July 22, 2003. In addition, the secretary of the Working Group stopped actively participating. Because timely minutes were critical, I looked for a volunteer to take notes at the meetings.

At this point in the project, a private citizen who lost his son in a horrific motor vehicle crash volunteered. Tony Huffman of Austin, Texas did an incredible job of providing material to the members. He traveled to meetings at his own personal expense and contributed greatly to the success of the project. When I asked what motivated him, his simple answer was the memory of son.

It was meeting people like Tony that kept me and others motivated.

On the same day (2/10/03) that Barbara E. Wendling sent the letter to the IEEE MVEDR working group, Stephen R. Kratze, the Associate Administrator for Safety Performance Standards of NHTSA gave a speech at the Society of Plastics Engineers Global Automotive Safety Conference in Detroit, Michigan. [4] Kratze noted:

> Automobiles are both good news and bad news for
> American society. The good news is that we have ready
> mobility in a private and secure environment. Our
> society's mobility now is taken for granted, but would
> have seemed like science fiction a century ago. Our

cars make us feel good. For many people they are expressions of our personalities, and for others they are merely a practical means of transportation. In any case, everyone expects their vehicle to be nimble at avoiding crashes, and take care of him or her in the event of a crash.

The bad news is that vehicles are also a contributor to a serious public health problem. Motor vehicle crash is the leading cause of death for children greater than 3 years of age and is the leading cause of years of life lost in our country, far exceeding cancer and cardiovascular disease like stroke and heart attack. In the year 2000, there were 41,821 Americans killed in or by motor vehicles. If there were any other disease in this country that killed as many people, there should be no doubt that our political system would ensure that every possible resource were devoted towards its elimination. Unfortunately, there is a great risk tolerance for death and injury from motor vehicles because of its perception as being unavoidable "accidents."

In that light, it behooves everyone who has anything to do with this industry to take corporate and personal responsibility to help eliminate this scourge on American society. We spend a lot of time speaking with and seeking the help of the automobile manufacturers about this safety mission. Considering the safety protections offered by vehicles 30 years ago, there have been huge strides made in occupant safety. But there is still more to do.

When we look at causation, we know that human factors and driver errors are responsible for the majority of motor vehicle crashes. We also know that an unfortunately large number of crashes are caused by physiologic impairments such as alcohol, drugs, or other physical infirmities. But if we assume that those factors account for 90 to 95 percent of all crashes, that still leaves 5 to 10 percent of crashes that are caused by vehicle problems or defects. Since there are about 6.3 million crashes yearly, that means that about half a million of those crashes are the result of vehicle problems. And this is unacceptable.

It is certainly much easier to consider safety when you are engineering a vehicle or vehicle component than it is for NHTSA to try to think of every possible way a person can be harmed in a vehicle and write a regulation to prevent it. We much prefer voluntary brilliance to involuntary compliance.

What I would like to do is to challenge you to begin to think long-term about your niche in the system of motor vehicle safety. I want each of you in the room to ponder for yourself what the safe vehicle of the future should look like. We know that propulsion systems may be vastly different over the next two decades. I ask that all of you begin to contemplate how to incorporate ideal safety design into what will be a blank slate. At the same time you figure out where to put batteries and tanks of hydrogen or borax, please think freely about how to draw the picture with safety as your first priority. I would encourage you develop a vision and to work personally towards that end.

There are so many obvious ways in which engineering can solve common but very severe problems. Dr. Runge often asks why is it that his 5-foot, 1-inch mother must manually hold the shoulder belt away from her neck while driving or riding in her car? Can we not engineer safety belts that are comfortable for people of all sizes and shapes? Surely if we can design a cruise control that knows how fast the cars around you are going, we should be able to create seat belt designs that are not a disincentive to their use.

Another question Dr. Runge often asks is why does it take a 4-day course and a post graduate degree to properly install a child safety seat in a vehicle? Why is it that seat manufacturers and child safety seat manufacturers aren't working together to ensure a perfect fit every time by an average mom or dad installer? Why is it that we at NHTSA have to create regulations for seat strength or ask for comments about how to avoid roof crush? Aren't these things inherently obvious to anyone who manufactures components or automobiles?

Those of you who make interactive devices for vehicles, please know that you bear a particular responsibility to assess the distraction potential from such devices. The agency has not yet decided which way we will go in performance standards for distracting components. We would prefer that you designers and manufacturers would prospectively accumulate such data that may alert you to inappropriate distraction potential. These data may likewise reassure both you and us that this problem has been addressed and does not warrant new regulations by the agency.

NHTSA has an Administrator who believes heavily in personal responsibility. But he also believes in corporate responsibility, and that corporations should do all that they can to be good citizens, not only for the benefit of their customers, but those people with whom its customers share the road. For example, when vehicles collide, they should do so compatibly. The fact that you are a responsible driver is great. But you should not pay the price for someone else's lack of responsibility.

Safety is also good for the American economy. The most rapidly rising cost in our society is health care and it has been for the last 25 years. Everyone needs to do his or her part, whether you are the CEO of a corporation or a design engineer with a mouse in your hand all day long.

All of the automobile manufacturers now understand that safety sells. We would emphasize that it should not be the perception of safety that sells, but real safety. NHTSA is committed through its NCAP Program and its consumer information to disclose data to the public that they can use in selecting a vehicle. Dr. Runge's favorite motto is *Esse Quam Videri*, which means "to be rather than to seem." We are committed to being, not just bearing the title of, the public's agent for automobile safety. We would encourage each of you to be the guardians of the public safety through your undivided attention to safety in your design and manufacturing processes, rather than simply giving manufacturers a marketing point.

We will be unwavering in our mission, and we would challenge all manufacturers and all suppliers to join us in working toward the same goal. As Dr. Runge likes to inform me, if everyone always did the right thing, we wouldn't need performance standards. In that light, NHTSA wishes to applaud the industry for the innovations in safety engineering that you have already made. The improvements you have made in occupant crash protection, including more advanced seat belts and front and side air bags, and in braking and handling capabilities represent quantum leaps forward for safety. But now we challenge you to create your vision for the safety concept car of the next decade, which can and must become a reality, as opposed to a marketing concept.

While these activities were taken place, NHTSA continued to receive comments. The following comments are important because they reflect the group that was most active in setting the global standard for MVEDRs.

DOCKET # NHTSA-2002-13546-39 02/14/03 IEEE Motor Vehicle Event Data Recorder (MVEDR) Working Group[5]

The WG wrote:

Dear Dr. Runge,

The following comments are hereby transmitted on behalf of the IEEE Motor Vehicle Event Data Recorder (MVEDR) working group, heretofore known as P1616, in reference to NHTSA's request for comments, published in the *Federal Register* on October 11, 2002. The purpose of these comments is to describe the purpose, scope, organization, and schedule of P1616. These comments represent the summary and status of P1616 activities by the co-chairmen and the majority viewpoints of P1616 members. It does not necessarily represent the views of IEEE, which has not reviewed or approved this response.

MVEDRs collect, record, store and export data related to motor vehicle pre-defined events. P1616 was formed in early 2002, to draft an industry standard to define a protocol for MVEDR output data compatibility and export protocol of MVEDR data elements. This standard does not prescribe which specific data elements shall be recorded, or how the data are to be collected, recorded, and stored. It is applicable to event data recorders for all types of motor vehicles licensed to operate on public roadways, whether offered as original or aftermarket equipment, whether standalone or integrated within the vehicle.

Many light-duty motor vehicles and increasing numbers of heavy commercial vehicles are equipped with some form of MVEDR. These systems, which are designed and produced by individual motor vehicle manufacturers and component suppliers, are diverse in function, and proprietary in nature. The continuing implementation of MVEDR systems provides an opportunity to voluntarily standardize data output and retrieval protocols to facilitate analysis and promote compatibility of MVEDR data. Adoption of the standard will therefore make EDR data more accessible and useful to end-users.

P1616 consists of members from many interests including academia, government, automobile manu- facturers, safety advocates, aftermarket manufactur- ers, and other stakeholders. The membership is open to interested parties willing to provide input to the standard drafting process. Recently, P1616 formed three technical subcommittees tasked with EDR defi- nition, data definition, and output protocol determination.

P1616 looks forward to finalizing this standard by early 2004. The following milestones are envisioned (listed by their completion date):

February 2003—Identify and define terms and technical design,
May 2003—Identify data and system requirements,
July 2003—Gain consensus on draft standard,
November 2003— Initiate IEEE sponsor ballot, and
March 2004—Standards approval.*

*NOTE: These were merely estimates of the benchmarks stated on February 14, 2003. The March 2004 draft standard was approved by the working group and then sent for spon- sor ballot. It was approved on the first ballot but re-circulated to address negative and coordination comments, and successfully passed again three times by August 9, 2004.

DOCKET NHTSA-2002-13546-41 02/20/03 American Insurance Associations[6]

The American Insurance Association (AIA) represents more than 400 insurers that write nearly 20% of the commercial and personal automobile insurance in the United States, or $27 billion annually in premium. Each year, AIA members use that money to pay bil- lions of dollars in claims. And, they have always made motor vehicle crash, death and injury prevention among their highest priorities.

These comments are provided as our response to the notice and its specific questions. We also incorporate by reference the comments of Advocates for Highway and Auto Safety.

At the outset, we urge that the Event Data Recorders, hereinafter "black boxes", be made standard equip- ment on all large commercial vehicles. They would be a powerful tool for safety, providing the best method for enforcing hours of service rules. This is especially critical considering the statistics on the relationship of fatigue and crash involvement.

AIA also views the black boxes, if installed in all types of vehicles, as potentially a breakthrough in providing real world crash data to supplement existing information, to establish more effective vehicle standards, traffic laws, law enforcement, safety education, emergency response and highway design. Accordingly, we support efforts to collect, aggregate, analyze and use as much data from black boxes as is possible, for the purposes set forth above.

AIA members also believe that black box data could significantly assist in the rapid, fair and cost effective settlement of claims and in fighting fraud. These data would be useful, especially in high value claims, to determine fault, mechanisms and forces of injury, driver behavior and vehicle performance. Information that would be useful includes data on: windshield wipers, steering, braking, speed, air bag deployment, headlights, points of contact, all other lights and signals, sound equipment, cell phone, seatbelts, direction of movement and sequence, horns, throttle and even air conditioning or heater use.

Finally, we do not believe there are any insurmountable legal, privacy or technological obstacles to accomplishing these goals. In summary, we urge the agency to work through regulatory and voluntary efforts to achieve the collection, achieve analysis and widest use of as much data as possible from black boxes.

DOCKET # NHTSA-2002-13546-45 02/20/03 Virginia Tech Transportation Institute[7]

Virginia Tech Transportation Institute (VTTI) is an interdisciplinary, multidisciplinary, university research center of Virginia Tech. VTTI was established in August 1988 in response to the U.S. Department of Transportation's University Transportation Centers Program, and in cooperation with the Virginia Department of Transportation. VTTI pursues its mission by encouraging research, attracting a multidisciplinary core of researchers, and educating students in the latest transportation technologies through hands-on research and experience. The institute is both a FHWA/FTA ITS Research Center of Excellence and a Mid-Atlantic University Transportation Center.

VTTI commented to NHTSA:

> In response to NHTSA's request for comments regarding Event Data Recorders (EDRs), the Safety and Human Factors Engineering Group of the Virginia Tech Transportation Institute (VTTI) would like to submit the following in response the "safety issues" questions.

> VTTI has limited direct experience with the other issues provided. A. Safety Issues: (1) Safety potential. EDR data could potentially provide a considerable safety benefit. Information such as seat belt use, air bag deployment stages, vehicle speed, and vehicle accelerations are a few of the data elements that could lead to improvements in vehicle design for crashworthiness and crash avoidance. EDR data coupled with post-crash analysis may provide considerable road design information. Also, EDR systems designed to activate an Automatic Collision Notification system and connect to emergency assistance is another potential safety feature. However, EDRs will not be a panacea to allow researchers to address all driving issues. Many pre-crash actions of the driver can not be specified with an EDR, nor can all interactions with other vehicles, pedestrians, animals, or other obstacles. The safety potential of EDRs can be quantified once all the possible data elements that can be recorded are specified through research and collaboration with stakeholders.

> Common data elements on all vehicle types will facilitate the most comprehensive post-crash analysis; nonetheless, it is plausible that not all data elements apply to all vehicle types or those additional data elements may be valuable for another vehicle type. Determining the data elements that should be recorded for each vehicle type should be determined through empirical study.

> In and of itself, EDR data provide limited but valuable insight into the Human Factors issues associated with vehicle crashes. Issues of driver fatigue, distraction, inattention, vehicle interaction, and, for many instances, determining appropriate driver response cannot be determined solely by examining the EDR data. Therefore, while EDRs show a great safety potential, its benefits will be limited to specific applications.

Aside from the myriad of benefits listed in other responses to this docket, VTTI does not know of additional safety benefits of EDRs. In general, improvements in pre-crash data will lead to more informed development of crash countermeasures of all types.

We have only used these data on a limited basis to date. However, we plan to use them extensively in the future.

The reference for these studies are: Dingus, T.A., Neale, V.L., Garness, S.A, Hanowski, R.J., Keisler, A.S., Lee, S.E., Perez, M.A., Robinson, G.S., Belz, S.M., Casali, J.G., Pace-Schott, E.F., Stickgold, R.A., Hobson, J.A. (2001). Impact of sleeper berth usage on driver fatigue: final project report (PB2002-107930) Springfield, VA: NTIS. Hanowski, R.J, Wierwille, W.W, Garness, S.A, and Dingus, T. A, (2002). Impact of local short haul operations on driver fatigue: final report (DOT-MC-00-203) Springfield, VA: NTIS.

In both studies, VTTI researchers instrumented commercial vehicles with systems that were far more capable than EDRs and included video data. Due to behavior of the drivers, and the number of critical incidents, near-crashes, and crashes recorded, VTTI researchers were of the opinion that the commercial drivers drove as they normally would. The data collected in the studies were confidential. If EDRs collected data that were regularly reviewed by an employer and could be used as grounds for dismissal, drivers may drive more cautiously. Eaton-Vorad has discovered such effects associated with the deployment of their forward collision warning system as reported on their website.

Possible new databases? Such a question should be answered by a committee of all involved stakeholders; however, several database "sites," perhaps at universities, would be a logical solution. Standards for collecting data must be set for all vehicle types in order to maximize the utility of the data. The data elements for EDRs should be standardized for maximum value of the data, to encourage ease of use, and to discourage misuse of the data.[8]

DOCKET # NHTSA-2002-13546-46 02/21/03 National Association of EMS Physicians[9]

This letter is in response to your request for comments on the Event Data Recorders. The National Association of EMS Physicians is quite pleased that NHTSA is investigating these issues and has asked for comment from a broad range of issuers. EMS physicians use motor vehicle crash information every day in order to make important EMS system and clinical decisions. Our comments are enclosed. We look forward to the opportunity to discuss this with you in the future.

The National Association of EMS Physicians (NAEMSP) is a national organized of physicians and other professionals who provide leadership and foster excellence in out-of-hospital emergency medical services. As part of its ongoing commitment to improving out-of-hospital emergency medical care, the NAEMSP promotes meetings, research, publications and products that connect, serve, and educate its members. In addition, the association acts as a resource and advocate of EMS-related decisions in cooperation with organizations throughout the country and the community at large, including agencies of the federal government. It is the continuing role of the NAEMSP to coordinate and focus advances in medical care, research and training related to EMS.

NAEMSP was a leader in creating the EMS Agenda for the Future, which proposed a vision for EMS in the years ahead. EMS of the future will be a community-based health management system that is interrogated with the overall health care system, public health and public safety agencies. EMS will remain the public's emergency medical safety net.

To realize the vision of the EMS for the Future, 14 required EMS attributes were identified. Among them are prevention, communication systems, information system, evaluation, and integration of health services. Implementation of this vision falls into three major categories: building bridges among isolated disciplines; creating tools and resources to facilitate growth and development of new roles and competencies; and developing infrastructure to facilitate access, communications, and information sharing, and interoperability between EMS and its new partners.

The emergence of digital crash data plays a strong role in the development and integration of emergency medical service systems. For example, there is heightened awareness of the complexity of managing the response of a crash involving hazardous materials. Yet, in today's world, the links between transportation, public safety, and emergency responders are hampered by lack of communications, information sharing and interoperability. The development of a standardized minimum data set of crash information dramatically affects the ability to enhance and integrate a coordinated response.

A few months ago, the Medical Subcommittee of the ITS America Public Safety Advisory Group released Recommendations for ITS Technologies and Emergency Medicine Services to provide recommendations regarding future emergency medical service-related activities to the Department of Transportation and ITS providers.

While ITS Technologies have the potential to provide more efficient, effective emergency medical services and safer, more secure highways, these benefits cannot be realized without immediate and substantive input from prospective end users in the EMS community. The EMS community currently is facing serious challenges because new telecommunications and end vehicle technologies have entered the market without sufficient development input from EMS physicians and other professionals. Several of the report recommendations are germane to this request for comments.

In particular, emergency medical services systems require:

1. A minimum data set of accurate crash data that provides location and severity of the crash information;

2. Real time, cross agency voice and data communications networks that allow responders from different departments and units to communicate with each other more effectively.

3. Traffic signal priority and route guidance systems that help move emergency vehicles through evermore congested roadways;

4. Automatic Collision Notification systems that contact emergency centers immediately upon a vehicles impact, providing instant location information, evolving to provide data related to crash severity and likely passenger injuries to both emergency responders and hospital or trauma centers.

5. Development of technical standards for telematics procedures and automatic crash notification systems to promote rapid implementation for life saving technologies;

6. Integration of ITS technology into existing emergency management systems and formation of ongoing operational partnerships and real-time communications network connecting a emergency responders-including at a minimum EMS providers, transportation agencies aw enforcement agencies and emergency management agencies.

In the aftermath of September 11, and with the growth of intelligent transportation systems, this integration and coordination between pubic safety, EMS, and ITS technologies is more important than ever before.

Comments on Even Data Recorders and Crash Data: The NAEMSP strongly believes that EDRs have the potential to improve highway safety and emergency medical services. Crash information is used on a daily basis by emergency responders in order to make important life saving decisions regarding emergency response and transport, clinical care, and systems resources. Most of this information is subjective in nature, creating inefficiencies in a system that is already over-worked. In addition, the lack of accurate crash information inhibits the ability of EMS administration and researchers to continuously improve injury prediction criteria.

Motor vehicle injuries are the leading cause of death to Americans under the age of 34 years old. Each and every day, EMS responders must care for thousands of injured crash victims. Each year over 42,000 people die and 3 million are injured on our nation's highways. Many of those who are injured suffer a lifelong disability. In fact, motor vehicle injuries are the leading cause of Long-term serious head and spinal injury.

EMS systems strive to rapidly respond to serious crashes, matching the resources needed with the severity of the crash. Unfortunately, this goal is often undermined by the delays in notification and the subjectivity of the information regarding the crash. Replacing subjective crash information with measured vehicle data will have numerous positive effects on the safety of the public. Better information will mean better decisions made by EMS administrators, clinical providers, researchers, and policymakers.

The crash severity and direction of impact plays a major role in the allocation of emergency medical service resources. This information together is used to predict the likelihood of injury and the patterns of injury received by crash victims. Dispatchers use initially reported crash information to determine dispatch priority and number and level of resources sent to the scene. Once at the scene of a crash, EMS providers must rapidly determine the likelihood for serious injury based upon injury criteria: anatomic injury, physiologic criteria, or mechanism of injury (MOI). Mechanism of injury essentially is the best guess of what happened in a crash. That "mechanism" is utilized as a proxy for the energy release of a crash. In today's world, by far the vast majority of patients sent to specialized trauma resources are based upon mechanism of injury criteria. Unfortunately, due to the subjective nature of that information, most patients are over triaged; that is, the patients are sent to higher lever resources than are actually necessary in order to ensure that the patient's needs are met. In contrast, because some crashes are more severe than observation of the vehicle or patient reveals, other patients may be under-triaged.

Better, more accurate, objective information will allow EMS systems, trauma systems and EMS providers to continually improve injury prediction criteria. This will make the system more efficient, less costly and better-matched system resources with patient needs. In addition, epidemiologists, traffic safety professionals, health care providers, and EMS researchers will have improved vehicle data for analysis. This will allow the creation and development of new insights and knowledge regarding motor vehicle injuries.

Location information is critically important to a rapid response. In essence, emergency responders cannot find you if they do not know where you are. For this reason, NAEMSP and others have strongly supported the move toward location-based wireless E911. A future safety benefit of a standardized minimum data set of crash information, including location, is that automatic collision notification may become widely deployed by simply adding vehicle location information into the existing EMS infrastructure.

Vehicle crash information can and should be readily available by EMS responders at the scene through a simple plug-in or electronic interrogation of the memory module Our understanding is that the Institute of Electrical and Electronic Engineers (IEEE) is currently defining protocols for EDR crash data collection, storage and export protocols. While that standards process does not prescribe which specific data element should be recorded, it is important to note that a standardization of a minimum data set as noted above would provide great benefits in a short period of time. Minimum data for EMS response includes location crash severity, direction of forces, and use/deployment of safety devices.

The Crash Outcome Data Estimate System (CODES) has now linked police crash records with hospital records in over 26 states to better look at how crashes may be prevented, what the severity of these crashes are, and their associated heath care costs. EMS data has been found to be a critical component of data linkage. Objective measured vehicle data will improve both data linkage and the quality of data systems created at the local and state levels. As EMS providers implement better diagnostic and therapeutic decisions in less amounts of time, more rapidly locate and respond to severe auto crashes, and improve the trauma triage and transport of patients, mortality rates and the severity of injury to victims of motor vehicles crashes will be decreased.

The National Highway Traffic Safety Administration is to be commended for its leadership both in emergency medical services and in vehicle safety. The growth of technology in the vehicle fleet has provided significant opportunities for improving safety, saving lives and preventing injuries. The National Association of EMS Physicians strongly believes that NHTSA should maintain its leadership role on these issues and strongly supports the development of EDRs and data elements with NHTSA and other partners in traffic safety.[10]

DOCKET # NHTSA-2002-13546-48 02/24/03 Anthony Huffman[11]

I am, Anthony A. Huffman, a private individual. I would like to offer my comments, as applicable, to the questions asked in the referenced document.

EDRs have the potential to improve highway safety by providing: a. Accurate and rapid crash site location to rescue and law enforcement personnel. b. Meaningful crash related data that would be available to the ambulance and rescue crews. Rapid accessibility of the data, by the authorized personnel, will result in better triage at the crash site. Time saved by the medical crews getting to the crash site and getting the crash victims to the proper medical facility as soon as possible will result in lives being saved. c. Much more accurate traffic crash reports that will be made public in a timely manner.

There must be different classes of EDRs for different types of moving vehicles. The environmental conditions found in passenger cars, SUVs, off-road and long-distance heavy load trucks vary greatly as do the potential impact conditions. However, the configuration of the device(s) used extract the data from the different types of EDRs must be limited to one basic configuration for all moving vehicles; otherwise, the crash data will not be fully accessible at all times in all conditions by all authorized personnel.

Regarding use of EDR data, I can only comment on my family's tragic experience. Our only son was killed in a collision between the vehicle that he was driving and a heavy load vehicle. The crash happened in southern Texas. Crash data was extracted from the EDR that was legally removed from the heavy load vehicle, six weeks after the fatal crash, by our attorney. The data revealed that the heavy load vehicle was going in excess of 75 mph; no brakes were applied prior to the collision and the rpm of the truck's engine greatly varied prior to the crash. The variable rpm indicated that the truck was inattentive, at best. The truck crossed the double yellow line in order to impact my son's vehicle. My son had pulled completely off of the highway in vain attempt to avoid the collision. The impact between the vehicles was of such magnitude that parts of my son's vehicle were found hundreds of feet from the point of impact. The collision ignited a fire and it took rescue workers over one hour to remove my son's body from the wreckage.

We were fortunate that we able to use the EDR data to obtain an indictment of Criminal Negligent Homicide against the truck driver. It is a fact that the truck driver would not have been indicted without the crash data. The crash data was instrumental in resolving both the criminal and the civil cases against the trucking firm and the driver. It is also a sad but true fact that no one at the crash site knew about the EDR device or how to extract the usable data. It took the Highway Patrol six weeks to produce an incomplete accident report. If any of the investigators at the crash scene had known anything about EDRs, the legal proceedings would have been concluded a year ago at less cost to the taxpayers and much less stress to my family. The data downloaded from EDRs will be a cost effective tool for the city, county, state, and federal government agencies. Many cases similar to ours can be processed through the various court systems faster and at much lower costs due to the standardized EDR. Out of court agreements between insurance companies and crash victims' families will be reached, therefore reducing the number of federal court civil court proceedings.

Future safety benefits: Suggested addition of an "out of tolerance driving conditions" trigger. Therefore, audio and/or mechanical crash avoidance systems will be activated, via the EDR, when certain critical engine performance parameters exceed the pre-set safe operating abort limits. This feature will aid in preventing collisions by vehicle that are being driven by an inattentive (asleep) drivers.

Standards: I just attended my first IEEE meeting concerning MVEDRs that was held in Colorado Springs, CO on February 10 and 11, 2003. It was apparent to me that the some of the representatives of the light vehicle OEMs want the SAE organization to produce the EDR standard. These same representatives indicate that the IEEE would work on the standard for other types of vehicles. This scheme, to divide the work between two Standard Development Organizations, at this late date, will have disastrous results. No viable standard will ever be developed under these adverse conditions. It appears to me that the three (3) IEEE, MVEDR working groups have completed a lot of excellent technical work. Much more progress is required by the IEEE in order to meet their goal of issuing the MVEDR standard in early 2004. In my opinion, the IEEE group must maintain focus on generating the MVEDR standard. A few people with different agendas should not impede the MVEDR project. The U.S. is many years behind the Europeans in implementing MVEDRs. Today, the U.S. has the technology to have standardized EDRs installed in all moving vehicles. Lives are being lost every day on the U.S. highways. It is critical that the EDR data output be standardized so that ambulance crews can quickly read the pertinent data no matter the type of EDR.

Data Element: The critical data elements in which I have limited experience are: vehicle velocity, application of brakes, engine rpm and time of day. These data elements give a history of the vehicles performance with respect to time. Other elements that are of concern are: seat belt(s) and air bag activation(s). Medical parameters are important; therefore, change in velocities when impact occurred will be critical. It cannot be emphasized enough that the final output of data MUST BE USABLE by the medical and rescue community immediately upon demand.

Training: Proper training for all authorized law enforcement and medical personnel is mandatory. It does no good to have a fully operational EDR system installed on moving vehicles when no one has the proper training to be able to extract the data from any of the EDRs in a timely manner.

Survivability: The EDR must be able to be function properly following exposure to the Insurance Institute "off-set" crash test. With the solid-state devices available today this is a realistic requirement. The solid-state can withstand high impacts and do have a high degree of reliability. The EDR must meet fairly substantial EMI requirements and be designed to be mechanically tamper proof. Present odometer laws could be amended in order to prosecute violators that tamper with the EDRs. The EDR could be designed so that the vehicle fails to start after the EDR has been tampered with.

Privacy: There are a lot of elements associated with EDR data privacy issues. Assuming that the crash site is on a public highway, the EDR data is much like skid marks on a highway. The data becomes part of the

crash site investigation report and available to all concerned parties.

Role of NHTSA: Requesting comments such as this
docket remains a good way to gather ideas. The U.S.
badly needs an effective MVEDR. Presently we have
the technology to make the EDR a reality but we do
not seem to have the will or commitment to make the
MVEDR a reality. The NHTSA has the expertise and is
in the position to help make the standard MVEDR a
reality. But it seems that we also need the support and
the commitment from some of our elected officials to
make the EDR a reality.

DOCKET # NHTSA-2002-13546-50 02/21/03 National Association of Independent Insurers (NAII)[12]

NAII is the nation's largest full-service property/casu-
alty trade association, representing more than 715
members. Its purpose is to advocate its members'
public policy positions at four levels: federal government, state legislatures, federal and state courthouses
and the public arena, and to provide its members with
targeted industry information. NAII stated that, "The
National Association of Independent Insurers (NAII) is
pleased to offer its observations and comments on the
role of the National Highway Traffic Safety Administration (NHTSA) regarding the standardization and use of
the event data recorders (EDRs) in motor vehicles.
NAII is offering several recommendations. First, we
encourage your agency to take whatever steps are necessary to see that the full potential of EDRs is analyzed
and understood, and that an appropriate inter-gov-
ernment initiative is then outlined to facilitate the
realizations of as many EDR benefits as possible. Second, we urge your agency to move forward on the task
of developing data element and data retrieval and
standards of EDRs. Third, we believe that EDR data
elements should be set with proper attention given to
data uses beyond the scope of safety engineering or
research. Fourth, in the interest of achieving greater
flexibility, it would appear preferable to investigate a
state role in setting standards or collections, storage
and access rights to the EDR data. Our letter will
expand on these and other related thoughts.

For the record, the National Association of Independent Insurers is a national trade group representing

over 715 property/casualty insurance companies in the U.S. that write at least $112 billion in premiums throughout the nation. NAII member companies account for a 33 percent share of the property-casualty insurance market. NAII affiliated insurers underwrite approximately $62 billion in personal auto insurance premiums (46 per cent on the U.S. market) and $8 billion in commercial auto insurance premiums (32 per cent on the U.S. market). Our client companies transact business in all 50 states, and they are a cross-section of the insurance industry. They include national firms that write insurance throughout the country, small single state or niche market providers, mutual companies, stock companies, reinsures, as well as surplus lines carriers. They insure accounts ranging from single, private passenger vehicles to interstate trucking firms or other commercial fleets so their interest in federal motor vehicle regulation are quite diverse.

Safety Perspective: Let me underscore that NAII supports the points that were made in the written comments of the Insurance Institute for Highway Safety (reference their January 13, 2003, letter to the Administrator Runge from IIHS Senior Vice President–Research Susan Ferguson). It is clear that IIHS believes that EDRs can play a major role advancing motor vehicle safety research. Like the Institute, NAII sees merit in moving the private market toward standardization, at least in the context of basic data collection and retrieval performance. On the other hand, NAII suggests that any regulatory standards should be flexible enough to address manufacturer design concerns, encourage private sector competition, and facilitate cost-effective training. Our technician should be able to retrieve and interpret EDR data regardless of the class of vehicle or the make or model. NAII recommends that EDR engineering be "mainstreamed" in a sense so that information systems technicians, regardless of their industry or field, all have the requisite skill sets to service an EDR, or at least to download and interpret its data. Our observation is predicated on the assumption that there could be broader use of the EDR data in the future, and accordingly, a demand for greater access.

The IIHS also made a number of useful recommendations in relation to the data elements that could play a greater role in advancing vehicular crash research. We support the Institute's suggestions. That organization's extensive experience in crash-testing motor vehicles over the last several decades has given them, like NHTSA, considerable insight on the factors and dynamics that affect driving, vehicle performance and in many cases, crash outcome. We think the Institute has identified a number of relevant measures that should be recorded and collected from EDRs in the future.

Another potential safety use of EDR data output that was touched on in the agency's Request for Comments release in the October 11, 2002, *Federal Register* relates to the automatic crash notification (ACN) systems. As we understand it, the data captured by the EDR could be conveyed to hardware in the vehicle that in turn reports a collision or the likelihood of an injury to an outside emergency service unit or network. If data elements collected by the EDR could help a rescue team or paramedics better assess crash-related injuries or the likelihood of injuries, we trust the agency will be receptive to guidance from the ANC and related medical community on what data elements are critical.

NAII is not aware of research or studies that have conclusively proven what the societal payoffs that are derivable under an ACN network; however it is our understanding that we are at a very early stage in the development of the ACN infrastructure. Until there is a more robust network and more subscribers to OnStar® and similar service providers, it may be premature to make definitive conclusions on the merits on ACN. Having EDRs that collect useful data for emergency responders seems logical; therefore we trust the agency will see that data element standards incorporate the relevant measures.

Other Applications for EDR Data: NAII respectfully suggests one agency priority should be to see that other potential applications of the event data recorder are thoroughly analyzed. The focus of the *Federal Register* notice was decidedly safety oriented in nature. In addition to safety and vehicle engineering applications, however, other uses for the data unrelated to highway or vehicle safety may also be relevant. We encourage NHTSA as well as the vehicle manufacturer community to remain as open minded as possible. It may take some time before the other stakeholder interests understand the potential utility of EDR data.

We will mention several of the possible areas in which EDR data might be of value to the property-casualty insurance sector. The observations that follow are offered as hypothetical illustrations of how EDR databases could be used. They are intended as a starting point for discussion, hopefully leading to more in depth analysis and evaluation.

One of the EDR applications of interest to NAII is in improving accident reconstruction and the allocation of negligence or responsibility among the drivers and other parties involved in a collision. In each of the 50 states, the accident reparations system is responsible for compensation of injured parties and in many cases, the assignment of fault. Under the tort liability system rules, it is often difficult to settle issues of fact relating to an accident. In some instances, the report from the police officer on the scene might be helpful, but in many cases the officer did not witness the crash so details are often sketchy or ambiguous. There may not be any witnesses, or if there are, the witness' observations may conflict or be unreliable. In short, the limitations of current fact-finding hampers efficient insurance claim administrations, creates expensive and often flawed discovery procedures during litigation petitions, and unduly complicates the resolution of disputes arising out of motor vehicle crashes.

If the EDR recorded data elements that could be useful in analyzing an accident and the actions of the various drivers involved in the moments before the collision, the data it yielded could be extremely beneficial. It could help accident investigators and claim adjusters separate fact from fiction and lead to more equitable decisions that affect litigation and liability arising from the case. A big payoff could be in the reduction in lawsuits arising out of personal injuries sustained when using a motor vehicle. In the same vein, it would be our expectations that the availability of data downloaded from the EDR could conceivably reduce the expenses of litigations associated with auto injury cases. The stakeholders that could benefit from the use of EDR data in accident reconstruction and dispute resolutions settings are many. They would include law enforcement agencies, insurance companies, accident investigation and engineers, municipalities, lawyers, and those charged with administering the court systems (both civil and criminal/traffic offenses). There are a handful of EDR data elements that could be of value to those involved in accident reconstruction and the injury claiming process.

Injury compensation determinations can be affected in some states depending on whether a motorist was wearing a seatbelt at the time of collision. In a number of states, the failure to wear a seatbelt could result in a reduction of compensation. In addition to seatbelt usage, there are other data elements that could be influential factors in ascertaining whether driver negligence contributed to a collision. There are many kinds of factual issues that become relevant when a personal injury suit is filed. Many cases could be resolved more expeditiously and efficiently if key facts were known. Illustrations of EDR data elements that could provide factual detail relating circumstances immediately prior to a crash include: Seatbelt usage at time of crash; Speed or direction of vehicle at the time of collision; Braking action or lack thereof preceding a collision; Whether vehicle navigation (headlights, taillamps, etc.) lights were illuminated; Position of the gear selector (*e.g.*, whether a special power distribution feature such as traction control, an active handling system, stability control, four wheel drive, etc. had been engaged before the crash); Whether a turn signal indicator had been activated immediately prior to crash; Exact time at which a collision/occurrence took place; and Last action/position of the driver before the collision.

If data element standards for EDRs are set by NHTSA, NAII suggests that they include as many of the above measures as possible. The data captured could be of value to both the engineering and research community as well as to those responsible for dealing with the economic consequences of vehicle collisions.

We would anticipate that those charged with enforcing traffic laws and prosecuting offenders might be interested in evaluating the use of EDR data for reasons similar to those outlined in the above paragraphs. Assuming it recorded the appropriate data elements; event data recorders could provide the information necessary to corroborate circumstantial evidence of traffic law violations.

There may be other insurance-related applications associated with the EDR in the future. One of those relates to tracking how many miles a vehicle is driven over some span of time and where or how the vehicle is used (*e.g.*, whether more miles driven are at a steady speed or whether driving patterns are more stop-and-go). In the future, it is conceivable that some insurance companies will be interested in having access to vehicle mileage records that could independently verify information supplied by a policyholder. "Miles driven" is a common rating factor but it does not carry much weight because the reported miles driven are often difficult to verify. Conventional wisdom suggests that those who drive more miles have a greater statistical chance of being involved in a collision. In some instances, however, greater miles driven may be offset by factors such as where a vehicle is driven and how a vehicle is driven. For example, a motorist making extended daily commutes at high speeds might nevertheless present a lower risk profile than a motorist that drove significantly fewer miles, but drove on busy city streets in stop-and-go conditions rather than on limited access roads or interstates in rural areas. The convergence of EDR technology and on-board GPS navigation systems could bring more reliable mileage-based insurance rating programs closer to reality.

If there is to be a future in mileage-based auto insurance rating alternatives, it may take several significant developments. First, the use of EDR and GPS technology on a standardized basis would seem desirable. Second, there will have to be greater market acceptance and consumer demand for these products/services. Lastly, access to the EDR and GPS data by third party users will have to be accommodated in some manner. The access issue is probably twofold in nature, one dimension relating to the ease of physical access to the EDR unit and the procedure of retrieving (*i.e.*, downloading) data. The second factor may require a paradigm shift in privacy concepts and privacy expectations. Our association is hopeful that agencies with jurisdiction over EDRs such as NHTSA will adopt regulatory policies that do not foreclose these developments from taking place.

The expanded use of EDR data (outside the safety engineering field) depend on there being broader access to the recorder, and NAII offers several observations later with respect to the privacy interests that may be a complicating factor to that access.

Other Issues: There were several topics raised in the *Federal Register* Notice in October that merit a comment. One concerns the question of whether all EDRs should be designed alike and meets the same set of standards. While not offering a recommendation per se, NAII believes that NHTSA has ample reason to evaluate whether separate EDR standards for these vehicle classes are warranted. There are significant differences in size, weight, handling characteristics, driver licensing-vehicle registration standards, financial responsibility requirements, and in the regulatory objectives in general etc. between large trucks and private passenger automobiles. In view of the safety risk posed by large trucks, or at least the public perception of that risk, it would seem reasonable that they have to meet a more rigorous EDR standard.

Another subject that was highlighted in NHTSA's *Federal Register* Notice related to complications that could arise as a result of privacy interests. Before formulating our thoughts on this issue, NAII reviewed the Summary of Findings from the NHTSA EDR Working Group. A variety of views appear to prevail on the privacy interests linked to the data collected in the event data recorder. Some legal barriers may exist that may not be easily circumvented. The federal Privacy Act of 1974 appears to be one such obstacle, barring a federal agency from releasing EDR data without the consent of the person to whom the information applies. In our view, the potential applications associated with event data recorder information are so far-reaching and the payoffs so promising, a thorough public policy analysis and legal review should be undertaken before any summary conclusions are reached that limit the use of EDR output to safety research. The framers of the U.S. Constitution could not possibly have foreseen today's information technology, nor could they have foreseen the complexity and inefficiency of our institutions today either. Accordingly, it would be too easy in our minds to summarily conclude that the Fourth and Fifth Amendments to the U.S. Constitution block more innovative uses of EDR information. NAII believes that a more expansive and fresh review of privacy issues associated with the event data recorder.

NAII is concerned that privacy concerns are being overstated. For example, is it reasonable for a motorist to expect that his privacy interest in how he drives is superior to the interest of the community at large and its elected officials to protect the public safety? There has to be a balancing of interests. When citizens use the court system, they are bound by its rules. One of those rules requires that each party in the litigation is entitled to discover all pertinent information relating to the controversy or dispute in the possession of the other party. Are discovery rules to be waived in the context of EDR data? On what legal ground? Consider the information that is purportedly cloaked with a privacy interest. EDRs do not record or monitor speech, that is, the conversation of a driver or other vehicle occupants. Furthermore, EDRs do not continuously record or monitor vehicle functions. At present, they are engineered to record only several seconds of information prior to an event and then the data correlating to the actual crash pulse. Under the standard insurance contract, when a policyholder is sued, the insurance company is contractually bound to defend their insured. The insured however is also contractually bound to cooperate with the insurer in the defense of the claim. This contractual stipulation is well grounded in the common law of most states. NAII believes that this contractual consideration may also contemplate that an insured consent to the release of EDR data in order to fulfill this obligation. These may seem like unrelated concepts, but our point is simple. We fear that privacy is becoming a red herring issue that nevertheless may persuade policy makers to prohibit the full use of event data recorders.

If anything, a new technology often demands new law and new legal concepts. The Internet is one recent illustration. A decade ago, few people realized that in the near future legal obligations could be created through digital signatures. NHTSA and other peacemakers should refrain from making premature and closed-minded conclusions about EDR data, EDR ownership rights, and corresponding privacy interests. There may be a compelling reason for legislative bodies, at the state if not the federal level, to hash out a new statutory scheme that would address the privacy issues in innovative yet respectful ways that balance the competing interests of the individual and broader societal institutions. NAII would encourage NHTSA to continue its dialogue with EDR stakeholders and to expand the participants if possible to get an even broader viewpoint. NHTSA's goal should be to promote more inclusion of state level governmental and private sector interests (*e.g.*, motor vehicle regulators, law enforcement authorities, the legal profession, the private insurance sector, the trauma car-automatic crash notification stakeholders, after-market as well as OEM manufacturer-suppliers) so that debate on these issues is as well rounded and uninhibited as possible. In our view, a legislative solution to EDR data collection and access should be crafted at the state level, the governmental tier where so many fundamental motor vehicle and driver compliance programs are currently administered. NHTSA should therefore assert limited jurisdiction in the context of EDR rulemaking actions. NAII commends the NHTSA for proceeding slowly but deliberately in moving forward on the regulation of event data recorders. Thank you for permitting interested parties the opportunity to share their views on this subject.

DOCKET # NHTSA-2002-13546-47 02/24/03 Harris Technical Services[13]

Harris Technical Services of Port St. Lucie, Florida, provides internationally recognized, published and court qualified traffic accident reconstruction experts for the analysis of auto, truck, motorcycle, and pedestrian accidents. Harris Technical Services noted:

> Event data recorders in motor vehicles have already proven themselves to be of tremendous value. Many proponents of privacy concerns fail to realize, or fail to mention, that the information obtained from an EDR can not only be used against the individual but in their interest when a claim of negligence or criminal action is levied against them.

> A case on point was Florida v. John Walker, 20th Judicial Circuit, Lee County, Case No. 00-002866CF RTC (2003). This was a criminal case with a two vehicle, head-on collision. The defendant was charged with two counts of Vehicular Homicide with a possible penalty of more than 20 years in prison if found guilty. At issue was the defendant's speed and in which lane the collision occurred.

> As with many co-linear collisions at high speeds, it is difficult, if not impossible, to accurately determine a speed for either vehicle. The prosecution presented a speed of 70+ mph for the defendant and 60 mph (the speed limit) for the other vehicle.

Over the course of the case, the defendant's vehicle was impounded by the police, then sold for salvage. The prosecution was not aware the defendant's vehicle had an EDR on board that could contain information that would be vital in accurately determining a speed for the defendant in this type of a collision.

After a motion was filed by the defense for the production of the vehicle, it was located and the EDR downloaded. In this case, the EDR only provided the velocity change of the vehicle through the collision but it was sufficient to contradict the prosecution's claim of the high speed for the defendant.

The defendant's speed was an integral part of the prosecution's case. The jury's verdict was not guilty on both counts. Without the EDR evidence, a guilty verdict would have been far more likely.

EDRs have already proven they can be a vital component of the analysis of a traffic crash. All auto and truck manufacturers should be required to include EDRs in new vehicles with a standardized system of data recording and retrieval for elemental data.

Automotive black boxes speak for the victims—they speak the truth.

DOCKET # NHTSA-2002-13546-52 02/25/03 Garthe Associates[14]

Garthe Associates of Marblehead, Massachusetts commented:

> Every year NHTSA expends about 250 million dollars on programs with states and organizations in an effort to reduce traffic deaths and disabilities, primarily by increasing restraint use and decreasing drunk driving. Requiring vehicle EDRs with the capability to rapidly download data to EMS personnel at the scene of the crash can produce survivability benefits similar to these current NHTSA programs at relatively low cost.

> A statewide, population-based study we completed in conjunction with the Massachusetts Department of Public Health and the Governor's Highway Safety Bureau showed that triaging seriously injured persons to the most appropriate medical facility by the most appropriate method of transport made a 2:1 difference in survivability. This result puts correct triage approximately equal to restraint use in importance to survival.

> A key to this survivability improvement is making scene triage more objective and therefore less prone to errors and external influence. Crash severity metrics computed from data downloaded from EDRs at the scene of the crash can provide the "on the spot" objective triage data EMS currently lacks. This capability can be added to future EDR's for about $1 per unit. Attached are our three slide presentations to the current IEEE P1616 EDR Working Group describing the details of the scene use of EDR data for EMS triage decision-making. We also presented similar information to NHTSA's earlier EDR workgroup.

> Opportunities to make a survivability improvement of this magnitude are rare. We request the NHTSA include the necessary technical specifications (as outlined in our presentations) to permit on-scene EMS use of EDR data in any standard NHTSA creates.

DOCKET # NHTSA-2002-13546-55 02/27/03 American Bar Association[15]

> The American Bus Association ("ABA") represents over 3,400 members; of those, approximately 850 member companies are bus operators, offering a variety of bus services: regular route intercity between fixed points on set schedules; charter service, where a group of passengers (such as a company or corporation)

purchase all of the seats on a bus for exclusive use on a particular trip; commuter bus services, generally from the suburbs into urban areas, and special operations, which are scheduled to enhance public transportation systems (such as a bus from a city to an airport), or which may be connected with a special event or attraction at the destination.

The rest of ABA's members include representatives and suppliers of products and services used by the bus industry. The intercity bus industry carries over 774 million passengers a year with a safety record that is unparalleled on the highways. As the national trade association of the intercity bus industry, ABA is submitting the following comments to Docket No. NHTSA-02-13546, in which NHTSA requests comments concerning the future role they should take in the continued development of Event Data Recorders (EDRs) in motor vehicles.

The American Bus Association believes that the development and installation of Event Data Recorders (EDR) should be market driven. As a result we believe that further efforts by NHTSA in this area are unnecessary. A considerable amount of data is available via electronic components currently available on trucks and buses. As example of this technology is the Detroit Diesel Electronic Controls IV Electronic Control Module (commonly referred to as DDEC IV) that is currently installed on Detroit Diesel Engines. This module which provides operational data for a vehicle and engine including trip activity, speed versus rpm, engine load vs. rpm, periodic maintenance, engine usage, and hard braking activity is not only a valuable tool for commercial vehicle operators, but has been used to successfully supplement traditional investigative techniques by the National Transportation Safety Board (NTSB) and others to determine crash causation. It is important to note that data from a DDEC IV helped NTSB determine the cause of the motor coach run-off-the-road accident near Canon City, Colorado on December 21, 1999 (NTSB Accident No. HWY-00-FH011). Other engine manufacturers have similar modules as well. In ABA's view, this technology is sufficient to accomplish both company and agency post-accident analysis.

Further, adding additional and unnecessary EDR capabilities will certainly place more demands on vehicle electrical systems, making them more complex, less reliable, and more difficult to maintain and repair.

It is important to note that "NHTSA has already twice rejected petitions for rulemaking to mandate the use of EDRs on motor vehicles, finding that the issues raised are "best addressed in a non-regulatory context." 67 *Federal Register* 63493, 63494 (October 11, 2002). Further, NHTSA has now received a third petition for rulemaking from a manufacturer of such devices, who is clearly looking to generate a market share through regulation. NHTSA should reject the petition out of hand.

In sum, the NTSB has never failed to determine a causation factor for an accident because current engine and vehicle technology effectively supplement traditional investigative methods. Additional detailed data may be interesting to engineers and accident investigators, but is it needed when the cause of the accident can already be determined by current methods and evaluation, possibly at the cost of system reliability? We believe that a more efficient use of resources would be to conduct crash testing of motor coaches in order to evaluate their capabilities during a crash.

The ABA believes that market-based solutions work best in this area. This approach has already been proven successful by the use of modules such as Detroit Diesel's DDEC IV and others.

DOCKET # NHTSA-2002-13546-56 02/27/03 American Automobile Association (AAA)[16]

On behalf of AAA, that nation's largest organization representing motorists, we are pleased to respond to the agency's request for comments on event data recorders (EDRs). For more than a century AAA has worked to improve the quality and safety of the nation's roadways and continually improve safety for the traveling public. Automobile crashes claim roughly 42,000 lives annually. They are the leading cause of death for people ages 1–34 and are one of the top 10 causes of death for all ages. Without question, traffic safety is among the most serious public health challenges of the 21st Century.

In our efforts to reduce the number and severity of crashes, we have fallen short in one critical area—the collection and analysis of scientific data to fully understand the dynamics and trends in crash causation. Comprehensive nationwide crash causation data are more than 25 years old.

Crucial pieces of information already exist within the body of the vehicle. Technology has made the cars we drive smarter. More than 80% of the systems in some cars are monitored and controlled by computer. Today's vehicles are highly sophisticated machines that make driving smoother, safer and more adaptable to existing travel conditions.

For some time many of these vehicles have been equipped to capture crash data through EDRs. While the data may vary by manufacturer, there is growing recognition that such data have the potential to shed more light on crash causation than existing efforts, which rely primarily on reconstructing evidence after the crash, or eyewitness reports.

AAA is encouraged that NHTSA recognizes the potential by seeking public input on EDRs. A signal from the lead government traffic safety agency may not only spur this technology along, but also the analysis of data it generates may contribute to a much improved national data base from which informed decisions can be made. AAA directs our comments to several areas mentioned in the rulemaking.

Safety Potential: Improving safety for the millions of people who travel is the single most important reason to promote the expansion and use the EDR technology. It represents the best opportunity to capture electronically what actually occurs in the vehicle during those last seconds before a crash. These data provide an objective measurement which, when combined with other crash investigation details, can provide a reasonably accurate picture of the actual crash event.

Taking the "crash pulse" should yield important benefits in vehicle design by identifying the types of changes that manufacturers could pursue to build more crash-friendly vehicles. This is an area where EDR technology can yield relatively short-term, high-payoff benefits for motorist safety.

There are additional long-term uses for the technology that should not be overlooked. The safety potential of EDRs reaches beyond road and vehicle design and can be directly applied to vulnerable populations that are at risk in any public heath crisis. Of note, children, teenagers and seniors each face special dangers on the road. Senior road fatality rates have climbed while overall fatality rates have remained fairly stable since 1991. Analysis of aggregated, non-identifiable data collected by EDRs and other crash investigation techniques can contribute to a better understanding of the causes of crashes involving these vulnerable populations. Taking EDR technology to this next step requires careful consideration to avoid compromising the rights of any one individual.

The key to enhancing traffic safety is striving to prevent crashes before they occur. Yet, studying crashes can educate experts on preventing future crashes. Details that can be gleaned from data recorded at the accident can provide "at the scene" information that will teach traffic safety professionals and engineers how to prevent future crashes. EDR crash data can be employed to improve dangerous roads, intersections or other trouble spots discovered in regular road safety reviews.

Standards: Standardization of EDR data will provide immeasurable public safety benefits. With non-standardized EDRs, critical time can be lost in analyzing each unique set of data elements. Further, different data elements with non-standardized EDRs can create different "pictures" of what occurred before a crash and could potentially lead to bias during the interpretation of events.

AAA supports the development of a national traffic safety database for crash causation because it offers the greatest potential for transportation safety professionals to learn about the factors contributing to crashes and affecting the severity of those crashes. Crash data from all sources are often not uniform across states or even within states. This also leads to difficulty in performing comparisons of state data. Creating a single repository capsule of storing universally recognized real-world crash data would be a critical step in enhancing primary crash prevention. Because EDRs continue to emerge in varying levels of sophistication, regular and systematic updates of the types of data elements included should be built into the development of traffic safety databases to ensure the information continues to be timely and useful.

Application: NHTSA raises an important question in asking whether different vehicle types would benefit from different EDR applications, particularly heavy trucks, which AAA believes should meet a higher standard of public trust.

In 1999, AAA called on Congress to authorize a national study to determine the factors contributing to car-truck crashes. That study is underway and is scheduled to conclude by the end of this year. Previously, our efforts to prevent car-truck crashes have been hampered because we lack scientific data to document where problems exist. The results from this study should help guide our efforts to improve truck safety and reduce fatalities from car-truck crashes.

Use of an EDR specifically designed to operate in a heavy truck would also add important information to our understanding of these crashes. The stark reality is we lack sufficient crash data for all vehicle types, including cars and light trucks. EDRs in heavy trucks have the potential to collect critical data to better understand truck safety factors as an ever-increasing number of cars and trucks vie for space on our nation's highways.

Future Safety Benefits: One of the clear future benefits of increased EDR use lies in enhancing the ability of engineers to design effective safety systems and improve overall vehicle design. Even small design decisions can be a critical factor in saving lives and preventing injury. One important example was the development and design of air bags. The development and implementation of air bag regulations would have been greatly simplified and expedited, while leading to higher safety benefits, if more data on real-world crash experience had been used. Data, such as that provided by existing EDR systems presently in use, would have proven invaluable. Analysis of collected data parameters could lead to engineering improvements with regard to vehicle rollover, braking, traction control and seat belts.

EDRs, combined with the development of automatic crash notification (ACN) technologies, also offer great potential for reducing response time to crashes, which not only saves lives but also reduces the severity and likelihood of long-term injuries.

Technology cannot only enhance the timeliness of emergency responders but can provide personnel with an advanced "picture" of the crash scene and the likely severity of injuries. Finally, post-crash data gleaned from EDRs could be utilized in safety reviews to gain a greater evaluation of improvements that will enhance primary crash prevention and lead to implementation of performance-based safety management systems.

Additional Issues: While the safety benefits of EDR data are widely accepted in the traffic safety community, making these data available to commercial interests warrants careful consideration to ensure consumers have adequate protection from intrusion of privacy and a fair degree of control of the use of these data by third parties. These issues reach beyond NHTSA's domain. For example, access to EDR crash data could provide useful information to insurance companies to help determine liability and damages with customer claims, possibly reducing the need for costly and time-consuming litigation. Crash data extracted to verify insurance claims can obviously lead to more accurate claim adjustments. But, legitimate privacy considerations should not be compromised in the process. The goal, while difficult, is to seek agreement from all parties to strike a reasonable balance between data access and appropriate use of the data once it is captured. If use of the technology is to reach its potential, it is also essential that consumers accept it by becoming better informed as to its existence and being assured of its fair use. Motorists are probably unaware that crash data are already being gathered in many vehicles, and most have not contemplated the consequence—good and bad—of their use in the context of automotive insurance. If those data become a crucial factor in determining liability in a crash, they could be equally unaware.

Privacy Concerns: AAA acknowledges that privacy and data ownership are very complex issues requiring careful scrutiny to ensure that motorists' and vehicle owner property rights are not compromised. Protecting personal privacy and confidentiality are major considerations that demand careful review. AAA believes that vehicle owners need to be thoroughly aware of all the advanced safety components featured in today's technically sophisticated motor vehicles, including EDRs, and educated as to how the data collected from this technology are used.

NHTSA's request for comment raises an important question as to who owns the data collected by the EDR. AAA strongly believes that the owner of the vehicle owns the data as they have purchased the technology when they bought their vehicle. When consumers drive off the lot with a new car, they own more than just the vehicle; they own the information that their vehicle generates. While collection of crash data through EDRs presents significant potential for crash prevention, protecting the confidentially of personally identifiable information is not only an important privacy consideration, as NHTSA correctly points out, it may be key to public acceptance.

While NHTSA states that its role and statutory authority is limited with respect to privacy, the issue remains a critical one if we are to gain the full benefits of EDRs. With any technology that can be used to track or monitor the movements of people in a free society, there is a substantial risk of public concern and possibly rejection. AAA believes the public trust must be fostered throughout the process, whether it is by NHTSA or another agency that is the arbitrator of this confidence.

DOCKET # NHTSA-2002-13546-57 02/27/03 Forensic Accident Investigations Inc.[17]

FAI is an accident reconstruction firm specializing in the scientific use of forensic evidence to reconstruct transportation and industrial accidents. FAI commented to NHTSA that:

In response to your request for comments I am writing to support Dr. Ricardo Martinez's petition to NHTSA for the mandated collection and storage of onboard vehicle crash event data in a standardized data and content format and in a way that is retrievable after the crash. Standardized EDR data will result in improvements to vehicle safety, occupant safety, and highway design and collision analysis. NHTSA should mandate collection and storage of onboard vehicle crash event data in a standardized data and content format and in a way that is retrievable after the crash.

This important standard will continue to develop, mature and evolve with time. This standard should not be considered as static. Safety systems and vehicle systems will evolve and EDR will evolve. Standardized EDR data will reduce collisions, save lives and improve public confidence in the transportation system.

CHAPTER 8

Turn Signals: February to March 2003

The global road trauma epidemic is real. Every day 3000 people die and 30,000 are seriously injured on the world's roads.

In the blink of a turn signal motor vehicles crash. The emotional cost of a car crash is hard to measure. It varies from individual to individual, from victim to survivor. Motor vehicle crashes are a leading cause of post-traumatic stress disorder (PTSD). People with this disorder can have a variety of symptoms, including psychological numbing of emotions, anxiety, poor concentration, and insomnia. Formal advisory bodies, such as the National Transportation Safety Board and the Transportation Research Board (part of the United States National Academy of Sciences) were set up to provide independent advice and guidance. The combination of new, dedicated institutes on road safety and greater scientific research has in many cases produced major changes in thinking about traffic safety and its interventions. However, at the same time, there is often a real conflict between the aims of traffic safety lobbies and those of campaigners for increased mobility or for environmental concerns. In such cases, the lobby for mobility has frequently been the dominant one. In the long term, increases in mobility, without the corresponding necessary increases in safety levels, will have a negative effect on public health.

—World Report on Road Traffic Injury Prevention, 2004.

The NTSB is an independent Federal agency charged by Congress with investigating every civil aviation accident in the United States and significant accidents in other modes of transportation—railroad, highway, marine and pipeline—and issuing safety recommendations aimed at preventing future accidents.

DOCKET # NHTSA-2002-13546-59 02/28/03 National Transportation Safety Board (NTSB)[1]

On February 28, 2003 the National Transportation Safety Board (NTSB) commented to NHTSA:

> The National Transportation Safety Board (NTSB) has reviewed the National Highway Traffic Safety Administration's (NHTSA's) request for comment title, "Event Data Recorders," published in the *Federal Register*, Volume 67, Number 198, on October 11, 2002. The Safety Board appreciates the opportunity to comment on this important issue affecting highway safety.
>
> The Safety Board has a long history of promoting on-board recorders in all modes of transportation. In fact, the issue of automatic information recording devise for all modes of transportation has been on the Board's "Most Wanted" list of Transportation Safety Improvements since 1997. On-board recording devices have proven themselves to be extremely valuable in the other modes of transportation, particularly aviation. The Board believes effective implementation of on-

board recording in highway vehicles can have a similar, positive impact on highway safety. Accurate recorded data can provide key information to help individual accident investigations. Also, the collection of record data in aggregate from can offer a wealth of information when addressing important safety issues, such as air bag system effectiveness. Even further, standardization of event data recorder (EDR) data could increase the benefit of collecting data in aggregate form.

The Safety Board's first recommendation to NHTSA regarding the on-board recording of data resulted from the 1997 NTSB Air Bag Forum. Safety Recommendation H-97-18 recommended that NHTSA work with the automobile industry to use data stored in current crash parameters. The subsequent initiative by NHTSA and automobiles manufacturers, such as General Motors, went beyond the Board's expectations. The Board is further encouraged by the efforts of some members of the automobile industry to continue to expand the recording capabilities of air bag modules.

As mentioned in the request for comment, the Safety Board's most recent Safety Recommendations to NHTSA regarding on-board recorders were H-99-53 and -54. These recommendations requested that installation of EDRs be required on newly manufactured school buses and motor coaches and that NHTSA work with industry to develop an EDR standard. For the development of a standard, the Board focused on key issues that have been critical to on-board recording standards in other modes of transportation: parameters to be recorded, data sampling rates, duration of recording, and survivability factors.

The Safety Board believes NHTSA's ongoing leadership in response to the Board's recommendations has been invaluable to the implementation of EDRs. NHTSA's formation of two working groups, the Event Data Recorder Working Group and the Truck and Bus Event Data Recorder Working Group, brought together a number of different manufacturers, user groups, and other stakeholders. The Board participated in both of the NHTSA EDR working groups and found that the efforts of these groups were an important first step in the effective implementation of EDRs.

The Safety Board's investigations have already bene-
fited from the availability of on-board recorded data
from such devices as engine electronic control modules. The impact that recorded data can have on an
accident investigation is no more evident than in the
Board's investigation of the Sierra Trailways motor
coach accident that occurred in Canon City, Colorado,
on December 21, 1999. Investigators were fortunate
that the motor coach engine's electronic control module recorded information, though limited, that
assisted the investigation. However, despite the
important contribution the data made to the investigation, investigators were still left with many unan-
swered questions regarding the status of the vehicle
systems and the actions of the driver. If the motor
coach had been equipped with an EDR that recorded
the parameters recommended by the Board on H-99-
54, investigators would have been able to more clearly
identify the status of the vehicles and the driver's
actions, and more focused safety recommendations
may have resulted.

In 1998, the Safety Board issued Safety Recommendations H-98-23 and -26 to trucking industry organizations, recommending that they advise their members
to equip their vehicles with on-board recording
devices to gather information on both driver and vehicle operating characteristics. These recommendations focused on two issues. The first issue was the
recording of driver data for the purpose of hours of
service compliance, which was outside the scope of an
EDR. The second issue focused on the recording of
vehicle information, as with an EDR. Unfortunately,
the trucking industry groups did not choose to take
the recommended action, and the recommendations
have been classified as "Closed-Unacceptable Action."
The Safety Board is disappointed that no action was
taken for the recording of vehicle data and continues
to believe the recording of data on board heavy vehicles remains an important issue.

The formation this past year of the P1616 Working
Group by the Institute of Electrical and Electronic
Engineers, Inc. (IEEE) is another promising step in the
implementation of EDRs. The Safety Board fully supports the development of a voluntary standard, and
the Safety Board staff is participating in the working
group. The Board hopes that the standard developed

through IEE will provide a framework to promote consistency and availability of EDR data for both light and heavy vehicles.

The Safety Board believes EDRs continue to be a key component in the process to improve highway safety. The issues associated with EDR development and deployment will be an important part of a Topical Technical workshop (TOPTEC) being co-sponsored by the Safety Board and the Society of Automotive Engineers. The TOPTEC will focus on recorders in all modes of transportation and is expected to take place June 4–5, 2003, in Alexandria, Virginia. The Board will keep NHTSA informed as the details are finalized. We hope that NHTSA will actively participate in this effort.

We appreciate the opportunity to comment on this important issue and urge NHTSA to continue pursuing the effective implementation of EDRs on highway vehicles.

DOCKET # NHTSA-2002-13546-60 02/28/03 Alliance of Automobile Manufacturers (AAM)[2]

The Alliance of Automobile Manufacturers, Inc. website noted, on the date of this submission, that it is a trade association composed of 10 car and light truck manufacturers with about 600,000 employees at more than 250 facilities in 35 states. Alliance members account for more than 90 percent of U.S. vehicle sales. Formed in 1999, the Alliance serves as a leading advocacy group for the automobile industry on a range of public policy issues. Open to all new car and light truck manufacturers, this industry association is especially committed to improving the environment and motor vehicle safety. Through the Alliance, members are able to convey this commitment as well as the industry's accomplishments and its positions on issues to the public, government, media and other interested parties.

AAM commented to NHTSA:

> The Alliance of Automobile Manufacturers (the "Alliance"), whose members are: BMW Group, DaimlerChrysler, Ford Motor Company, General Motors, Mazda, Mitsubishi Motors, Nissan, Porsche, Toyota, and Volkswagen, offers the following comments in response to the notice referenced above which requests comments on what future role the National Highway Traffic Safety Administration (NHTSA) should take related to the development and installation of event data recorders (EDRs) on motor vehicles.

Alliance members recognize that EDRs have the potential of field performance data, roadway designs, and emergency response systems. It is possible that EDRs could improve existing safety databases both with respect to the accuracy of existing data elements, and through the addition of select new data elements that are not currently available.

While recognizing the potential benefits of EDRs, the Alliance believes it is important for NHTSA to address the following points in any evaluation of possible rule making to these devices. Although it is reasonable to anticipate that improved databases could lead to enhanced motor vehicle safety, such safety benefits have not yet been quantified or verified.

A number of manufacturers are voluntarily expanding the installation of EDRs in motor vehicles. The development and introduction of EDRs is proceeding even without regulatory intervention.

Alliance members, NHTSA personnel, and others are actively involved in EDR standards-writing activities. For example, an SAE task group has recently been initiated to develop an EDR standard for data output compatibility and download protocols. ISO is also working on standardizing impact severity recording. These standards-setting efforts should be allowed to continue as NHTSA considers regulatory action.

EDR technology and applications are continuing to evolve to allow collection of vehicle related parameters before, during, and after a crash event. The desire for more detailed medical diagnostic information goes well beyond the EDR data recording function.

Concerns over consumer and public acceptance of EDRs need to be taken into consideration, whether EDRs are provided voluntarily by vehicle manufacturers or required by the government. The recording of information by EDRs may raise a number of potential privacy issues when the information can be associated with an identified or an identifiable person. These could include the questions of who may have rights to the information that has been recorded, the circumstances under which various parties may obtain that information, and the purposes for which that information may be used. We believe that de-identified or aggregate information that is not associated with specific individuals, such as used to augment public safety databases, does not raise the same kinds of potential privacy concerns. Ultimately the resolution of these issues will likely not be in at the purview of the vehicle manufacturer. Voluntary implementation of EDRs may facilitate the development of consensus on how best to manage the legal and privacy implications of EDRs by allowing the policies to evolve through public debate and judicial precedent on the basis of real-world experience.

The alliance encourages NHTSA to continue to monitor the implementation of EDRs, and to remain active in the ongoing standards-writing activities. We also believe that the agency has an important role to play in defining how to incorporate EDR data into FARS, NASS, and state crash databases.

DOCKET # NHTSA-2002-13546-62 02/28/03 Advocates for Highway and Auto Safety (Advocates)[3]

The Advocates for Highway and Auto Safety is an alliance of consumer, health and safety groups and insurance companies and agents working together to make America's roads safer.

Advocates commented:

Advocates for Highway and Auto Safety (Advocates) files these comments in response to the National Highway Traffic Safety Administration's (NHTSA) notice and request for comments regarding Event Data Recorders. The main issue addressed in the notice is the "future role the agency should take related to the continued development and installation of EDRs in motor vehicles." 67 FR 63493 (Oct. 11, 2002). In addition, the agency notice poses a series of questions on EDR-related topics.

NHTSA is a regulatory agency with statutory authority to protect the public by mandating safety standards, conducting research, and engaging in data collection that will enhance safety and reduce traffic crashes on America's roads and highways. It is the agency's duty to advance public safety by requiring performance-oriented standards, and the use of equipment, that reflect the state-of-the-art in vehicle engineering or design. The agency's authority extends to requiring the use of established or emerging technology. Advocates is convinced that NHTSA should exercise its authority

to require the installation of EDRs in all newly manufactured motor vehicles. The agency must exercise its
discretion and authority to bring uniformity and standardization to real-world crash data collection in
order to promote the public's interest in improved
highway safety. To fulfill that end, the agency should
determine the use or uses for which EDR data can provide a safety benefit, *e.g.*, vehicle and highway
research, emergency medical response, heavy truck
operation, and establish appropriate requirements for
the collection of a set or sets of data elements to further each purpose. The agency will also have to adopt
technical standards for data recordation, storage, and
accessibility that provide a simple, universal means for
EDR data retrieval. Finally, the agency will have to
ensure that EDR data is collected by state and local law
enforcement agencies for transmission to a national
database.

Potential Value of Event Data Recorders: Research literature and practical experience make it abundantly
clear that data obtained from EDRs after crashes and
near-crash events can be used to substantially
improve traffic safety. This type of data is already being
used by vehicle manufacturers to confirm the proper
functioning or failure of specific items of vehicle
equipment, and to adjust the performance of air bags
and restraint systems.

Data obtained from EDRs in individual crashes also
provides detailed, objective information for crash
reconstruction investigations. Critical information
such as vehicle speed and change in velocity, seat belt
use, braking, and timing of air bag deployment can be
ascertained from EDR data rather than estimated
based on post-crash eyewitness accounts or conclusions derived from circumstantial evidence. The specificity of EDR data will increase the capacity for
detailed research and investigation of particular types
of crashes and special investigations. EDR data would
enhance the specificity of the information available
for analysis, for example, as part of the National Automotive Sampling System—Crashworthiness Data System (NASS/CDS). NHTSA has already begun collecting
EDR data in NASS/CDS investigations.

On a broader scale, a nationwide database of crash-
specific vehicle performance and driver input information can provide a real-world basis for future safety

initiatives and vehicle design improvements. Aggregate data collected from EDRs across a large number of crashes and near-crash incidents can support research for safer vehicle and highway and improved crashworthiness countermeasures, and to guide the future development of crash avoidance technology. Such data can augment the Fatal Analysis Reporting System (FARS) by providing greater specificity and detail about fatal crashes to complement the data already collected in FARS.

In more immediate terms, availability of EDR data will improve emergency medical response for crash victims. EDR data for certain crash parameters, such as vehicle speed at time of impact, occupant belt use, crash mode or angle of impact, yaw, etc., provide critical information that can expedite emergency medical response to a serious crash. Immediate access to such crash information facilitates appropriate medical response at an earlier point in time. Treatment-related information should be accessible to medical responders at the scene, or by non-medical law enforcement personnel who can transmit the data to medical response dispatchers. Eventually, treatment-related EDR data could be transmitted by Automatic Crash Notification (ACN) systems in order to ensure that the emergency medical responders are aware of pertinent crash-specific information.

The potential safety benefits of EDR technology were summarized in the findings of the NHTSA EDR Working Group on light vehicle data recording: *Event Data Recorders, Summary Findings by the NHTSA EDR Working Group, Final Report, Findings, Section 11*, p. 67, U.S. DOT (Aug. 2001) (NHTSA EDR Report). The report states that EDRs hold promise of potential safety applications for all types of motor vehicles. According to the report, real-world crash data would be extremely helpful in conducting research to evaluate crash avoidance countermeasures, improving occupant protection systems, and monitoring safety equipment and safety systems. The report further states that by providing accurate data about the vehicle systems, the driver and occupants during pre-crash, crash, and post-crash events, EDR data will be provide a basis for developing safer vehicles and reducing crash-related deaths and injuries.

Event Data Recorders Are Already In Widespread Use: In one form or another, EDRs are already in widespread use in transportation. EDRs are currently mandated or are otherwise used in several modes of transportation. Flight data recorders, a type of EDR specifically designed for use in commercial and passenger aircraft, have been mandated for use for over four decades. Likewise, "black box" recording instruments are required on railroad locomotives and are now used in merchant and passenger shipping. In addition, the National Aeronautics and Space Administration (NASA) uses multiple sensors and flight and data recorders on the Space Shuttle. These recording instruments have two purposes: 1) in accident reconstruction to assist in identifying the cause or causes and contributing factors in a catastrophic failure and, 2) in research to identify trends or patterns that could improve operation and design to avoid future crashes. Thus, EDR-type technology is being used to improve safety for transportation modes that involve high levels of financial investment, but comparatively low levels of fatalities and personal injuries. Installation of EDRs has not been required on motor vehicles even though highway traffic crashes annually account for almost 95 percent of transportation fatalities and 98 percent of transportation injuries in the U.S. From a safety perspective, the use of EDRs on both light and heavy motor vehicles holds far more potential to save lives and prevent crashes than all other transportation modes combined.

In fact, millions of motor vehicles are already equipped with sensors, diagnostics, electronic control modules and computer processing hardware/software that record one or more aspects of vehicle and equipment operation including air bag deployment, seat belt engagement, vehicle speed, engine revolutions, brake status and use, as well as other vehicle equipment conditions and operational status. Many of the data recording devices currently in use are component-specific, *e.g.*, air bag and traction control system status sensors and diagnostic modules. However, vehicle manufacturers are also linking equipment sensors located throughout the vehicle to a central memory to more comprehensively record and store a wide range of vehicle component status and operating data. As a result, and to varying extents, EDR technology has been installed in some form in all current production vehicle models, either with or without the knowledge of the original purchaser. Manufacturers of light vehicles have access to the EDR data with the agreement of the owner when repairing vehicles, after repurchasing a vehicle post-crash, or through litigation. Thus, without any federal requirement, EDRs have become ubiquitous on a voluntary basis, and the data is being accessed and used by vehicle manufacturers for research, repair and other purposes.

At the present time, the collection and use of EDR data serves private interests. The use of EDR data to advance public safety is only a secondary consideration. NHTSA collects EDR data, when and where available, as part of the NASS/CDS investigations and the NHTSA Special Crash Investigations (SCI). EDR data could also support research being conducted under the auspices of NHTSA as part of the Crash Injury Research and Engineering Network (CIREN). However, the agency privately contracts with each vehicle owner to obtain access to the EDR data.

In addition, manufacturers of heavy trucks and other commercial motor vehicles voluntarily install on-board recorders, which are EDR-type technology commonly referred to as "black boxes," to serve numerous purposes of their customers, especially fleet operators, including real-time tracking of shipments, routing, recordation of hours-of-service requirements, engine on/off time, and to record and monitor other aspects of vehicle and driver operation. For commercial vehicles the potential use and data collection includes crash-related information such as would be collected by an EDR. But the application for data collection in commercial vehicles is broader than passenger vehicles, extending to the type of real-time information that is collected by an on-board recorder, which is not being considered for non-commercial vehicles. Since EDRs are required in several modes of transportation, and are being voluntarily introduced in light and heavy motor vehicles, the benefits of this technology should be harnessed to improve traffic safety for the public.

Since EDRs have been voluntarily installed and are being used in many vehicles, the data acquired from EDRs has already been used in crash investigation and reconstruction, as well as in litigation. Societal concerns regarding EDR data ownership and privacy rights have already been raised. Thus, regardless of whether EDRs are voluntarily introduced in greater numbers by manufacturers, or are required by federal regulation, the issues of privacy, ownership, control and use of EDR data have already been broached. Although NHTSA cannot resolve contentious legal issues and general privacy concerns that the existence of EDR data has spawned, the agency can ensure that EDR data can be collected to serve the public interest in improving traffic safety.

The Role of the National Highway Traffic Safety Administration: In the National Motor Vehicle and Traffic Safety Act of 1966 (Safety Act), Pub. L. 89-563 (Sept. 9, 1966), Congress directed the Secretary of Transportation, and by designation NHTSA, to conduct research and collect information to carry out the purposes of the Safety Act. 49 U.S.C. _ 30168(a). The law specifically states that the research shall include collecting data to determine the relationship between the performance of motor vehicles and motor vehicle equipment and vehicle crashes, deaths, and injuries that occur in motor vehicle crashes. This research authority was intended to permit gathering information that can improve traffic safety. NHTSA's role is to accomplish that purpose. Thus, since EDR data can provide real-world data on the precise question of the relationship between motor vehicle performance and crashes, deaths and injuries, requiring the installation of EDRs in all new light and heavy motor vehicles, and the collection of EDR data in a national database, will further the research goals of the statute.

The National Transportation Safety Board (NTSB) evidently believes that NHTSA should take the initiative to collect EDR data. NTSB has issued a recommendation to NHTSA to "develop and implement, in conjunction with the domestic and international manufacturers, a plan to gather better information on crash pulses and other crash parameters in actual crashes, utilizing current or augmented sensing and recording devices." NTSB Recommendation H-97-18 (1997). NTSB has since issued similar recommendations to NHTSA regarding data collection for buses. NTSB Recommendation H-99-53 and H-99-54 (1999).

In addition, NHTSA has issued two reports regarding the use and potential benefits of EDRs in both light vehicles, NHTSA EDR Report, and heavy motor vehicles, *Event Data Recorders, Summary Findings by the NHTSA EDR Working Group, Volume II, Supplemental Findings for Trucks, Motor coaches, and School Buses, Final Report*, NHTSA, DOT HS 809 432 (May 2002). The reports made clear that, at current levels of installation, there is an insufficient number of EDRs in vehicles that are involved in crashes to provide enough data for analysis. The report on light vehicles also found that the lack of uniformity in the kinds of data that are being recorded in different EDR systems, as well as the lack of a common means of access to the data, are significant obstacles to the acquisition and use of real-world crash data to improve safety.

In addition, the Federal Motor Carrier Safety Administration has issued a report on data collection by electronic control modules in commercial vehicles. *A Report to Congress on Electronic Control Module Technology for Use in Recording Vehicle Parameters During a Crash*, FMCSA, DOT-MC-01-110 (Sept. 2001).

NHTSA's role as the federal traffic safety agency must be to ensure that EDR technology can be used to further national safety goals. Federal regulation requiring vehicle manufacturers to install EDRs that will record and collect crash data is well within the jurisdiction and authority of the agency provided by its enabling legislation. Agency regulations would have to address a number of issues and different aspects of EDR technology. NHTSA must require that all new vehicles, including light vehicles, heavy trucks and buses, be equipped with EDRs that collect a specified minimum set of data elements. Furthermore, agency rules must require that access to EDR data must be by a uniform technology or interface.

Requiring EDRs in all new vehicles is both necessary and practicable to ensure the type and amount of real-world crash data that would be useful for safety research purposes. The NHTSA EDR Report points out, however, that "[t]he degree of benefit is directly related to the number of vehicles operating with an EDR and the current infrastructure's ability to use and assimilate these date." NHTSA EDR Report at 67. NHTSA's leadership is needed for precisely these reasons, to ensure that all vehicles are equipped with EDRs, and that minimum, uniform data is collected for analysis. If the goal of a national crash database is appropriate to assist NHTSA in fulfilling its safety mission, then a regulation requiring EDRs is within the authority of the agency. Indeed, this is precisely the type of minimum performance requirement Congress expects NHTSA to take on its own initiative in order to ensure sufficient information is being recorded and collected for a national crash research database to fulfill the agency's research mission and to improve vehicle and highway safety.

Regulatory action by NHTSA is necessary to bring uniformity and coherence to the technological issues mentioned above. The agency must establish a performance standard for the set of data elements that all EDRs should collect. At present, each manufacturer collects different types and amounts of data, some only bring bag modules, others from a wide variety of vehicle sensors and equipment. There is currently no agreement among manufacturers regarding what data should be recorded by EDRs.

In order to ensure that each vehicle records information that would be valuable for crash research and analysis, it is necessary for NHTSA to prescribe, by rule, a minimum required EDR data set.

In addition, each vehicle manufacturer currently installing EDRs uses a different proprietary system for accessing the data installed in vehicles produced by that manufacturer. The need to use unique hardware and/or software to download the data prevents others from acquiring that data without the specific proprietary technology. While such a system may serve the needs of individual manufacturers and to some extent may protect individual ownership and privacy interests, it presents an obstacle to general data collection and analysis for public safety research. The specifications for a uniform interface are best developed by technical experts in an open forum. NHTSA can either consider adopting a pre-existing technical standard for a uniform interface, or develop its own standard. In either case, the standard should be promulgated through public meetings and informal rulemaking.

NHTSA already has sufficient authority to require installation and collection of EDR data and no additional legal authority is needed. Neither should NHTSA have to seek permission of the owner to obtain EDR data in each and every crash or investigation. Currently, state and local law enforcement officers have the legal authority to investigate crashes, obtain information, conduct test, and obtain physical evidence. Police reports, and FARS data, are generated in this manner based on existing state and local police powers and authority. It is likely that existing authority would also cover access to EDR data, although in some states further legislation or clarification of police investigative authority may be required. NHTSA can require, by regulation and through existing cooperative agreements, that state and local officials provide EDR data for a national crash information database. Just as with FARS data, the agency can ensure that state and local authorities will transmit EDR data to the agency for deposit into the national database. However, in order to accomplish this end, the agency must first ensure that EDR data can be retrieved easily and conveniently through uniform technology, and that state and local officers can, for example, download EDR data into a laptop computer or a hand-help personal digital assistant (PDA), where it can be stored for later transmission to the national database.

Finally, as is discussed in the following section, NHTSA must assure the public that the use of EDR data by the agency will, as with FARS, be coded in order to protect individual privacy, and to assure the public that EDR data will be subject to the restrictions of the federal Privacy Act. While the issues involved in requiring EDR technology and data collection raise complex technical and legal concerns, agency regulations to accomplish these goals are well within its research authority for collection of crash data.

Advocates suggests that NHTSA begin rulemaking in the near future to:

1. Determine what information, at a minimum, should be collected;

2. Obtain information regarding a uniform EDR interface;

3. Determine technical specifications for recording and storing EDR data;

4. Gather information regarding the creation of a national EDR database that would be available to the public.

Any national database should contain only coded information that protects the privacy of the individuals involved in reported crashes.

Initially, NHTSA should collect EDR data in all fatal crashes, which would also be used to augment existing FARS data. However, since fatal crashes are not necessarily representative of all crashes in which serious injuries occur, the agency should collect EDR data in all crashes that result in the hospitalization of one or more persons, including non-occupants. Ultimately, collection of EDR data would, if possible, extend to all tow-away crashes. Such a national database would provide the agency, the public, and the research community with a rich source of real-world data and information concerning crash events.

Concerns Regarding Privacy Are Not An Impediment To NHTSA Regulation: The voluntary introduction EDR technology has given rise to privacy concerns regarding the use of vehicle data. Public concern about the use, or misuse, of such data arises in a number of contexts, including access to and use of the data by emergency medical responder, police investigators, the judicial system, auto manufacturers, motor carriers, insurers, and others. These privacy concerns derive from the existence of EDR data which, to date, have been developed and collected as the result of voluntary actions by motor vehicle manufacturers. Regardless of federal regulation, privacy issues will remain a concern as more and more diagnostic and event data are recorded by vehicle systems and EDRs. Thus, the privacy concerns regarding EDRs are separate from, and independent of, any eventual federal regulation requiring the collection of EDR data.

Assuming that vehicle owners also own any data in the vehicle, their privacy right in the EDR data may be protected in everyday situations. However, vehicle owners may have to give up the right in order to obtain vehicle service or repair, if access to EDR or equipment diagnostic is necessary to provide service. Moreover, privacy rights will likely have to yield in the event of a crash investigation or litigation. A federal regulation requiring installation of EDRs in every motor vehicle, and requiring that certain data be collected, may increase the amount of data that is available and contested, but it will not affect the legal rights of parties to obtain available information.

NHTSA has no jurisdiction or authority to prevent the use of EDR data for many purposes that the public may consider intrusive. For example, NHTSA has no authority to prohibit the use of EDR data from use as evidence by the police during a crash investigation. Neither can NHTSA restrict the use of the EDR information required by legal process or judicial order. The agency can only ensure that EDR data it obtains for use as part of a national data base will be coded to protect individual privacy, and the privacy will be protected when the data is deposited in a national database maintained by the agency.

Ownership of EDR Data Is Not An Impediment to NTSA Regulation: Finally, although NHTSA needs to address the legal question regarding the ownership of EDR data, the agency only needs to clarify the existing legal right of state and local authorities to acquire relevant information in the course of a traffic crash investigation. Since NHTSA will acquire EDR data for the purpose of population to a national crash information database, the agency will generally receive information that has previously been obtained by state and local law enforcement officials pursuant to an investigation following a crash, or other near-crash event (*e.g.*, air bag deployment), to which the police have responded. As with any other information currently collected for police traffic incident reports, state and local law enforcement personnel will obtain EDR data in the course of pursuing a traffic investigation. In such cases, law enforcement officials have the authority to inspect and photograph the vehicle or vehicles involved, interview witnesses, conduct tests of the vehicle or equipment, and take other actions, including impounding the vehicle, either at the crash scene or afterwards. Thus, the information received by NHTSA for FARS is obtained by state and local officials pursuant to state law providing those officials with investigative powers at traffic crash scenes. EDR evidence would be obtained in the same manner, and requires no new federal legislation or agency authority.

Federal law authorizes federal grants to the states and interstate authorities to carry out motor vehicle safety research. 49 U.S.C. § 30168(a)(2).

EDR Data Collection for All Vehicles and Emergency Medical Treatment: Advocates supports collecting

EDR data for all types of vehicles, with the understanding that the data collected to provide a sufficient amount of information needed for safety analysis of crashes involving passenger vehicles will differ from the data collection need for buses and medium and heavy trucks. As mentioned above, far more data is already being collected voluntarily in commercial vehicles to increase efficiency and allow for greater monitoring and management of commercial fleets. The two NHTSA EDR Working Groups for light vehicles and for buses and trucks, both concluded that universal installation of EDRs that collect a minimum set of data and that can be accessed by a uniform technology is reasonable, practicable, and appropriate for each type of vehicles. Therefore, requiring similar EDR technology that collects different types and amounts of data is within the authority of the agency.

The data collection required to expedite emergency medical response after a crash, however, presents a different set of issues. This type of data would have to be required in all vehicles, including potentially, motorcycles. NHTSA can and should ensure that treatment-related EDR data is collected so that it can be available for emergency medical response. This presents both a technological issue of whether data useful for medical treatment should be physically separated from other, non-medically useful data, and what legal and privacy rights are involved in divulging such data to emergency responders.

With respect to the technological issue, it may be possible to cordon off a subset of data that is relevant to medical treatment from other data, so that emergency responders only access the data they need to provide treatment. Alternatively, it may not be feasible or, more likely, cost effective, to devise separate EDR data storage for limited purposes. In that event, if emergency medical responders are given access to EDR data they will obtain data that is medically relevant and data that is not. This creates concerns regarding data that is not relevant for medical treatment but that has been divulged to third parties. NHTSA can require separate collection of medically relevant data, however, the legal issues and privacy rights regarding access to medical and other data is the jurisdiction of state legislatures. States could by law explicitly sanction the release of such data for emergency treatment and provide appropriate privacy safeguards for the use of the data in that manner, *e.g.*, confidentiality or subject protection as privileged information. The legal issues and privacy concerns regarding EDR data in general are already being raised as a result of the voluntary introduction of this type of data. The use treatment-related data becomes available and accessible. These issues will be sorted out by society and are not within the purview or authority of NHTSA to decide by regulation. NHTSA's role is to ensure that technology and data that are otherwise being used for other purposes are harnessed to improve traffic and public safety.

DOCKET # NHTSA-2002-13546-63 02/28/03 Association of International Automobile Manufacturers, Inc. (AIAM)[4]

AIAM is the voice of international automobile manufacturers in Washington, D.C. It is unique, respected and independent. The motor vehicle industry is one of the most highly regulated sectors of the United States economy. International automakers designing, producing, marketing and selling high-quality products in the United States must ensure that they are full partners with legislative and executive branches of government as they deliberate on important issues affecting the industry. AIAM assures that strategic representation by providing its members with timely information, thorough analysis, expert guidance and forceful advocacy on those issues that vitally affect the motor vehicle sector.

AIAM commented:

The Technical Affairs Committee of the Association of International Automobile Manufactures Inc. (AIAM) provides the following comments in response to the NHTSA Request for comments on Event Data Recorders (EDRs) (67 FR 63493). Some vehicle manufactures currently install what may be characterized as EDRs on certain of their vehicles. Aftermarket versions of EDRs are also available for installation by fleet owners and private vehicle owners. At present, the term "EDR" is imprecise when applied to motor vehicles, as it may refer to basic systems that collect data only from the air bag module or to complex systems that might include data on the status of other vehicle systems (*e.g.*, braking, steering, transmission), global positioning system information, and audio and video recordings. Whether a vehicle has a form of EDR, what and how much data an EDR collects, and how that data is stored and downloaded varies by vehicle, vehicle type, and vehicle application, due to differences in vehicle technology, use, and owner interests.

Some EDR data may provide incremental safety benefits by improving the data available for highway safety research, but it is not necessary that all vehicles be equipped with an EDR to accomplish that goal. As noted in the Request for Comment, NHTSA currently supplements the National Analysis Sampling System/ Crashworthiness Data System (NASS/CDS), Special Crash Investigations (SCI), and Crash Injury Research and Engineering Network (CIREN) crash investigations with EDR data when it is available and when granted permission by the vehicle's owner.

EDR data can best inform research when it is considered along with other data collected as part of a crash investigation. NHTSA currently makes this type of comprehension data available to researchers through the NASS/CDS, SCI, and CIREN programs. These NHTSA databases have defined protocols for case selection and data collection, which provide the comprehensive information needed for an analysis of specific crashes within the context of motor vehicle safety. In an October 17, 2002, letter to NHTSA, AIAM and others encouraged the Agency to fully fund the NASS/ CDS program to ensure that sufficient data can be collected to produce a statistically valid, nationally representative sample.

Regarding Automatic Collision Notification (ACN) systems, EDR data can only supplement the most critical information for emergency responders, which is that a crash has occurred and the location of the vehicle. Any incremental benefit provided by EDR data to whatever benefits may exist for ACN systems will depend on whether the infrastructure is in place to receive, interpret, and communicate this data, and the ability of emergency responders to use this information to make meaningful adjustments to the manner and time in which they respond to a motor vehicle crash. The application of EDRs in the automotive environment is evolving, and it is simply primitive for NHTSA to undertake regulation of EDRs at this time. As more EDRs are placed in the vehicle fleet, more will be learned that could help evaluate alternative applications (*e.g.*, choice of data elements, data collection period). Rather than regulate this emerging application now and locking in specific technical requirements, manufacturers should be allowed to develop systems of their own and work with voluntary standards organizations as a means of achieving a consensus on at least some of the many unresolved technical issues.

Because it is not clear today which direction(s) in the development of event data recorders and automatic crash notification will yield significant and realistic benefits, AIAM recommends that NHTSA take a leadership role to establish a strategic direction in these areas. AIAM believes that NHTSA is in a unique position to bring together key participants in this area: crash data investigators (including the NHTSA SCI and CIREN programs), the emergency medical response and treatment community, the vehicle and component manufacturers, the legal professions, the general public and city/state/Federal agencies. Given the maturity of the technologies available to provide affordable and high volumes EDR and ACN features, everyone could benefit from more coordination between beneficiaries and providers and more standardization. Without a clearer vision on the best direction for EDR and ACN development and implementation over the next 10–15 years, gaps and incompatibilities between vehicle features, inadequate emergency medical infrastructure, and conflicts among the users of EDR information will likely negate much, if not all, of the promise of better post-collision treatment and improved future designs driven by real-world data.

On February 28, 2003, the Electronic Privacy Information Center (EPIC) respectfully submitted these comments on the National Highway Traffic Safety Administration (NHTSA)'s role in the development and installation of Event Data Recorders (EDRs), or "black boxes," in motor vehicles. Our comments focus on the privacy implications of EDR technology. [5]

EPIC noted:

We recommend that, in order to respect the privacy interests of drivers, the collection of driving-related information through EDRs must follow Fair Information Practices, including obtaining unambiguous or "opt-in" choice from drivers to collect such data. With respect to the proposed EDR database compiled by NHTSA, we recommend that in addition to complying with the letter and spirit of the Privacy Act of 1974, any such database be constructed with the goal of preserving the privacy of drivers so that only aggregate information is collected and made available to third parties.

EPIC is a non-profit research and educational organization that examines the privacy and civil liberties implications of emerging technologies. Our experience in the field has shown that the most effective way to tackle emerging threats to privacy posed by new technology is to craft strong, technologically neutral standards to protect the privacy interests at stake. The comments focus on the key privacy issues implicated by the use of EDR technology and suggests a policy framework of Fair Information Practices to effectively protect the important interests at stake.

Event Data Recorders: Event Data Recorders (EDRs) are electronic "black boxes" that collect and store information about the operation of a motor vehicle. The data recorded might include the date, time, velocity, direction, number of occupants, air bag data, and seat belt use. The devices might even include location data, which would raise additional significant privacy issues. In addition, there are open questions about how the data can be accessed, recorded and transmitted. There are several different types of EDRs in the market ranging from the Vetronix system, which is installed in cars produced by General Motors, to the more elaborate MacBox system currently being tested by the Drive Atlanta project at the Georgia Institute of Technology. Each type of device collects different kinds of data for different purposes. The NHTSA has attempted to limit the definition of EDRs in the request for comments, but this does not address the public concern about these devices, as the different types of EDRs are available in the market. Any limitation of the purpose of EDRs must be part of a broader privacy protection framework, as we argue below.

Advocates of EDR technology suggest that the information might be useful in accident reconstruction and developing safer vehicles through "real world" testing. Insurance companies want the data to settle claims expeditiously. These companies, along with car rental agencies, and others have also demonstrated interest in obtaining this data in support of efforts to control driving behavior through surveillance.

The former head of NHTSA, Dr. Ricardo Martinez, who now runs Safety Intelligence Systems Corporation (SISC), wrote a letter asking NHTSA to consider mandating the use of EDRs. SISC, which was formerly Loss Management Systems, Inc., aims to find cost effective ways to service insurance claims and simplify investigation and litigation procedures. It has entered into a partnership with IBM and the Insurance Services Office, Inc. to promote a global auto-crash database that envisions a point where "information can be automatically and instantly transmitted from cars" to a centralized database.

SISC supports the MacBox technology behind the Drive Atlanta program at Georgia Institute of Technology. The project also receives funding from NHTSA. The MacBox records location data, voice, and video images of a vehicle in addition to information about the vehicle's operation. It also uses Global Positioning System (GPS) and cellular technology to transmit information about the car back to a central command center. One of the principle researchers behind the project, Dr. Jennifer Ogle, co-authored a paper discussing the potential use of EDR technology in insurance. The study examined the use of variable insurance premiums designed to "discourage risky driving behavior." The report says that, "For example, premiums may increase significantly for vehicle activity above 65 mph, accelerations over 8 mph/second, etc. "The system also tracks how much a person travels to adjust insurance premiums accordingly. The aim is to both punish driving patterns that are considered to be "risky" and to modify driving behavior through the constant surveillance enabled by EDR technology.

Clearly the privacy implications of such a monitoring program are significant. The project currently is being tested with volunteers who are fully apprised of the technology and have consented to being monitored. If EDR technology is mandated by the NHTSA or becomes required through the coercive pricing of insurance rates, there needs to be a very strong set of privacy safeguards established to protect the interests of drivers before any such technology becomes widely deployed.

Significance of Automobile Privacy in American Culture. Over the past century, ownership of automobiles has expanded from being the privilege of the very elite to becoming essential to the transportation and liveli-hood of most Americans. Regulating the use of auto-mobiles has become crucial to safety on roads, but along with these regulations, the regulation of private uses of automobiles presents risks to individual pri-vacy. Any intrusion on automobile use has been predi-cated on the government articulating public safety goals. The use of EDR technology must follow a similar model where any public safety goal is first proven to be necessary and effective and then must be used only in a manner that minimally infringe on the rights of individuals.

Creating A Fair Information Practices Privacy Architec-ture: The privacy issue concerns not just who "owns," *i.e.* controls the use of the data (which should be the operator of the vehicle), but the entire set of informa-tion practices, including how the data is collected, processed, transmitted and stored. The Organization for Economic Cooperation and Development (OECD) developed robust Privacy Guidelines in 1980, which have been adopted by several countries, government agencies, and corporations. These guidelines would provide an effective framework for addressing the pri-vacy issues surrounding EDRs as they provide strong, technologically neutral privacy rules.

The Guidelines incorporate eight core principles:

1. Collection Limitation Principle: There should be limits to the collection of personal data and any such data should be obtained by lawful and fair means and, where appropriate, with the knowledge or consent of the data subject.

2. Data Quality Principle: Personal data should be rele-vant to the purposes for which they are to be used, and, to the extent necessary for those purposes, should be accurate, complete and kept up-to-date.

3. Purpose Specification Principle: The purposes for which personal data are collected should be specified not later than at the time of collection and the subse-quent use limited to the fulfillment of those purposes or such others as are not incompatible with those pur-poses and as are specified on each occasion of change of purpose.

4. Use Limitation Principle: Personal data should not be disclosed, made available or otherwise used for

purposes other than those specified in accordance with Purpose Specification Principle except:

(a) with the consent of the data subject; or

(b) by the authority of law.

5. Security Safeguards Principle: Personal data should be protected by reasonable security safeguards against such risks as loss or unauthorized access, destruction, use, modification or disclosure of data.

6. Openness Principle: There should be a general policy of openness about developments, practices and policies with respect to personal data. Means should be readily available of establishing the existence and nature of personal data, and the main purposes of their use, as well as the identity and usual residence of the data controller.

7. Individual Participation Principle: An individual should have the right:

(a) to obtain from the a data controller, or otherwise, confirmation of whether or not the data controller has data relating to him;

(b) to have communicated to him, data relating to him within a reasonable time; at a charge, if any, that is not excessive; in a reasonable manner; and in a form that is readily intelligible to him;

(c) to be given reasons if a request made under sub-paragraphs (a) and (b) if denied, and to be able to challenge such denial; and

(d) to challenge data relating to him and, if the challenge is successful, to have the data erased, rectified, completed or amended.

8. Accountability Principle: A data controller should be accountable for complying with measures, which give effect to the principles stated above.

There are different EDRs in the market that collect varying amounts of information. A clear purpose specification, for example, would determine what information needs to be collected and would limit the uses of the data for surveillance. There need to be clear guidelines for how the data can be accessed and processed by third parties following the use limitation and openness or transparency principles. Similarly the data quality principle and the security principle provides guidance on the standards for protecting transmission of the information from the vehicle and how the data should be handled to ensure that there is a robust audit trail. The NHTSA needs to conduct further analysis to develop appropriate guidelines following the Fair Information Practices framework. Even if the NHTSA were not to mandate the use of EDR technology, it should consider developing such a framework for vehicles that do have EDR installed.

The Deutsche Akademie Fuer Verkehrswissen-schaften (German Academy of Traffic Science) has recently released a report on the use of black box data in German courts. The document proposes limits on the collection of information for purposes such as reconstruction of accidents in civil and severe criminal cases, and grants control of the data to the vehicle operator. NHTSA might consider the German approach, and also build on the experience and expertise of the international community in EDR technology while building its own privacy framework.

Proposed NHTSA Auto-Crash Database. The NHTSA should follow the letter and spirit of the Privacy Act of 1974 in developing the auto-crash database. We plan to submit comments on the Privacy Act notice, if and when it becomes available. The proposed nationwide database of auto-crash data should respect the privacy interests of drivers by containing only de-personalized information about automobiles in the event of an accident. The privacy of drivers should be protected using Privacy Enhancing Technologies (PETs) during the collection stage, rather than later at the processing stage where although a database administrator might choose to withhold that information, it might still be subject to disclosure. If a particular person is identified with a record, there must be strict guidelines for giving access to that information following the Privacy Act.

NHTSA must not mandate the use of black box technology without ensuring that strong privacy safeguards are in place to protect the interests of drivers. Indeed, strong privacy safeguards might further any public safety interests the agency has in EDR technology, by promoting adoption of the technology by drivers who do not feel the presence of these devices are a risk. EPIC encourages the agency to engage in further public discussions to develop a Fair Information Practices framework to cover the use of automobile black boxes.[6]

DOCKET # NHTSA-2002-13546-71 Society of Automotive Engineers (SAE)[7]

The Society of Automotive Engineers is a non-profit educational and scientific organization dedicated to advancing mobility technology.

SAE commented:

> As you are aware, the Society of Automotive Engineers (SAE), in addition to its many other functions related to serving the automotive and aerospace industries, continually is involved with the development of Standards and Recommended Practices that are referenced by suppliers and manufacturers during the design, development, manufacturing and testing of products in those industries. I would like to inform you of a new standards development effort within SAE—specifically, the formation of the Vehicle Event Data Interface (VEDI) Committee.
>
> This Committee seeks to develop common data output formats and definitions for selected data elements that may be useful for analyzing vehicle "events," most notably crashes. Further, the Committee seeks to specify common connectors and network communications protocols to facilitate the extraction of such data from the vehicle. The initial emphasis of the Committee's work will be on the development of standards or recommended practices as applicable for light duty vehicles, but similar efforts focused on other vehicle classes is not precluded.

DOCKET # NHTSA-2002-13546-68 Art Kellermann[8]

On March 3, 2003, Art Kellerman, Professor and Chairman of the Department of Emergency Medicine of the Emory School of Medicine and Director for Injury Control of the Rollins School of Public Health of Emory University, Atlanta, Georgia commented to NHTSA.

Kellerman noted:

> Event data recorders are a highly promising and important advance in highway safety. If the capacity is developed for EMTs and paramedics to download EDR data from a vehicle into a PDA (not difficult technically) EDRs could generate accurate information

about the physics and biomechanics of a crash that would be HIGHLY relevant to clinical assessment of trauma patients brought from the scene. Such data could help emergency physicians and trauma surgeons determine when costly diagnostic tests are needed, and equally important, when they are not. This would save health care costs and reduce the risk of missed diagnoses because physicians received inaccurate or imperfect data on the amount of energy, direction of forces, and safety belt use in a crash.

By correlating the final diagnoses, including both patterns of injury and severity, the EDR data, controlling for age, gender, size, weight, comordidites, pretence, etc.—we have the potential to revolutionize biomechanics research to improve vehicle safety. In effect, every crash victim will become a "Vince" or "Larry."

Physician input is essential to this process. Engineering expertise is vital, but clinical input on these standards and their interpretation is EQUALLY vital.

DOCKET # NHTSA-2002-13546-67 03/03/03 Colorado State Patrol[9]

For the last two years, the Colorado State Patrol Accident Reconstruction Team, in cooperation with the National Highway Traffic Safety Administration (NHTSA), has compiled crash data retrieved from event data recorders (EDRs). Prior to 1999, many Reconstructionists were relatively unaware of the integral role EDRs would eventually play in technical accident reconstruction. Recently, however, EDRs have become indispensable tools for Reconstructionists, which has invariably enabled them to derive more precise conclusions following technical accident investigations.

Practical Application: Currently, our accident reconstruction team downloads data from EDRs—if available—following most serious-injury or fatal crashes. To facilitate the successful retrieval of data, Reconstructionists principally rely upon the Vetronix CDR. Following retrieval, data from EDR is then compared to manual crash reconstruction speed estimates and delta v calculations to objectively determine if they are consistent with one another. Comparisons that yield similar findings validate mathematical formulae commonly used by Reconstructionists in technical accident reconstruction. Data which is captured by EDRs are a useful tool for automobile manufacturers, governmental agencies, and accident Reconstructionists alike. Increasing integration of EDRs can only enhance the reliability of technical accident investigations in the future, and perhaps provide crucial data for establishing vehicle safety standards. Expanded use of EDRs also bolsters the inherent reliability of the devices themselves—a frequent source of contention during criminal proceedings. At any rate, EDR data will improve our understanding of accident reconstruction concepts, assist in determining whether additional vehicle safety design standards are needed, and determine whether there are inherent safety defects that may have contributed to a crash.

Education and Training for Reconstructionists and Prosecutors: Education and training on the subject of EDRs is a necessity, particularly for Reconstructionists and prosecutors. As EDRs become more technologically advanced, and perhaps capture increasingly complex data, the need for training and standardization become more pronounced. Without a true understanding of the data—method of capture, for example—and basic principles of crash reconstruction, any intended use of EDR data would only complicate efforts to arrive at a reliable conclusion following a technical accident investigation. Education and standardized procedures for the design and retrieval of EDR data would ensure admissibility in court proceedings and expand our knowledge of a vehicle's crashworthiness. It is imperative that Reconstructionists use EDR data to supplement their technical accident investigations—instead of using the data as a convenient substitute for conventional reconstruction methods.

EDRs and the Right to Privacy: The issue of "Big Brother" using this unit as a means to monitor an individual's driving habits will always be a concern. The issue of ownership and right to privacy is something we continue to research. Since accident Reconstructionists are aware of the useful information these units can provide, we either obtain voluntary consent from the owner or obtain a search warrant to both remove the EDR from the vehicle and retrieve the data. The ends justify the means when looking at the information the unit records if it will increase occupant safety and survivability in real-world crashes. As our world evolves and technology advances, the crashworthiness and inherent design characteristics of motor vehicles require constant evaluation, and EDRs demonstrate the capability of improving our understanding of factors that lead to serious-injury and fatal crashes.

DOCKET # NHTSA-2002-13546-69 03/03/03 Sandhills Community College[10]

On March 3, 2003, the Event Data Recorder (EDR) Survey of 437 College-Age Individuals at Sandhills Community College, Pinehurst, North Carolina was submitted to NHTSA:

> The following Event Data Recorder (EDR) survey, conducted on the Internet in the Sandhills region of North Carolina is hereby transmitted on behalf of Tom Kowalick, a professor at Sandhills Community College in Pinehurst, North Carolina in reference to NHTSA's request for comments, published in the *Federal Register* on October 11, 2002.

> The survey was conducted from November 22, 2002, to February 26, 2003 at [http://198.85.71.78/transpo/]. It represents the views of 437 individuals on utilizing EDRs in motor vehicles. Although there were some respondents from outside the Sandhills geographical region, the vast majority of respondents were college students with an average age of 25 years.

> The survey was designed within recommendations from the White House Office of Management and Budget (OMB) [http://www.whitehouse.gov/omb/] in conjunction with the initiative titled E-government.

> Details of how it was conducted are contained on the cover sheet of the survey. The rationale for the survey is that "vehicle and highway safety is the shared responsibility of government, industry and the public." According to the National Safety Council (NSC) approximately 21 million Americans were involved in a motor vehicle crashes in 2002.

> Thus, the purpose of this survey was to provide the "college-age public" with an opportunity to comment on the pros and cons of EDRs. A secondary purpose is to provide NHSTA with this perception of EDRs.

There were twelve questions. The survey and its results were available for public review as the data was accumulated. Respondents were informed that the survey would be submitted to NHTSA Docket-2002-13456 as a matter of public record. A review list of individual survey takers (alphabetized by last name) and their comments is also provided.

The results, a general report listing the number of respondents who picked each of the different answers offers NHTSA a better understanding of views on the issues of safety potential, vehicle applications, privacy vs. safety, additional safety benefits, role of NHTSA, mandating, access to data, and public vs. proprietary crash data codes. The overall finding of the survey indicates: 1. A willingness amongst the college-age public to implement EDRs within limitations and 2. A strong sense that the college age public perceives the greatest safety potential thru public safety benefits of crash data via open access to public crash data.

Finally, it is fitting to submit this survey in memory of Ashley Lynn Suits of Bear Creek, NC, a SCC student who participated in the survey on 1/15/03 and who perished in a motor vehicle crash on Friday, February 28, 2003. Today would have been her 19th birthday."

The debate continued. There would be choices and speed bumps in the road ahead.

CHAPTER 9

Speed Bumps: March to November 2003

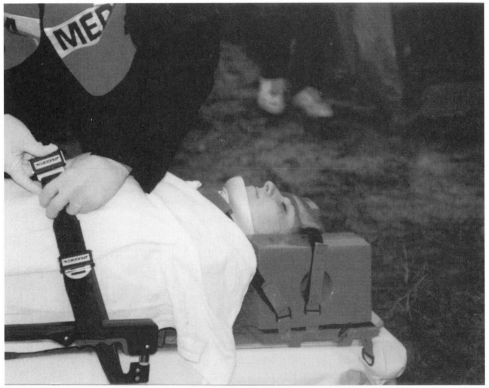

No more important challenge exists than finding ways to improve the safety of everyone while riding in motor vehicles. This occupant suffered permanent brain damage.

There are human tragedies behind road crash statistics. Here is an example.

It has been almost a year now since my life was turned upside down, literally and figuratively. And it will be a long time until I forget my boyfriend and the two good friends who were in the car with me that cold December night. We were returning from a night of partying. My boyfriend Matt was driving and with him up front was my friend Kyle. I was in the back seat with my other friend Jeremy. Yes, we were speeding and drinking—a lot of things we shouldn't have been doing that night. The most important thing we didn't do though was buckle up.

As we came over a hill going extremely fast, I knew this was going to be it. Jeremy and I were holding hands and he said to me, "I love you, you're my sweetheart." And, in that instant, I made a decision that saved my life—I buckled my safety belt. I don't know exactly what made me do it, but I was the only one of us who did.

Matt lost control of the car and we started rolling, over and over again. When the car finally came to a stop I unbuckled myself. I was alone in the car and the engine was still running. I reached up and turned the car off and discovered we had plowed into a ditch in a frozen field on the side of the road. Out in the field I saw my boyfriend Matt. He was lying there like he was sleeping, but he wasn't. I called to him, but he didn't answer. I walked over and knelt beside his head and lifted it up into my hands—I could feel Matt's life pouring away. His blood and brains collected in a pool in my palms. He was gone.

My friend Jeremy was nearby—his neck was broken. It was broken so badly it was tucked underneath his body. And finally there was Kyle, laying there covered in his own blood—gurgling in it. His eyes opened and he was staring straight at me. I put my coat over him and told him to hang on while I got help. But he wouldn't make it either. Strangely, in the midst of this horrific scene, I noticed that Kyle's shoes were no longer on. They had been literally ripped off his feet by the violence of the crash. That was something that really bothered me— one of those flashbulb images you never forget. I ran to a nearby farmhouse covered in my friends' blood and called 911. But there was nothing they could do. Matt and Jeremy both died instantly—Kyle passed away soon after that at the hospital. There were three funerals that week, one after the other. A 17 year-old kid shouldn't have to be planning her boyfriend's funeral—but I did. We were together for five years. The sound of that crash will be with me a lot longer though.

When I tell this story to kids my age, I wish they could get inside my head for just one moment—just one sleepless night when the nightmares come, the nightmare that was the last time I saw my friends alive. If they could get inside my head they'd understand how terribly much I just want to go back and live that night again and make them buckle in those last moments— before the car rolled and everything went black. But I can't go back.

But there are hundreds of people out there—a lot of them just kids that still have a chance to change their fate. They don't have to end up in a ditch bleeding and gasping for their last breaths. I have to live with the memory of seeing my friends like that for the rest of my life. There's not a day that there are still a lot of questions I need to answer for myself, but I come to the realization that the best I can do is live as an example for other kids like me. Kids my age don't like to hear what's good for them, but I'm trying. I'm thinking that maybe if they see me and hear me out that they'll think twice before going for a drive without putting their safety belt on.

Taken from Mary Reinhart's speech before the Wisconsin Department of Health and Family Services—Statewide Trauma Care System Forum, September 2003.

On March 4, 2003, the American College of Emergency Physicians (ACEP) commented to NHTSA:

DOCKET # NHTSA-2002-13546-72 03/04/03 American College of Emergency Physicians (ACEP)[1]

The American College of Emergency Physicians (ACEP) is committed to improving the quality of emergency care through continuing education, research and public education. Founded in 1968, ACEP represents 21,000 members and is the national emergency medicine specialty society recognized by organized medicine. ACEP's 53 local chapters, 21 membership sections and multiple national committees provide expertise and guidance for critical issues in emergency medicine.

Emergency physicians provide professional services on the front lines of emergency medical services and hospital care. Injuries comprise between 20 and 30 percent of hospital emergency department visits, and motor vehicle injuries represent the largest category of fatal and life-threatening injuries, often leading to long-term disability.

The American College of Emergency Physicians has approved policy statements supporting the need for enhanced motor vehicle safety, focused and improved injury control efforts based on national and geographic specific data, and the development and implementation of population-based data systems for injury and trauma care.

ACEP welcomes the opportunity to comment on the issues of crash data and event data recorders (EDRs). These comments will briefly outline the safety benefits, technical issues, and role of NHTSA in this area. While these comments are brief and concentrate on broader issues, ACEP members work diligently in many of these areas in-depth and may add additional insights and guidance for the agency.

Safety Benefits: Motor vehicle injuries are a leading cause of death in America to children, teenagers, young adults, and in the workplace. Motor vehicle injuries are the leading cause of major head and spinal injury, and a major cause of long-term disability.

Understanding the root cause of any disease process is essential to effective prevention and treatment. Unfortunately, despite their societal impact, relatively little is understood about motor vehicle crashes simply because investigations rely on physical evidence observed after a crash. Replacing this process of derived data with measured information captured during a crash will provide tremendous safety benefits to policy makers, vehicles designers, researchers, medical professionals, public safety officials, and others. ACEP's comments will be limited to some of the medical benefits reaped from EDR data.

Injuries to victims are a result of the energy of a crash. Because these victims are predominately younger, their body's ability to compensate means that many serious injuries may be undiscovered until compensatory mechanisms reach their limit. More accurate crash information would allow better injury prediction and help guide diagnostic and therapeutic decisions.

In addition, patterns of injury are associated with different types of crashes and better knowledge of crash kinematics allows the physicians to anticipate possible injuries.

Injury Research: While information about victim's injuries, treatment, and hospital stay are often recorded in depth, the underlying crash severity and type is highly subjective. Measurement of crash severity and direction of force data can markedly improve injury prediction, algorithms, biomechanics research, cost of injury research, and identification of problem injuries.

Injury control research and trauma data banks require improved crash data in order to assess crash causation and epidemiology, evaluate trauma care guidelines, and study long-term consequences and rehabilitation of crash injuries. The Crash Outcome Data Evaluation Systems (CODES) projects have been successful in providing insights to traffic safety, health care, and injury control by linking traffic records with EMS and hospital records. Replacing subjective traffic records data with measured crash severity data will greatly enhance the quality and value of these data sets.

Future Safety Benefits: Recently, ACEP members participated on the Medical Subcommittee of the ITS America Public Safety Advisory Group which released Recommendations for ITS Technology In Emergency Medical Services. That document provides ample examples of future safety benefits of improved crash data including: rapid deployment of Advanced Automatic Collision Notification systems (AACN), improved integration of crash information within emergency medical services and hospitals, improved trauma center and hospital emergency department care, and development of ITS technologies that help prevent crashes.

One strong recommendation of this report called for greater involvement by medical professionals into the development of ITS technologies such as traffic management, automatic crash notification, public safety operations and communications, emergency vehicle routing systems, and data communications networks. Early effective input and involvement by emergency physicians and other medical professionals assures seamless integration of new technology into the existing EMS and health care systems and enhances their lifesaving benefits. ACEP strongly supports NHTSA's leadership in bringing the various stakeholders together on new issues.

Research Databases: Knowledge is power. As stated by NHTSA EDR Working Group, EDRs have the potential to greatly improve highway safety. Crash information will be available as single cases and in the aggregate. The aggregation of crash data, similar to that of medical cases, provides the power of epidemiology, problem identification and effectiveness studies. In order to aggregate information, a minimal data set must be established.

NHTSA currently collects crash information for a variety of databases (FARS, NASS, SIREN, and SCI) that are readily used by our members for research. NHTSA is currently collecting EDR information for some crash investigations on a limited set of vehicles. These efforts should be expanded as vehicle and data availability allow.

Standards, and Data Elements: Everyday in emergency departments around the country, emergency physicians use basic crash information such as crash severity, restraint use, and direction of impact to make critical decisions about patient care. A minimal data set of data elements required for clinical and research applications should be captured by the EDR. Again, a minimum data set allows aggregation of data in meaningful ways.

Location information is critical to a timely emergency medical response. Location information minimizes the time of response, ensures the allocation of appropriate resources, and helps direct response and transport protocols so that the call goes to the right PSAP, EMS system, and hospital facilities.

Many vehicles, especially trucks, may be carrying hazardous materials. When so loaded, information about the type of substance is vital for emergency responders to both protect themselves and manage the consequences of any such crash.

Both the NHTSA EDR Working Group and the ITS Medical Subcommittee have recommended a minimal data set of crash data that would support improved medical response operations and research.

A minimum data set allows the widespread deployment of Automatic Collision Notification Systems as wireless capabilities are added to the vehicle fleet. In addition, on scene response personnel can more easily interrogate a vehicle for crash severity and make trauma decisions based upon measured vehicle forces.

In order to maximize the utility of EDRs, these devices must be crash-survivable and located in vehicles in such a way as to provide easy access to data for emergency responders. Since 1996, U.S. domestic vehicles are mandated to have a universal connector port for download of vehicle power train and emissions data (on board diagnostic-11 protocols). Crash information needed for emergency response and clinical decisions should be as easily accessible.

The Role of NHTSA: NHTSA has been the leader in promoting motor vehicle safety through crash research and vehicle standards. As technologies rapidly enter the vehicle, the NHTSA has the opportunity to enhance motor vehicle safety beyond crash worthiness into the crash avoidance and post-crash arena. A minimum data set of crash information is required for capture across the vehicle fleet. NHTSA needs to develop standards for the types of data that are being recorded and require standard means for accessing the data. The American College of Emergency Physicians looks forward to working with the agency as it brings this vision to reality.[2]

DOCKET # NHTSA-2002-13546-73 03/04/03 University of Miami[3]

The William Lehman Injury Research Center welcomes the opportunity to comment on the value of event data recorders (EDRs) in improving the emergency care of people with crash injuries and on the role of NHTSA in advancing this technology to improve public safety. As director of the William Lehman Injury Research Center and as a practicing trauma surgeon, I am acutely aware of the difficulties that presently exist in providing rapid and appropriate care for crash survivors with time-critical injuries. My staff and I have published a number of papers in which we describe how the difficulty in quickly recognizing injuries has increased with improved occupant protection features in motor vehicles (Augenstein 1992, Augenstein 1995). The EDR technology offers the promise of providing valuable crash data that could save lives by identifying the crashes in which occupants are most likely to be injured, and even predicting the kind of injury to expect.

Based on studies of crashes in the CIREN database and in NASS, my staff and I have developed several tools to assist triage decisions and to predict specific injuries. The first tool was called the "SCENE SCALE". Based on this research, in 1993, NHTSA published a Research Note and distributed SCENE SCALE posters to the emergency rescue community [Lombardo, 1993]. The posters describe crash attributes that predict the presence of time-critical injuries that might be present in crash involved occupants who "look ok" at the scene.

More recently we have been working to validate and improve the URGENCY algorithm that was developed for NHTSA by Malliaris [Augenstein 2001, Augenstein 2002, Augenstein 2003]. The URGENCY algorithm uses information available from the EDR to predict the risk of severe injury. The application of the technology in conjunction with the Automatic Crash Notification System offers tremendous promise of saving lives by improving post-crash rescue treatment.

The safety potential: The National Highway Traffic Safety Administration (NHTSA) has reported that 27 million vehicles were involved in over 17 million crash events on US roadways in 2000. During these events, an estimated 2 million occupants sustained injuries requiring medical care, but only 1 in 8 sustained injuries that were considered life threatening. Although these 250,000 seriously injured occupants require the most urgent medical attention, they are not easily distinguished from the less severely injured using current rescue protocols. This inability to distinguish occupants at high risk for severe injury results in costly delays in treatment and poor allocation of medical resources.

A number of crash attributes have been recognized as important indicators of injury potential, yet the use of this information to improve rescue care has been limited to date. In the event of a motor vehicle crash, potentially injured occupants rely on passing motorists or accessible cellular technology to initiate a call for help. Once this call has been made, rescue services verbally gather location and crash severity data from callers in order to select and deploy rescue services to the crash site. A study by Evanco estimates a potential reduction of 3,069 rural fatalities if notification times within one minute of the crash are achieved [Evanco 1999]. Clark and Cushing estimate this potential fatality reduction to be 1,697 for the 1997 fatally injured population [Clark, 2002].

Upon arrival to the crash, first care providers rely on anatomical, physiological and mechanism criteria to distinguish occupants who require trauma center care from those who do not. In many cases, evidence of severe internal injury is difficult to discern in the field. A large number of crash involved occupants are improperly transported to non-trauma center care before the true severity of their injuries is recognized.

Conversely, many occupants are triaged to trauma centers based on "High Suspicion of Injury" criteria in the absence of definitive evidence of injury. In this case, first care providers may choose trauma center care based on their overall impression of an occupant's condition even if they do not meet any established trauma criteria. This use of paramedic judgment greatly improves the chance that an occupant who has sustained non-obvious or occult injuries will receive necessary trauma center care. In many cases, this practice taxes rescue and in-hospital resources. In Miami, Florida 60% of occupants triaged to the Ryder Trauma Center under "High Suspicion of Injury" criteria are discharged within 24 hours of hospital arrival. This suggests that better methods to discern the seriously injured from uninjured in the field may help to reduce the unnecessary use of valuable medical resources.

In 1997, Malliaris conducted research that was the basis for the URGENCY algorithm to predict the risk of serious injury in the event of a motor vehicle crash [Malliaris, 1997]. The algorithm processed crash conditions using logistic regression models to predict the likelihood of AIS3 or higher injury for crash involved occupants. A single regression model was developed to predict injury risk for all crash modes based on characteristics known to be influential for injury outcome.

Our 2003 ESV paper supports further implementation and enhancement of Automatic Collision Notification technology to improve crash rescue care. Further development of the URGENCY algorithm is described and its predictive ability is documented through an analysis of real world crash cases. Four independent injury models by crash mode were developed. Each algorithm was created in two levels of complexity and tested for its accuracy. Model performance is also compared with the use of delta V alone as an independent predictor of injury. The paper provides information on the crash parameters that if record in the EDR world assist in predicting serious injuries. In addition, the relative benefit of collecting each additional data element is indicated.

Study Summary: Predictive models have been developed which utilize a series of crash attributes to estimate the likelihood of MAIS3+ injury. These models were created in three levels of complexity to understand the relative benefit of additional variables during the estimation of injury likelihood. The study indicated that the accuracy of injury predictions improves significantly with the addition of selected variables; however, this improvement depends heavily on crash mode. The most influential variables are those shown below: 1 Lateral Delta V (for each impact event), 2 Longitudinal Delta V (for each impact event), 3 Lateral Acceleration Profile, 4 Longitudinal Acceleration Profile, 5 Three Point Belt Usage (all occupied positions), 6 air bag Deployment (in occupied positions), 7 Intrusion Extents, 8 Occupant Age, 9 Occupant Height/Weight, 10 Rollover Occurrences, 11 Occupant Ejections.

During the study, logistic regression models were created based on historical crash data to process crash attributes readily available through on-board vehicle sensor systems and on-scene observations to predict MAIS3+ injury risk for frontal, nearside, far side, and rear impacts. These models were subsequently evaluated using an independent set of crash cases to understand the ability of each model to distinguish seriously injured occupants within the total population of crash involved occupants.

The predictive ability of models constructed in three levels of complexity was compared. The first model utilized crash mode and delta V threshold as a simple criterion for Automatic Crash Notification (ACN) devices. Sensing, processing and transmission of this data alone would adequately detect close to 64% of the MAIS3+ injured population. With the addition of information regarding 3-point restraints usage and air bag deployment, the accuracy of injury recognition improves only slightly to 67% but the occurrence of miss-classifications declines nearly 10%. With the addition of crash attributes shown in the list above, proposed models correctly identified 74.2% of the MAIS3+ injury occupants involved in tow-away crash events for NASS/CDS cases from 2000 and 2001. 12.5% of the uninjured population was incorrectly classified as injured for this population.

The usefulness of each crash parameter shown above varied significantly between crash modes. For each optimized combination of variables by crash mode, the overall accuracy of each model at a given threshold level are shown. Within the following text, a mode-detailed description of the model development and validation process is presented. Further information regarding the predictive accuracy of models is given in the paper.

It is well understood that rapid notification of rescue services and appropriate administration medical care will reduce the likelihood of secondary injury or death of crash-involved occupants. Methods to process crash conditions in order to estimate the likelihood of injury have been established and the accuracy of these methods has been reported. When compared with injury prediction based on delta V alone, proposed models were shown to improve accuracy of injury estimates based on crash attributes available at the time of the crash. Additional crash attributes must be recorded for subsequent processing by predictive models and transmission.

For the NASS/CDS populations tested, the sensitivity of models predicting the likelihood of MAIS3 and higher injuries is 74.2% with an overall specificity of 87.5%. When compared with predictions based on delta V alone, the use of proposed models offers more accurate estimate of injury potential based on readily available crash information for frontal crashes and far-side crashes. This improved accuracy is not readily observed for nearside and rear crashes.

In order to make use of any injury model including those based only on delta v, methods to automatically collect store and deliver crash information to the most appropriate individuals must to implemented. This effort will require continued cooperation between auto manufacturers, rescue providers and in hospital clinicians to collectively agree upon the most appropriate methods to reach this goal.

NHTS's Role in EDR development: NHTSA needs to encourage the technology to save lives by improving post-cash emergency care. To this end, NHTSA needs to sanction an injury-predicting algorithm that incorporates the data from the EDR. NHTSA should set minimum standards for the data to be transmitted and should standardize the format. Finally, NHTSA needs to provide training and training materials to all segments of the post-crash service and care professionals so that the critically injured crash victims will benefit from the ACN/EDR technology.[4]

DOCKET # NHTSA-2002-13546-75 03/12/03 *Public Citizen*[5]

Public Citizen is a national, nonprofit consumer advocacy organization founded by Ralph Nader in 1971 to represent consumer interests in Congress, the executive branch and the courts. They fight for openness and democratic accountability in government, for the right of consumers to seek redress in the courts; for clean, safe and sustainable energy sources; for social and economic justice in trade policies; for strong health, safety and environmental protections; and for safe, effective and affordable prescription drugs and health care.

Public Citizen is a good example of how the nongovernmental sector can play a major role in road casualty reduction. Nongovernmental agencies are called NGO's in the World Report on Road Traffic Injury Prevention (2004). NGO's serve road safety most effectively when they:

- publicize the true scale of the road safety problem;
- provide impartial information for use by policy-makers;
- identify and promote demonstrably-effective and publicly-acceptable solutions, with consideration as well of their cost;
- challenge ineffective policy options;

- form effective coalitions of organizations with strong interest in casualty reduction;
- measure their success by their ability to influence the implementation of effective road casualty reduction measures.

Public Citizen commented:

Public Citizen offers these comments to the agency's Request for Comments on Event Data Recorders (EDRs). We praise the agency's attention to this issue. An increasingly large number of vehicles manufactured each year have Event Data Recorder-type systems installed, and consumer demand for these systems remains high.

Given the widespread and growing use of EDRs, it would be a grave mistake for the agency to allow this opportunity for better data collection to go unrealized. In the absence of agency action, manufacturers will act opportunistically and at whim, taking away or adding categories for data collection as it suits their interest in addressing defects or in litigation. To protect consumer interests in this new technological landscape, and to enable the public to rely upon these data as a developing and crucial supplement to the agency's current data collection efforts, standardization is essential and would allow National Highway Traffic Safety Administration (NHTSA) to maximize the benefits and utility of the data collected.

We urge the agency to use the voluntary standard proceedings at Institute of Electrical and Electronics Engineers, Inc. (IEEE) as an outline of possibilities, but not to be limited in scope or requirement by decisions made by this industry-dominated panel. The list of potential attributes being developed by the group serves as a useful guide, or "menu" of options, for a binding agency standard. Yet the flaws in a voluntary approach are legion: variability in the categories would do much to denigrate EDRs value for traffic research purposes and would place consumers at the mercy of capricious industry decisions. For these reasons, we would strongly support the development of a uniform standard that includes reporting obligations for private or government parties utilizing the data for other purposes. This historic opportunity to widen the scope of NHTSA's inquiries and data collection capacity should not be left to chance.

Below are Public Citizen's responses to the questions posed by the agency in its request for comments.

Safety: (1) Safety potential. While we agree that Event Data Recorders have the potential to greatly improve highway and automotive safety, Public Citizen believes that realizing the safety potential of EDR data requires the mandatory installation of the devices in all vehicles. The utility of EDR data will depend largely on NHTSA's ability to analyze a sample of crashes that accurately represents the real-world crash landscape. In addition to (1) mandating the installation of EDRs in all new vehicles, NHTSA must (2) take regulatory action to standardize a minimum set of data elements, (3) develop a template for uniform data output and (4) enable uninhibited data access.

The most significant safety potential of EDR data is to prevent needless deaths and injuries by helping NHTSA identify gaps and holes in safety performance, develop or upgrade Federal Motor Vehicle Safety Standards and better target its crash and defect investigations. A 2002, report published by the Office of the Inspector General criticizes the Office of Defect Investigation (ODI) for its unstructured method of analyzing data to determine if a potential defect exists, and for the limits upon, and poor quality of, the data used by ODI to identify defect trends. Public Citizen believes that EDR data may help ODI to reduce the apparent randomness of defect investigations by providing ODI with a sounder analytical basis for conducting defect investigations.

(2) Application. In order to maximize the safety benefits of EDRs, NHTSA must require their installation in all new vehicles, including heavy commercial trucks. We presume that the decision to install a specific type of EDR will vary according to the data set being collected and the number of sensors needed to collect such elements. While some data elements, such as the crash pulse, changes in longitudinal and lateral velocity, stages of air bag deployment and restraint use are germane to all vehicle classes, a factor such as loading condition would be more pertinent to a vehicle class that may regularly carry heavy loads, such as commercial or light trucks. Once NHTSA clearly defines the information it wants to collect from particular vehicle classes, EDRs that meet the agency's relative performance requirements should be permitted. NHTSA

should, as logic dictates, establish minimum data sets that reflect differences across vehicle classes. However, within a specific vehicle class, such as light duty vehicles, the data collected by EDRs should be very similar, if not identical and should contain a minimum set of uniform elements to enable easy comparison. Lastly, the agency should initiate different rulemaking processes for passenger vehicles and commercial trucks as the data collection, privacy and political factors vary greatly across these groups.

(3) Use of EDR data. NHTSA currently employs EDR data as part of both its routine and special crash investigation efforts. Also, the agency is already working to incorporate EDR data into the National Automotive Sampling System database. We note that the agency's current policy guidelines require the vehicle owner's written permission in order to obtain any EDR data. NHTSA must develop a better protocol which shields the privacy of individuals involved in crashes while better facilitation access to EDR data for studies using only aggregated data. Compartments within the EDR itself or partitions in the data collected, should be developed and employed. Presently, the usefulness of EDR data is limited, but as the technology improves and the number of EDRs increases, EDRs will be of increasing utility to researchers studying the dynamics of motor vehicle crashes.

Unfortunately, the consistent lack of funding for NHTSA's data collection systems, namely the National Automotive Sampling System—Crashworthiness Data System (NASS-CDS) and the Fatality Analysis Reporting System (FARS), restricts the ability of researchers to uncover trends in motor vehicle crashes. Although originally designed to investigate 18,000 crashes a year, NASS presently investigates only 5000 crashes annually.

Because EDR data are currently only collected as a part of NASS, this severe lack of funding likewise restricts the availability of EDR data from this important new source. Police and municipal officials must receive the training and funding necessary to collect accurate and complete EDR data to be included in the FARS database. If the EDR program is structured to increase the scope and depth of crash investigations, EDRs will help researchers collect representative crash samples to supplement the agency's emaciated data collection systems. This is critical for effective agency decisions, and for priority-setting within the agency and by Congress.

(4) Future safety benefits. Public Citizen is anxious for NHTSA to harness the potential for EDR data to improve highway design, the administration of emergency medical services and to assist NHTSA in research and rulemaking efforts. In addition, the Federal Highway Administration (FHA) can use EDR data to identify and amend aspects of highway design that either cause or fail to prevent crashes, such as the inability of highway dividers to impede large sport utility vehicles. The Federal Motor Carrier Safety Administration can use EDR technology to improve commercial vehicle safety. Automatic Crash Notification Systems provide emergency medical teams with information that is critical in treating injured occupants. As a supplement to NHTSA's crash investigation and reconstruction efforts, EDR data will be critical in understanding crash causation factors and injury mechanisms at a reduced cost to the agency. When used in conjunction with the early warning database, the data may also help to uncover emerging defects, although the penetration of EDRs in the vehicle fleet would have to be substantial, and the scope of data collected extensive to reveal most defects.

The agency must harness the ability of EDRs to provide researchers with solid data on the performance of emerging technologies such as side impact and window curtain air bags in real-world crashes. Due to the relatively low penetration of emerging safety technologies into the overall vehicle fleet, NHTSA should conduct special data collection efforts to track and compare the safety performance of emerging technologies as a basis for future rulemaking.

(5) Research databases. The agency must clearly flag the EDR data when they appear in agency databases, standardize the coding and data presentation formats, and ensure that the data are internet accessible. Clearly identifying data obtained from EDRs within agency databases will make the utility and benefits of EDR data more evident. The homogeneity of EDR coding and data presentation is critical for crash comparisons and identifying trends in motor vehicle safety. Rather than collecting years of discordant data that may or may not be salvageable for future use, NHTSA

must act quickly to protect the utility of EDR data. In order to make the data available to any interested parties, we urge NHTSA to make EDR data available on the NHTSA or National Center for Statistical Analysis Web Site.

(6) Prevention of crashes. Public Citizen hopes that the mandatory installation of EDRs in the commercial vehicle fleet will yield crash reductions comparable to the decreases observed in the European studies. We anticipate that, as in the studies, the biggest safety gains from influences upon driver behavior would similarly occur in the commercial trucking fleet. As some of the comments in this docket clearly indicate, the commercial trucking industry is troubled by the ability of EDRs to corroborate the figures recorded in hours of service log books. Judging by the trucking industry's intense opposition to the mandatory installation of EDRs in commercial trucks, Public Citizen suspects that the presence of EDRs in commercial vehicles could substantially improve both industry practices and commercial driver behavior, particularly as to the rampant problem of drivers who continue driving while fatigued.

(7) Possible new databases. The EDR program should strive to collect and analyze data from every motor vehicle crash resulting in a death or injury, and of tow away severity. The data collected will undoubtedly be used to complement the data collected in the NASS and FARS databases, but Public Citizen believes that it will be useful to store EDR data in a database independent of those currently employed by the agency. For example, the utility of NASS is constrained by the limited number of crashes investigated annually. Due to inconsistencies in coding and collection procedures, much of the data contained in FARS reveals little about trends in motor vehicle safety, and Public Citizen urges the agency to review the FARS criteria. We believe that maintaining a separate storehouse of EDR data, in addition to data linked to FARS and NASS, might help corroborate conclusions drawn from other databases.

(8) Standards. The EDR activities of The Society of Automotive Engineers (SAE) and IEEE should not delay NHTSA in taking the regulatory action necessary to standardize procedures for EDR data collection, storage, and output. Although the findings of the SAE

and IEE may guide the agency in developing a government standard, relying on an industry standard alone will not yield safety benefits that are remotely comparable to those flowing from a federal requirement. Any voluntary standard will be subject to the caprice of the automobile manufacturers, and will be tailored to benefit the industry interests rather than consumers or NHTSA's interest in data collection. The agency must not relinquish this historic opportunity to protect the future of crash data collection from the predictable, self-interested whim of the automobile manufacturers. Also, it is imperative that the agency require installation of EDRs in all new vehicles to collect enough reliable data to effectively analyze trends in vehicle crash dynamics for FMVSS improvements.

(9) Standardization. The ability of NTHSA to standardize data elements, procedures for data collection, data organization, data storage, and data output will be the primary determinants of the program's effectiveness. The failure to homogenize data categories and elements within the multiple facets of the EDR program will unquestionably result in a flawed system and a squandered opportunity. In addition, standardization will prove highly cost-effective for consumer and the industry, as supplier costs are driven down by uniform requirements. Standardization would also, as NHTSA suggests, greatly enhance efforts to monitor new crash protection technologies as they emerge. Technical Issues:

(10) Data elements. NHTSA must establish a minimum set of data elements that would apply to all light duty vehicles, and, separately, to all heavy trucks. Furthermore, the agency should determine these data elements that offer the greatest potential to inform researchers of developing safety issues.

(11) Amount of data. NHTSA must be mindful that establishing too brief of a pre-crash recording period will limit the ability of crash investigation teams to identify the entire sequence of destabilizing events leading up to a crash. The amount of data collected prior to a crash should correspond with the maximum amount of valuable information needed and available to determine crash causation and crash injury factors. Public Citizen recommends that NHTSA ascertain the pre-crash, crash, and post-crash time window that will be most helpful and inclusive for the purposes of crash

causation and research based on an evaluation of real-world crash events, including driver maneuvers, cross-highway swerving and other time-consuming event scenarios.

(12) Storage and collection. Once EDRs are standardized and widely used, the breadth of the data collected will provide the agency with a panoramic portrait of the nation's motor vehicle crashes. The current databases provide researchers with only a peephole through which they can observe the milieu of real-world crashes. In order to effectively analyze EDR data on a national scale, the agency should standardize the methods for data collection and storage. Ideally, the information would be automatically loaded to a regional database through a wireless connection and then integrated into a national system.

It is critical that the agency protect the identity of vehicle owners, but we believe that location and date stamping, in addition to the Vehicle Identification Numbers (VIN), is necessary for crash reconstruction and investigation efforts. Partitioning of data released routinely to the public can adequately address any privacy concerns that may arise. The agency should strive for the cleanest (and most comprehensive) data set for use in its summary analyses, as it does under other current systems of data collection.

(13) Training. The agency must strive to minimize the collection procedure's vulnerability to human error. The variability in collection and coding procedures remains one the primary deficiencies of FARS. NHTSA should simplify the system by mitigating the potential for inadvertent data corruption by collection officials. To insulate the data collection process from human variability requires maximizing the role of EDRs in data collection, and minimizing the amount of data coded and recorded by collection officials. We recognize, however, that collection officials would be needed to authenticate the data collected by EDRs, and such officials would need regular, ongoing training.

Privacy Issues: The use of EDR data for statistical analysis does not involve privacy concerns, and Public Citizen believes that NHTSA should not be deterred from proceeding with rulemaking by privacy issues at this time. We recommend that the agency remain focused on the potential contributions of EDR data to the advancement of motor vehicle safety. The tension between privacy and safety is an important issue that requires initial agency proposals and can be adequately addressed by partitioning technology and other means best evaluated as a part of the rulemaking process.

Role of NHTSA: (17) Role of NHTSA. NHTSA must take action that will optimize the benefits of EDR data and obviate industry attempts to displace the agency's role in laying the ground rules for this crucial new source of public data. Public Citizen recommends that NHTSA:[6]

- Mandate the installation of EDRs in all new vehicles;
- Standardize a minimum set of data elements according to vehicle class (*i.e.* light duty passenger vehicles and commercial trucks);
- Conduct a study to ascertain an accurate pre-crash, crash, and post-crash time window;
- Develop a model database and strict guidelines for data collection for individual states or regions;
- Train state and local government officials in the collection of EDR data for research purposes;
- Create a template for uniform data output;
- Enable uninhibited data access via the Internet;
- and Develop privacy protections as needed without blocking to defend themselves in such lawsuits.

DOCKET # NHTSA-2002-13546-76 04/02/03 Mitsubishi Motors Research and Development, Inc.[7]

Mitsubishi Motors Japan, (MMC) would like to submit the following comments to the above-mentioned request for comments. Further, Mitsubishi Motors has participated in the preparation of the comments submitted to this notice by the Alliance of Automobile Manufacturers, and supports those comments.

MMC general comments: 1) Although EDRs may, by providing additional field performance data, have the potential to improve overall vehicle and roadway safety, MMC believes that fundamental legal, societal and technical issues must be resolved prior to a mass deployment of EDRs in the vehicle fleet. 2) MMC believes that the standardization of EDR technologies will provide the necessary backbone for future EDR technology development and deployment. MMC hopes unbiased, sufficient discussions take place during the standard development.

EDRs themselves do not directly lead to advances in vehicle safety. However, EDR data when it is properly linked to medical or other relevant data has the potential to improve vehicle safety by improving the accuracy and reliability of accident causation analysis. 2) If traffic data (road conditions, traffic situations, and environmental circumstances) and vehicle operation (driving behavior) can be linked, a better understanding of road-vehicle interaction can be realized.

There are at least four technical elements within an EDR (hardware, software, communication protocol, supporting infrastructure) and all of these elements should be designed to achieve the best performance on a particular vehicle. Since there are a wide variety of vehicles, based on how an EDR is intended to be used as well as EDR characteristics, MMC believes eventually there will be many different types of EDRs in use. In addition, even if the same technical elements are used for different vehicles, based on system and/or vehicle design, the same or compatible data may not be produced. For example, a G-sensor can produce a wide range of data based on the sensor location and vehicle structure. It is more important to focus on the compatibility of output data rather than applying the same box to different vehicles.

Large-sized vehicle EDRs may be different from light-vehicle EDRs because they can be integrated into truck/bus fleet management system (drive recorder). MMC believes proper care should be taken when handling EDR data because of privacy and data ownership issues. If the purpose of using this data is to improve safety, NHTSA should be allowed to collect and analyze EDR data, if it is used within the proper, agreed to social parameters, understood by vehicle owners from which the data would be collected and users of EDR data.

Although there is data to support EDRs reduced the accident incidence rate in Europe, is there any other such research? In Japan, the Institute for Traffic Accident Research and Data Analysis (ITARDA) published an investigation report, "Implementation of Traffic

Accident Countermeasures with Driving Recorders"
(June 2002), which may be relevant for NHTSA's study
of this issue. 2) Are the benefits applicable to the US
too? MMC cannot determine whether benefits also
apply to American drivers.

MMC cannot definitively determine whether there are
safety benefits directly related to EDR standardization,
but setting standards would make it possible to
improve the quality and accuracy of accident reconstruction studies. For that purpose, MMC believes that
the standardization of the data from EDRs (not EDR
itself) would be appropriate. MMC also believes hat
the industry should explore the possibility of implementing existing standards rather than developing
new standards.

Although there is a NHTSA EDR Committee proposal
for new EDR data elements, MMC believes the choices
in the proposal span too far a field and should be narrowed and focus on only safety-related elements.
MMC believes this is inevitable if the two issues "What
is an EDR?" and "How will EDRs be used?" are to be
discussed and resolved in the future.

Should the data recorded change by model or purpose? Since the data selected for recording depend
according to what the EDR is used for, MMC cannot
respond at this time. The data recording parameters
should only be stipulated after examining what applications an EDR will be used for. Remarks: MMC
requests the following regarding "Data," EDR (device)
and EDR data (anything recorded in the device)
should be clearly and separately distinguishable. 2)
Specify that EDR data does not include vehicle data
communicated via a bus and diagnostic data. 3) Leave
the possibility open for EDR data to be distributed
among and exist within multiple ECUs, rather than
only within the EDR. 4)Regarding the word EDR, we
request to make a clear distinction among device
(hardware), application (software) or recorded data
(contents) to avoid potential misinterpretations when
using the word "EDR."

Since the duration of EDR data collection should be
different for various models of events and collisions, it
is very difficult for us to recommend a particular
amount and length of appropriate data collection at
this moment. Since there is no perfect technical solu-

tion, the most effective technical measures should be implemented at the appropriate time.

MMC believes if the privacy issues are resolved, the above question will be consequently answered. 2) How should actual vehicle and accident tracking be handled? MMC cannot yet make a general comment on this issue since the objective for using EDRs is not yet clearly defined. However, MMC believes it is not necessary to track EDRs if the intent is to only use them for gathering collision statistics.

A procedure to download the data and an explanation of the legal responsibilities for those parties in charge of undertaking such work. MMC believes EDR collision durability is not so important as long as the EDR data remains intact, even if an EDR itself is damaged. Therefore, MMC believes it is sufficient if the device durability is left to the vehicle manufacturer. If NHTSA is considering setting durability requirements, EDR durability should be stipulated such that an EDR should not be catastrophically damaged in a collision. Accordingly, MMC believes a durability study based on collision analyses is needed. The most appropriate method would be to transmit data from the vehicle by an automatic event-reporting device, but MMC does not know how this could be implemented. MMC cannot make a cost determination before clarification of EDR usage is made.

MMC believes that the government should set regulatory parameters for EDR data usage to protect the privacy of our customers.

MMC believes that rather than focus on technical development, NHTSA should instigate activities to directly address privacy issues, data usage, and other non-technical issues. [8]

Appeal Is Denied

In April 2003, an appeals panel assembled by the IEEE Standards Association Standards Board heard the appeal of twelve members of the P1616 Working Group. The Alliance of Automobile Manufacturers (AAM) and the Association of International Automobile Manufacturers (AIAM) members involved in the project, along with two individuals appealed.[9]

The appellants based the appeal on the five reasons given below:

1. Substantial changes were made without P1616 knowledge and agreement.

2. These unapproved changes were obscured in a way that prevented most members from noticing them.

3. Some of the changes were not noted in any way.

4. Negative comments were not addressed.

5. The minimum ballot period was not observed.

The twelve appellants included representatives from DaimlerChrysler Corporation, Honda Research and Development, the Association of International Automobile Manufacturers, General Motors Corporation, Mercedes-Benz USA, Volkswagen of America, Mitsubishi Motors Corporation, Toyota North America, Nissan North America, and Booz-Allen & Hamilton of Tyson's Corner, Virginia, as well as a forensic analyst Roger L. Boyell, and Joseph Marsh of Ivy Consultancy.[10]

While in general, most of the population had little knowledge or understanding of the issues, articles did begin appearing in some of the world's most widely read and viewed media: the *Toronto Sun, New York Times, San Francisco Chronicle, Los Angeles Times, USA Today, Last Vegas Review-Journal, Wall Street Journal, Electronic Engineering Times, Toledo Ohio Blade, Milwaukee Journal-Sentinel, The Clarion Ledger, US News and World Report, Boston Globe, ABC News, Newsweek, Arizona Republic Online, Baltimore Sun, Philadelphia Enquirer* and *Forbes.com*. The IEEE standards development process also helped to increase public awareness by keeping the meetings open and placing the minutes of each meeting on a public website. (http://grouper.ieee.org/groups/1616/home.htm). Several chat rooms and discussion groups devoted to concerns about the issues of privacy and safety in the use of black box technologies also began to surface on the Internet.

Slowly but surely, the general public was becoming more aware of MVEDR technologies.

On April 18, 2003, at IEEE headquarters in Piscataway, New Jersey, the IEEE-SA denied the appeal.[11] In fact, all five charges were invalidated by unanimous decision of the IEEE panel.[12]

Automakers Begin Another Standards Initiative

On April 24, 2003. SAE International voted to create the Data-Extraction Sub-Working Group (Project J1698-1) based on the work of the Vehicle Event Data Interface Committee (VEDI), which was an ad hoc group formed in December 2002.[13]

Some members of IEEE P1616 believed that the work of VEDI was duplicative of the IEEE effort. Indeed, minutes of the SAE meeting on 24 April indicate that SAE had been approached by IEEE to coordinate efforts so that no duplicative, competing standard would be created. Members of the new SAE J1698-1 have stated their belief that their project was not duplicative. This question remained contentious throughout the process.

Some believed that the automakers had decided to do their work in SAE because they believed they would have more control over the content of the standard in that venue.

The SAE J1698-1 meeting minutes also note that, "after some discussion, the Committee agreed by unanimous consent to add a sentence in the 'scope' on the confidentiality of the conversations." Furthermore, three points were considered:

1. Nothing should stop an OEM from stating that the data should be this way to a supplier due to the high cost,

2. Vehicle system satisfy user requirements for common communication among vehicles need required resolutions,

3. When the data are reported, it must be at a specific resolution.

On another issue, it was noted by many that the automakers' desire to control the dissemination of the details of their discussions was matched only by their insistence that they also control the information that might be culled from MVEDRs. Some believed that under smokescreens of claiming to speak for the privacy rights of motorists, the automakers seemed to in fact be more concerned with their own privacy and protection. Otherwise, how could they reconcile their alleged concern for end-user privacy with the fact that sensor and memory capabilities are already being installed in automobiles? Safety advocates noted the fact they wanted to establish proprietary claims over the data derived from a crash for their own protection and profit. There were lingering questions. If a crash is a public event (which it is to many), then should the data be publicly available, particularly to those involved in the incident. Who's protecting whom remained the vexing question for safety advocates and investigative journalists.

DOCKET # NHTSA-2002-13546-81 Advocates for Highway and Auto Safety (Advocates) Statement from the Surgeon General of the United States

On April 28, 2003, the Surgeon General of the United States, Vice Admiral Richard H. Carmona, M.D., M.P.H., F.A.C.S. issued a Call for Better Data:

> Progress in motor vehicle injury prevention stands out as one of the most significant public health achievements of the 20th Century.[14]

> However, motor vehicle-related injury and death remains the nations' largest public health problem. Last year about 22 million Americans were involved in a motor vehicle crash.

Safety and injury prevention must be among our highest public health priorities as a nation. Our national commitment to reducing injuries and deaths from motor vehicle crashes is an important objective.

No more important challenge exists than finding ways to improve the safety of everyone while riding in motor vehicles.

Effective public health policy must be based on sound scientific evidence. The annual death rate for motor vehicle crashes has decreased 90% though six times as many Americans drive today as did in 1925, covering ten times as many miles in eleven times more vehicles.

Despite these improvements about 42,000 Americans of all ages still die each year as a result of motor vehicle trauma. Motor vehicle crashes are not "accidents," and much more can be done to prevent these and resulting injuries.

Assessing the problem can be difficult because of lack of consistency in the use of terminology and data elements for motor vehicle crashes. Without such consistency, we cannot monitor and track trends for motor vehicle crashes to determine the magnitude of the problem.

Thus, MVEDR (Motor Vehicle Event Data Recorder) recommendations are crucial for standardizing definitions and data elements for motor vehicle crash surveillance. Better quality and timely incidence and prevalence estimates can be useful for a wide audience, including policy makers, researchers, public health practitioners, victim advocates, service providers, and media professionals.

MVEDR technologies provide that evidence, thus expanding and recording the knowledge needed for informed decision making to improve the health of every citizen involved in a crash.

It is my pleasure to congratulate the Institute of Electrical and Electronics Engineers (IEEE) for sponsoring this initiative. The IEEE is helping to advance global prosperity by promoting the engineering process of creating, developing, integrating, sharing, and applying knowledge about electrical and information technologies and sciences for the benefit of humanity.

I would encourage decision makers to use the IEEE findings. I am confident that this initiative will be a major milestone toward the goal of motor vehicle occupant injury prevention and will build new opportunities for greater national success. [15]

On May 5–6, 2003, the eleventh IEEE Motor Vehicle Event Data Recorder meeting was conducted at the National Academies of Science, Washington, DC. There were twenty-nine individuals in attendance.[16]

During this meeting the Surgeon General's "Call for Better Data" statement was distributed. Within the document the Surgeon General of the United States noted that safety and injury prevention must be among our highest public health priorities as a nation and that our national commitments to reducing injuries and deaths from motor vehicles are important objectives and that no more important challenge exists than finding ways to improve the safety of everyone while riding in motor vehicles.

There was a presentation by David L. Karmol, Vice-President Public Policy and Government Affairs of the American National Standards Institute (ANSI). The American National Standards Institute (ANSI) is a private, non-profit organization (501(c)3) that administers and coordinates the U.S. voluntary standardization and conformity assessment system. The Institute's mission is to enhance both the global competitiveness of U.S. business and the U.S. quality of life by promoting and facilitating voluntary consensus standards and conformity assessment systems, and safeguarding their integrity. Karmol stated that, "A standard is a document not a regulation. He described U.S. compliance and enforcement, how ANSI works, and the structure of the U.S. Standardization System. ANSI believes inclusiveness and incorporating everyone in the process is important. The accreditation process was detailed. ANSI is a neutral policy forum. He described the ANSI relationship with the U.S. Government. He also described International and regional representation in standardization organizations. Mr. Karmol acknowledged the background issues of who should manage the MVEDR standards process. He offered similar examples of competing Standards Development Organizations (SDOs) and suggested that although there is no one way to create a standard, ANSI places high value on those that adhere to openness and widespread inclusiveness. Finally he described ANSI's role in International Standards Organization (ISO) and IEC and answered a few questions from the members. Mr. Karmol offered his continual support and guidance to the project.[17]

Dr. Susan Ferguson, Senior Vice-President of Research of the Insurance Institute of Highway Safety (IIHS) gave a presentation titled "MVEDRs and Crash Data." During the presentation Dr. Ferguson received several questions from the members as to the rationale for choosing recommended EDR data elements. She answered each question with a specific example of how EDRs can provide useful information to understand vehicle performance and injury mechanisms in crashes. She noted EDR usefulness in crash investigations and explained that currently estimated delta V is only a partial measurement of crash severity. There were questions about EDR usefulness in crash investigations and Dr. Ferguson stressed the value of determining seat belt usage via an EDR. She noted that seat belt use can be difficult to access in some crashes, and she implied that accurate usage could be determined by an EDR with the implication being that vehicle and highway safety would be greatly enhanced. There were questions about the current limitations of EDRs. Dr. Ferguson noted that information may only be available for frontal crashes, access to the EDR units were not always easy in crash investigations, and that an EDR is not infallible.

In summary, Dr. Ferguson stressed that EDR data elements and access should be standardized and she congratulated the members for proceeding towards that end. Dr. Ferguson offered her support and assistance toward the IEEE P1616 project goals.[18]

Finally, the topic of IEEE and SAE future co-operation was brought up by a Working Group member. He introduced himself as the Chair of the SAE VEDI Committee. He had been directed to come to the meeting and represent the SAE's position. He gave a presentation, the main focus was whether the SAE and IEEE would continue working separately with the likely outcome of duplication of work or whether IEEE P1616 could modify their focus to areas not already covered by SAE. This topic prompted a lively discussion. When pressed by some IEEE members if he was speaking on behalf of the SAE, he noted he was not—that he was offering personal advice.

After much discussion, the Working Group decided that the grassroots IEEE standards work would continue on with the scope and purpose and would include a wide variety of end users.[19]

DOCKET # NHTSA-2002-13546-79 05/07/03 Consumer's Union[20]

On May 7, 2003, Consumer's Union commented to NHTSA:

> Consumers Union submits these comments in response to the Administration's request for comments on the future role of the agency in the continued development and installation of event data recorders (EDRs) in motor vehicles.

> Consumers Union, publisher of Consumer Reports, believes that the installation of EDRs in motor vehicles provides enormous potential for increasing road safety. Accurate crash data, and a better understanding of which components of vehicles and of driver behavior are most associated with crashes, unquestionably serve an important societal purpose. However, the installation of EDRs, and more importantly, the collection and distribution of the information these devices record, raise several significant concerns for consumers. Consumers Union believes that NHTSA should take these concerns into account when developing its regulations for EDRs.

More than 10,000 people a year are killed in rollover crashes. This crash occurred in a rural area and was not discovered until the next day.

In addition, given that NHTSA's goal is to maximize the utility of this data in service of enhanced traffic safety, it is imperative that NHTSA play a central role in the collection, access, and management of EDR data. NHTSA's playing a prominent role will also help prevents improper access to or use of EDR data by third parties.

According to the August 2001 report produced by NHTSA's Working Group, the majority of vehicles on American roadways today contain some sort of data capture capabilities. According to NHTSA's website, the Working Group was made up of representatives from government, universities, the original manufacturer industry, the aftermarket products industry, and the general public.

Although the precise capabilities of the devices and the elements they capture vary considerably among manufacturers, most of them record air bag deployment, at minimum. But most consumers have no idea that such devices are active in their vehicles. If EDRs are to become a more widespread technology, and if they expand both in their prevalence and in the data elements they capture, it is critical that consumers be informed in a uniform and conspicuous way that their vehicles contain this technology.

In addition, given that there are still myriad unknown ways in which this device can and may be used in the future, it is critical that NHTSA consider ways to protect consumer privacy as EDR technology moves forward. Finally, in order to weigh properly the competing concerns regarding this technology, Consumers Union recommends that NHTSA create a commission to further examine the implications of the widespread installations of this technology in vehicles. This commission should include, among others, representatives from consumer advocacy groups.

We understand that NHTSA believes it has limited authority over privacy issues. However, we respectfully request that the agency not overlook its power as a federal agency to guide and set policy. By incorporating into the final EDR regulations standards concerning encryption and data access, NHTSA may speak to the protection of consumer privacy rights without straying from its legislative mandate and/or the bounds of authority. At a minimum, consumers have the right to know that EDRs are installed in their vehicles, that they are capable of collecting data recorded in a crash, and which parties may have access to this data.

We believe that different types of EDRs should be used for different applications. Primarily, different EDRs should be used for private automobiles that should be used for all other types of vehicles (such as commercial vehicles). The justification for this distinction is that while safety data is needed for both types of vehicles, the expectation of privacy is far different for the driver/owner of a commercial vehicle than it is for the driver/owner of a private automobile. In a commercial vehicle (which may be, for example, part of a large fleet of vehicles used for transport), the driver is a professional employee, often with a special license, driving the vehicle in the course of his or her employment, *e.g.*, a truck driver of a shipping company employee. Such individuals are aware that all of the actions they take related to their work are in furtherance of the job requirements, and not for their own personal needs. Put simply, they are aware that the vehicle space is not theirs, but their employer's.

On the other hand, private individuals in vehicles they have purchased for private use have a different relationship to their vehicles. It is true that from a Fourth Amendment standpoint, the Supreme Court has constricted the privacy that an individual can reasonably expect with regard to search and seizure of the car and its contents. But from the standpoint of consumer privacy rights, most individuals are not aware that their vehicles are recording data that not only may be used to aid traffic safety analyses, but has the potential of being used against them in a civil or criminal proceeding related to an auto crash, or by their insurer to increase rates.

Therefore, EDRs in privately-owned vehicles need to collect fewer data elements than those collected by commercial vehicles. They should record only those technical elements needed by NHTSA and other qualified parties for improving traffic safety in general; these elements are detailed in our answer to Question 10 of the request for comments, below. In addition, these data should be collected in a fashion that protects the anonymity of the owner. When transmitted, the data should be divorced from any information that identifies the individual, such as the name, address, or social security number (SSN) of the owner. These personal identifiers are not critical to NHTSA's ability to analyze effectively the cause(s) of the crash, and should therefore not be recorded. In a commercial vehicle, on the other hand, regulating the recording of such personal information is less of a concern, since the event most likely occurred during the employee's work day, where such activity is already recorded in some fashion.

Consumers Union believes that any information gathered for analysis from individual vehicular events should eventually rest in a database controlled by NHTSA, and that consumers and/or their legal representatives should have access to this data. Having NHTSA as the central, ultimate repository of this information will help to centralize and standardize any privacy protections or data encryption protocols developed to safeguard the data. Rather than have the data held by various parties, or even among various state or local governments, we believe that both the privacy interests of consumers and the analytical exigencies of improved traffic safety will be best served if NHTSA is the repository for all EDR data (both from private and commercial vehicles).

The following are the data points the EDR should record: 1) Longitudinal and lateral acceleration and principal direction of forces, 2) Seat belt status by seating location, 3) Number of occupants and location within/without the vehicle, 4) Pre-crash data, such as steering wheel angle, brake use, vehicle speed, 5) Time of crash, 6) Rollover sensor data, 7) Yaw data, 8) ABS, traction control, and stability control data, 9) Air bag operation data, 10) Tire pressure data, 11) VIN (alphanumeric portion, not 6-digit serial number).

Inclusion of only the first eleven digits of the VIN will ensure that only necessary information (country of origin, make, vehicle type, gross vehicle weight rating, car line, series, body style, engine, check digit, model year, and assembly plant) is associated with each EDR. The remaining six digits should not be included in the EDR, as they do not bear on crash analysis, and may implicate individual identifying information.

As NHTSA and the IEEE working group note, EDRs only need to record data within a very short window of time before a crash in order to be constructive. Therefore, we recommend permanent encoding of data for the 10 seconds prior to the crash. We also do not support the recording of data not related to actual crash events.

Deployment of the air bag in a vehicle should be the trigger for crash information becoming permanently recorded by the EDR; all other events that do not trigger air bag deployment should be overwritten by the EDR system. In addition, we believe that the technology that best maintains consumer privacy and minimizes any potential for abuse of EDR data is one that overwrites any recorded data after a certain set period, such as 250 ignition cycles (and which does not allow for access to non-crash-related data prior to overwriting).

The most important issue to consider regarding traceability of EDR data is the balance between protection of consumer (*i.e.*, vehicle owner) privacy and utility of the captured data. To that end, we support EDR IDs that do not contain any personal identification information. That is, the EDR should not collect as part of its data set an owner's name, address, phone number, social security number, license number, or any other information that makes it possible to connect the driving data contained in the EDR to an individual. However, we do believe that when the EDR is installed, it may have the first eleven digits of the vehicle identification number (VIN) of the vehicle coded into it, or alternatively, an EDR identification number which provides critical information, such as the make and model of the vehicle.

Regarding protection of consumer privacy, our concerns are twofold. First, as we have stated above, the most important function of EDRs is the information they provide to improve understanding of vehicle crashes, and how we can make both vehicles and highways safer. Therefore, personal identifiers are not germane to the goals of EDRs. Second, the potential for crash data linked to an individual to be used for other than safety purposes requires that the information reside only with NHTSA. We address potential abuses of EDR data below.

We understand that there may be a need for certain individual pieces of information to be included within the EDR data set to maximize data utility, *e.g.*, the VIN (which itself is linked to personal information about the vehicle owner). The information should be encrypted in a way that is only decipherable by individuals authorized to have access to the data, *e.g.*, NHTSA analysts or verified parties to litigation stemming from a specific crash incident.

As mentioned above, there are significant potential dangers that may result from inappropriate access to EDR data. We outline several below.

Auto Insurance Pricing-Out. One concern with the availability of extremely detailed crash data is its use by auto insurers. With the increased detail provided by EDRs, insurers may be able to obtain information such as *precisely* how many miles per hour a driver was going prior to a crash. While we do not deny that auto insurers are entitled to base their rates upon a driver's past risk experience in order to better spread risk, we are concerned that the availability of sub-level specific detailed information could be used by insurers to determine future rates in an unfair manner. In addition, there remains the possibility that EDRs may malfunction from time to time. Should an insurer continue to base its pricing decisions on incorrect EDR data, we are concerned that consumers will not have adequate means to ensure that the units within their vehicles are repaired, and that they do not suffer adverse pricing consequences as a result of such malfunction.

Insurers Requiring EDRs as a Condition of Coverage. It is foreseeable that in the near future, many more vehicle manufacturers will start putting more comprehensive EDRs in their vehicles. Since EDRs have the potential to provide detailed data about vehicle behavior (*e.g.*, vehicle speed prior to a crash), there is the possibility that auto insurers may begin, as EDRs become more prevalent, to require an insured to have an operational EDR in their vehicle as a condition of coverage—*i.e.*, auto insurers may refuse to issue coverage unless the consumer has purchased a vehicle with an EDR, or a certain type of EDR, installed. Therefore, we recommend that NHTSA include in its regulations or guidance that auto insurers not be allowed to force consumers to have EDRs installed in their vehicles as a condition of coverage.

Public Access to Private Crash Data. EDR data may well be of interest to engineers, auto safety experts, and analysts from various fields of expertise. Experts in private or academic circles may want to access EDR data in order to further automotive safety and engineering research. However, there is currently no one place where such an expert might go to access these data. The most inclusive existing databases on vehicle crashes reside with NHTSA, insurers, and state departments of motor vehicles. If NHTSA develops such a central repository and the data can be accessed by the public (which we support), it is important that NHTSA maintain the anonymity of drivers in the database.

It is our understanding that academics would be obtaining this data from NHTSA after it has been fully encrypted and any personal identifiers have been removed, since they would not be interested in the driving habits of a particular individual, but rather aggregated data.

It is also important that NHTSA create and enforce protocols governing how any outside parties may gain access to EDR data, and keep track of which parties have requested such data.

Use of EDR Data in Crash-Related Litigation. One of the most important factors creating a need for thorough consumer education and protection regarding EDRs is their potential use in litigation, both civil and criminal. It is foreseeable that in both criminal prosecution of automobile accidents, as well as in the civil litigation that may result from accidents, parties will seek to discover the data contained in these devices to aid in legal argument. What is more, if the data that EDRs collect become increasingly standardized, it is feasible that EDR data, and the information they may reveal about a driver's pre-crash behavior, will become a vital element in such litigation. Therefore, we believe that NHTSA should not neglect the legal implications of EDR data in considering what standard data elements should be collected. The data elements should advance only the cause of traffic safety, and should neither hinder nor help either plaintiffs or defendants in crash-related litigation. Therefore, NHTSA should ensure that the data elements capture information as close to the actual crash as possible, and not collect data on general driving habits or history. NHTSA should also ensure, through its guidelines, that both plaintiffs and defendants in any potential litigation have equal and easy access to specific crash data contained in an individual vehicle's EDR.

Use of GPS and other Locator Devices. Certain manufacturers, both original equipment and aftermarket, have discussed the possibility of including GPS technology within EDRs. In the event of a car accident, GPS can help to summon aid rapidly to injured individuals. However, absent such exigent circumstances, there exists the potential for abuse of the GPS and the potential to locate a vehicle at any given time. For example, an auto insurer could use a GPS to determine where an insured drives his or her car. We believe, in the interest of consumer privacy, that NHTSA should carefully examine and limit the use of GPS technology in EDRs, and include any such limitations in any rules, regulations, or guidelines it promulgates. If a GPS is functional in EDRs at all, it should only be activated to assist first responders in locating a vehicle after a crash; it should not be functioning other than in crash situations. In addition, we would like the police to be limited by law or regulation to using GPS systems for emergencies like vehicle crashes.

III. Recommendations for National Highway Traffic Safety Administration Action

a) *NHTSA Commission.* Given both the potential utility of EDR technology, as well as the numerous as-yet-unknown consequences of the implementation of these devices, we recommend that NHTSA establish a commission to study fully the practical, "real-world" consequences that will result from the widespread installation of these devices. This commission should include, among others, consumer advocates.

b) *NHTSA Management of Data Access.* Because of the myriad potential uses and abuses of EDR data, it is critical that NHTSA take a management role in over-seeing who may have access to the data. NHTSA should remain the central repository for this data, and ensure that if data is accessed by other parties, that it does not contain any personal identifying informa-tion. The data should be equally accessible by all par-ties, whether they be the vehicle owners/drivers themselves, academics, or parties to litigation.

The difference in access by these various parties, of course, is whether or not each would be allowed to view any information that identifies individual driv-ers/owners.

c) *Consumer Notification.* NHTSA states that the utility of EDR technology increases as the device becomes more prevalent in vehicles on the road. As these devices come into widespread use, effective and con-spicuous consumer notification about these devices is important.

To that end, we believe that NHTSA should include in any regulations, standards, or guidelines it promul-gates protocols for consumer notification. Automakers installing these devices—GM, for example—are dis-cussing EDRs in the owners' manuals that accompany each new vehicle sold. We recommend that new car sellers be required to notify consumers of the EDR when the vehicle is sold; this notification should be mailed separately one to two weeks after a consumer completes purchase of their vehicle. Notice and information to the consumer should also be clear, conspicuous, and comprehensible. This ensures that the information does not get lost either in the owner's manual, or in the flurry of paperwork given to a consumer at time of purchase.

d) *Downloading Capabilities and Access.* Determining the first party to download the data will determine, in large part, who has access to the data. NHTSA should provide that the data should be downloaded only by the local or state police at the scene of an accident, who will in turn be authorized to transmit the data to NHTSA for analysis and encrypted storage. The individual driver/vehicle owner involved in the accident and his or her legal representatives should also have access to the data. NHTSA's regulations should further stipulate that a vehicle manufacturer, or any private or academic party requiring access to the data can have access to aggregate data, but not to individual information; NHTSA may transmit these data to the requesting parties, leaving out all personally identifying information (*including* VIN or EDR ID).

e) *Protocols for Consumer Redress for Malfunctioning EDRs.* Consumers may suffer potentially serious consequences, economic and otherwise (*e.g.*, insurance pricing, crash-related litigation outcomes), should the EDRs in their vehicles malfunction. Therefore, NHTSA should include in the regulations protocols for evaluating the accuracy of EDR devices. These protocols should include the party or parties responsible (*e.g.*, vehicle manufacturers, NHTSA) for ensuring that the devices are functioning properly. Such protocols will also ensure that the crash-related data NHTSA receives are in fact correct and of use in advancing the cause of traffic safety.

IV. Conclusion

EDRs have an important role to play in enhancing our knowledge of the causes and effects of automobile accidents. However, because they can be designed to capture more detailed information than has previously been available at a crash event, the potential exists for the information to be used in ways that could violate a consumer's privacy interests. NHTSA should use its authority to set forth standards regulating the data elements to be captured by all EDRs. These data elements should be the minimum necessary to enhance traffic safety analyses.

Any information capable of identifying an individual should either not be recorded by the EDR, or should be encrypted. In addition, NHTSA should be the official repository for this information, and should be the source through which parties should seek access to the information. Parties other than the individual/driver involved in the crash and his or her legal representative or local or state police should be required to obtain the driver's permission before being allowed to access the data. These steps will aid in protection of this sensitive information.

During this time frame a major issue in the news centered around how safe or unsafe SUVs were. What follows is a open letter from Joan Claybrook, president of Public Citizen addressed to Josephine Cooper, president of the Alliance of Automobile Manufacturers (AIAM). There is a direct connection between automotive black boxes and SUVs. I include this letter to show the tension between safety advocates and motor vehicle manufacturers over issues directly related to enhancing vehicle and highway safety. Automotive black boxes will be used to determine if SUVs are safe as advertised.

One out of eight traffic fatalities in 2001 resulted from a collision involving a large truck.

Cease and Desist Letter

May 22, 2003

Ms. Josephine Cooper

President

Alliance of Automobile Manufacturers

1401 H Street, Suite 900

Washington, DC 20005

Dear Ms. Cooper,

I am writing to ask that the Alliance of Automobile Manufacturers cease and desist in its practice of blaming consumers for the manufacturers' poorly designed, unsafe sport utility vehicles and pickup trucks. Several weeks ago, the National Highway Traffic Safety Administration (NHTSA) published preliminary estimates for 2002 traffic fatalities, which showed that sport utility vehicle (SUV), pickup, and van rollover deaths account for more than half the total increase in traffic deaths for 2002.

In response to this appalling news, the Alliance blamed the victims whose lives were claimed by these disastrously unsafe SUVs and pickup trucks. Eron Shostek, your spokesman, claimed that "If every SUV driver wore their belt, we'd save 1000 lives a year." In the same breath, he also pointed out that alcohol-related deaths rose, overall, in 2002. Robert Strassburger repeated similar claims about belt use in *The Washington Post* on May 3, 2003.

This is just the "nut-behind-the-wheel" theory of the 1960s in updated clothes. It is far past time that the

Alliance acknowledge its members' responsibility to protect public safety, instead of maliciously changing the subject. NHTSA statistics show that SUV drivers have the same safety belt-use rates as car drivers in fatal rollover crashes, and use alcohol while driving at a rate slightly less than car drivers in fatal crashes. The 100 percent belt-use rate referred to by Mr. Shostek has not been achieved by any country in the world (even Canada, which ranks first, has a belt-use rate of 90 percent).

While the Alliance attempts to divert attention from the poor design of SUVs by damning consumers for lack of belt use, your companies know thousands of lives could be saved, and injuries mitigated, by better SUV design. The key problem is that the poor design of SUV s and pickups makes them wobbly and rollover-prone. When they do roll over in a crash, people are harmed by the weak roofs which crush vulnerable heads and spines of passengers, seat belts that do not tighten in rollovers, and weak door locks which open portals for ejection.

Consumer groups. including Public Citizen are avid supporters of increased belt usage. But the manufacturers focus on safety belts as the only remedy is profoundly hypocritical given their installation of inadequate safety belts that too often fail in rollovers. their failure to install effective belt reminder systems. the lack of sensors for air bags and belts to provide crash protection in rollovers. and their repeated refusal to support federal legislation mandating primary safety belt enforcement.

Moreover, SUV and passenger car belt-use rates are virtually identical in fatal rollovers, yet these crashes account for 61 percent of SUV occupant deaths and only 24 percent of car occupant deaths. This critical information about a death toll unique to SUVs places an even greater duty on manufacturers to make vehicles that do not rollover easily, yet when NHTSA considered a minimum rollover propensity standard in the early 1990s, a measure which could have helped to prevent rollover crashes from occurring, industry pressure caused the agency to cave.

Further, Alliance members have done little voluntarily to address these hazards or improve the survivability of rollovers, instead going on record against safety standards to improve roof strength for SUV s and pick-ups. And now, as deaths mount, the Alliance has the nerve to blame the public for the recent rise in fatalities.

Further, the Alliance avoids explaining why 30 percent of those killed in SUVs were killed while wearing a safety belt, and many other belt wearers are injured catastrophically. Ms. Sandy Turner of Little Rock, Ark., is one of the numerous people living with the devastating consequences of these perilous designs. She, like too many others, was not drinking and was wearing her safety belt, but nonetheless suffered paraplegia-inducing injuries when her SUV rolled over. What do you say to her?

Stop deflecting blame and start designing vehicles to save lives.

Sincerely,

Joan Claybrook

President, Public Citizen

On May 19–22, 2003, the 18th International Technical Conference on the Enhanced Safety of Vehicles (ESV) Proceedings was conducted at Nagoya, Japan. There were several papers presented specific to EDRs.[21]

A paper titled "Estimating Crash Severity: Can Event Data Recorders Replace Crash Reconstruction?" by Hampton C. Gabler, Carolyn Hampton and Thomas A. Roston noted that, "The Event Data Recorders (EDRs), now being installed as standard equipment by several automakers, have the potential to provide an independent measurement of crash severity, which avoids many of the difficulties of crash reconstruction techniques. This paper evaluates the feasibility of replacing delta-V estimates from crash reconstruction with the delta-V computed from EDRs. The potential of extracting manual seat belt use from EDRs is also discussed and compared with the corresponding results from NASS/CDS gathered by crash investigators. Although EDRs are expected to greatly enhance the investigation of a crash, it should be noted however that current EDRs are not perfect. The paper discusses the limitations of current EDR technology and the need for enhancement of future Event Data Recorders."

On June 4–5, 2003, the National Transportation Safety Board (NTSB), in conjunction with the Society of Automotive Engineers (SAE) hosted a symposium on vehicle recorders in Arlington, Virginia.[22] The symposium brought together a broad spectrum of manufacturers, operators, safety and regulatory officials, and other industry and government specialists to share the latest technical information and experiences in the use of vehicle recorders in all modes of transportation.

"Vehicle recorders are critical tools for providing information in an accident investigation," said NTSB Chairman Ellen Engleman. "These innovative technologies that will be discussed at the Symposium should provide us with vital information for future investigations."

Topics explored during the two-day meeting included: 1) State of the art in accident recorder technology, 2) Accident recorder survivability/crashworthiness requirements, 3) Video/imaging recorder technology, 4) Data privacy issues, 5) Acquiring data during regular commercial operations and 5) Proactive use of data in commercial operations to prevent accidents and improve efficiency.

On July 21, 2003, Charles J. Murray wrote an article in *EE Times*: The Industry Source for Engineers and Technical Managers Worldwide, titled "Automakers Face Standards Choice for Black Box Recorders".[23]

Murray wrote:

> After being stalled for months by squabbling, standards for automotive "black boxes" are moving forward again, as automakers, suppliers, crash experts and even insurance companies swing their collective weight behind two separate efforts.

> The IEEE plans to publish a vehicle data recorder standard by year's end, and the Society of Automotive Engineers (SAE) expects to release a separate recommended-practices document in the same time frame, thus providing two distinct solutions to the companies and government agencies awaiting a standard.

> The resulting documents, which take different approaches to defining the automotive black box, are expected to serve as a first step toward the creation of a standardized method for collecting accident data. Automakers and crash experts hope that by supplying that first step, they will pave the way for greater highway safety.

> "Standards would be absolutely invaluable for doing crash research," noted Hampton C. Gabler, associate professor of mechanical engineering at Rowan University (Glassboro, N.J.) and a former crash investigator. "There's no doubt that they would lead to the creation of safer cars."

> Observers warn, however, that neither the SAE nor the IEEE effort is a slam-dunk. Although virtually everyone involved recognizes that black boxes offer

tremendous potential for improving crash safety, their use in vehicles is nevertheless hotly debated, and fraught with public-policy questions.

Moreover, the existence of two standards raises questions of which set of specifications automakers will ultimately adopt. Experts said that during the next year, automakers will have to decide whether they want to build to an SAE standard, which would give them more control over black-box data, or an IEEE standard, which would prepare the data for use by a broader swath of industries.

The stakes in the black-box debate are high. Some claim that tens of thousands of lives could be saved if the vehicle-based data recorders were to be universally implemented. Others, however, warn that the data from such devices could end up in the hands of insurance companies or others that could, for example, use it to gather information on the driving habits of high-risk drivers.

Automakers, all of which want to improve vehicle safety, are nevertheless siding with the privacy advocates to some degree. Although the carmakers say they are squarely behind the idea of using crash data to improve safety, they argue that they must keep control of the data to prevent it from landing in the wrong hands.

"We are concerned that, depending on different parties' agendas, they may want us to record more information than is necessary to further the goal of automotive safety," said one automotive representative who spoke to EE Times on the condition of anonymity.

Automakers launched their effort to standardize black boxes, or "event data recorders," in November 2001 through the IEEE, with the goal of providing more useful data to crash investigation agencies like the National Highway Traffic Safety Administration (NHTSA).

Such data is considered a gold mine by crash investigators and automotive engineers who have spent decades trying to reconstruct accidents by employing hand measurements of skid marks and crumple zones. By replacing tape measures with on-board micro controllers, sensors and memories, these groups gain access to information gleaned by wheel speed sensors, air bag sensors, seat belt sensors, yaw sensors and a host of other on-board devices.

Goal: Safer Vehicles

With such data, many believe crash investigators could provide a "data vault" of crash information that would dwarf today's available information, and would eventually lead to the development of safer vehicles. A key area of improvement, they believe, would be in the arena of crash safety for children and the elderly, where available data is now very sparse.

Some automakers already employ data recorders. General Motors Corp., for example, uses its air bag sensing-and-diagnostics module in millions of GM vehicles to serve as a black box of sorts, and can draw pre-crash data from it by allowing investigators to plug into the on-board diagnostics (OBDII) connector, which is commonly used for emissions measurements.

Still, investigators say that obtaining the data isn't always easy. OBDII connectors, they say, are almost impossible to reach amid the crumpled steel of a severe crash. What's more, a recently published study from crash experts at NHTSA and Rowan University suggested that investigators often can't retrieve data because they don't have the right cables to hook up to the array of connectors that can be used on vehicles.

General Motors alone is said to employ more than 10 different cables, and other automakers are similarly unstandardized.

"Investigators end up having to carry a suitcase full of cables to the accident site," said Gabler of Rowan University.

Another source of difficulty is the fact that the information in the recorders is also unstandardized. While some data recorders measure vehicle velocity, others measure acceleration. And while some of those recorders store data in 1-millisecond increments, others store it in 10-ms increments.

Apples and Oranges

"The problem comes if you want to meld the data together," Gabler said. "It's like mixing apples and oranges."

Despite the obvious need for standards, however, the IEEE effort, which was slated to be finished late in 2002, ran into snags last year when automakers disagreed over key issues.

Automotive OEMs claim that the IEEE effort was geared toward determining which data elements would be recorded, and was aimed at a variety of vehicle types, ranging from passenger cars to trucks to ambulances. At least two automakers-General Motors and DaimlerChrysler-backed away from the effort, saying that they wanted the individual manufacturers to decide for themselves which data elements would be recorded. The automakers also said they wanted to limit the application of the standard to passenger cars only.

General Motors, DaimlerChrysler and other automakers, as well as some tier-one suppliers, ultimately decided to move their standards effort to SAE.

"There were a number of folks from the automotive industry that felt the IEEE effort was stumbling," said Bob Kreeb, chairman of the SAE's Vehicle Event Data Interface Committee.

"A lot of the [IEEE] meetings were spent arguing over policy," added an automotive supplier who asked not to be identified. "It was like a dysfunctional family."

Some found the new SAE effort more acceptable because it focused only on passenger cars and "data definitions." If an automaker wants to report acceleration, for example, the SAE document describes the units, level of resolution and level of accuracy for doing so.

"The standard provides a template that manufacturers can adhere to when they record certain data and report it," SAE's Kreeb explained.

A largely unspoken part of automakers' concerns with the IEEE effort, however, was that too many forms of data were being included, which made some of them nervous. Some were said to be concerned that the data would end up in the hands of insurance companies or even marketers who weren't concerned with automotive-design issues.

Some observers believe that the automakers moved to SAE because they could have greater control over the standards process there and, as a result, could keep the data closer to home. At SAE, they say, automakers also will work with familiar fellow automakers, whereas IEEE was notable for tighter ties to the computing industry.

"It's a natural tendency to be uncomfortable when a lot of new players are sitting at the table," said Dr. Ricardo Martinez, chief executive officer of Safety Intelligence Systems Corp. (Atlanta), who is also an emergency room physician and a former head of NHTSA. "But crash information can't be designed with any one group in mind. It has to be designed for the totality of society."

Many observers are convinced, however, that the SAE and IEEE standards are complementary. The SAE document, they say, codifies rudimentary forms of data, while the IEEE standard calls for the data to be richer and more usable by crash investigators, insurance companies and companies such as Safety Intelligence Systems, which intends to build a database of crash information.

"SAE is a good idea for the shorter term," Martinez said. "But IEEE is a longer-term solution that will produce a better quality of data, which will have broader use."

Security Concerns

Computing-industry representatives added last week that they can allay the concerns of automotive OEMs by ensuring that crash data is maintained in a secure fashion. IBM Corp., for example, announced two months ago that it was teaming with Celestica Inc., a provider of electronics manufacturing services, to offer a Java-based black box. Known as eDevices, the new black boxes are being employed by Norwich Union, the United Kingdom's largest auto insurer, and by American Transit Insurance Co., which is said to insure 80 percent of New York City's yellow taxis.

IBM said the insurance companies expect the boxes to provide "unimpeachable" accident data, thus reducing fraudulent claims.

IBM executives also said that such products can be designed to provide a security framework that protects privacy, even when those products use wireless communication techniques.

"We've anticipated the security requirements that the auto industry is going to have, and we're focusing our research and product development on that," said Jim Ruthven, director of IBM's Automotive Telematics Solutions (Detroit).

Many observers hope that as the security issues get resolved, SAE and IEEE will settle on a single standard.

"The idea would be to have close cooperation for the good of the industry," said Tom Kowalick, co-chair for IEEE's black-box standard. "There have been conflicting views, but the goal is still consensus, and we believe we will achieve that."

On August 11–12, 2003, the 12th IEEE Motor Vehicle Event Data Recorder (MVEDR) meeting was conducted at IEEE/USA, Washington, DC. There were thirty individuals in attendance.[24] This two-day meeting centered on technical issues and benchmarks towards completing the standards work. There was a presentation by IEEE-Standards Association staff explained the balloting process and voting procedures for an IEEE standard.

On September 4, 2003, the National Cooperative Highway Research Project 17-24 Intern Report Session was conducted at the National Academies of Science, Washington, DC.[25] There were sixteen in attendance.

The mission statement of the project reads:

There is a critical need to obtain accurate and reliable "real-world" crash data to improve vehicle and highway safety. The use of Event Data Recorder (EDR) information has the ability to profoundly affect roadside safety. EDRs are capable of capturing vehicle dynamics data, such as vehicle speed; lateral and longitudinal acceleration-time histories; principal direction of force on the vehicle; the status of braking, steering, seat belt usage, and air bag deployment; and other valuable crash information. This represents a new source of objective data for the highway and vehicle safety community because it will provide a "real world" connection between controlled test results and actual field performance of vehicles and highway design features.

EDRs have the potential to capture a large number of crash-related and other data elements for a wide range of users with different data needs. The data elements related to improving vehicle safety and driver performance are being used, but little has been done to apply the data elements to roadside safety analysis. Research can identify data elements relevant to roadside safety and improve methods to retrieve, store, and access these data.

Objective: The objectives of this research are to (1) recommend a minimum set of EDR data elements for roadside safety analysis and (2) recommend procedures for the retrieval, storage, and use of EDR data from vehicle crashes.

Tasks: Accomplishment of the project objectives will require at least the following tasks: (1.) Synthesize the current U.S. and international literature on collection, storage, and use of EDR data for roadside and vehicle safety. Meet with a data collection agency to assess current EDR data collection techniques (2.) Identify existing and potential EDR data elements that could be used to improve vehicle and roadside safety. The EDR data elements shall be prioritized based on roadside safety analysis needs. (3.) Review the data elements that are currently recommended for collection in "Model Minimum Uniform Crash Criteria" (MMUCC) and identify those that can be more accurately and effectively collected using EDRs. Identify and prioritize, based on roadside safety needs, data elements not included in MMUCC that could be provided accurately and effectively using EDRs. (4.) Investigate current methods for initial retrieval and storage of, as well as subsequent use of, EDR crash data for roadside safety analysis. Identify key issues, problems, and costs associated with these methods. (5.) Prepare an interim report documenting the findings of Tasks 2 through 4 and meet in Washington, D.C. with the project panel approximately 1 month after submittal of the interim report. (6.) Recommend procedures for improved retrieval, storage, and use of EDR crash data. The recommendations shall consider, as a minimum, resource requirements, cost-effectiveness, legal acceptability, and public acceptance. Identify possible obstacles to implementing the recommended procedures. (7.) Submit a final report that documents the entire research effort.

Status: The final draft "white paper" on the legal implications has been completed. The EDR consumer survey has been completed and the data analysis is
underway. The interim report will be submitted by the
end of July and a panel meeting was conducted on
September 4, 2003.

United Nations Initiative

On September 9, 2003, in a report issued ahead of the upcoming session of the
United Nations General Assembly, Secretary-General Kofi Annan recommended
that the UN's chief legislative body call on Member States to stimulate a new
level of commitment in tracking the problem of road traffic injuries, projected to
rank third among causes of death and disability by 2020.

A United Nations press release noted: [26]

> Improving road safety requires strong political will on
> the part of Governments," Mr. Annan says, recom-
> mending that countries be encouraged to develop and
> implement a national strategy on road traffic injury
> prevention and appropriate action plans.

> An estimated 1.26 million people worldwide died as a
> result of road traffic injuries in 2000 alone, represent-
> ing 25 per cent of all deaths due to injury. The UN
> World Health Organization (WHO) also estimates that
> by 2020, road traffic injuries could rank third among
> the cause of death, ahead of malaria, tuberculosis and
> AIDS.

> Mr. Annan notes that despite the widespread impact
> of traffic accidents, funding for research into the prob-
> lem has been limited. "A lack of research means that
> the magnitude of the problem, its impacts and the cost
> and the effectiveness of intervention are not fully
> understood, particularly in low and middle income
> countries," he says.

> Road crashes can indeed be prevented, but the histori-
> cal approach that places responsibility on the road
> user is inadequate, the Secretary-General says, advo-
> cating an approach that recognizes not only the falli-
> bility of road users but also the infrastructure,

> "In a systems approach, not only the driver, but also
> the environment (infrastructure) and the vehicle are
> seen as part of the system in which road traffic injuries
> occur," Mr. Annan says, adding that Member States
> should also "aim to ensure that sufficient resources are

available, commensurate with the size of the road safety problem in their country."

The Secretary-General also recommends a General Assembly call for efforts by the UN system to address the global road safety crisis. "Most United Nations agencies could integrate road safety into other policies, such as those related to sustainable development, the environment, gender, children or the elderly," he notes.

On September 22, 2003, *Automotive News* published an article titled "Technology : Guardian Angel or Big Brother? Debate over using black boxes intensifies after high-profile crashes" [27]
The articles noted:

> In June, Edwin Matos was sentenced by a Florida judge to 30 years in prison for double vehicular manslaughter. Evidence from a data recorder in his 2002, Pontiac Trans Am helped convict him. It showed that Matos was going more than 100 mph in a 30-mph zone before his car struck another vehicle and killed two teenagers. On the surface the case looks like a clear-cut victory for those who favor the use of "event data recorders"—similar to airliner black boxes—to advance vehicle safety.

> But the situation is not so simple. Some legal experts argue that the recorders violate a motorist's right to privacy and protection against self-incrimination. For the industry, there is a risk of backlash against car companies for putting the devices in their products.

> The government is studying the issue. Dr. Jeffrey Runge, the head of the National Highway Traffic Safety Administration (nhtsa.dot.gov), appears to be leaning in favor of wider use of the devices. Event data recorders "would make our data better," he said.

> Onboard witnesses: But in online discussions after the Florida sentencing, some motorists said they don't want to own vehicles that could wind up being witnesses against them. Even before the high-profile conviction of Matos, General Motors (gm.com) was the target of a class-action lawsuit claiming the recorders, which are part of air bag controls, violate motorists' rights. It subsequently was dismissed.

> Nevertheless, "if consumers won't buy cars with them, automakers aren't going to put them in," says Clay

Gabler, an associate professor of mechanical engineering at Rowan University (rowan.edu) in Glassboro, N.J. As a safety researcher, Gabler favors equipping vehicles with the devices. He says they would be "an absolute gold mine" of information for improving vehicles, safety devices and even roadways and roadsides. But he recognizes that there are technical, legal and consumer hurdles. He's leading a study for a branch of the National Academy of Sciences on those hurdles. Initial results are due late this year.

Behind the scenes, safety researchers and privacy advocates have been wrestling for years over the use of data recorders in privately owned motor vehicles. There is no question that technology is available to record what a vehicle is doing in the moments before a crash.

NHTSA says modules built by GM since 1999 are capable of collecting 16 major categories of information. They include vehicle speed, engine speed, throttle position and brake status. The main issues about the information are who can get it and how it can be used. Now, criminal prosecutors are forcing the debate into the open. In addition to the Florida case, South Dakota authorities say they expect to use data recorder information to make a case against Republican Rep. William Janklow, accused of speeding through a stop sign and killing a motorcyclist on Aug. 16. He was charged with manslaughter on Aug. 29.

Another closely watched case relying on data from both a defendant's vehicle and the one it struck is scheduled for trial in suburban Washington in October. Brian Shefferman, a public defender in that case, has filed a motion to have the data recorder information excluded on the grounds that the government hasn't proven the technology is reliable. Montgomery County, Md., State's Attorney Doug Gansler, the prosecutor, says, "There's no case law suggesting it's not reliable." NHTSA, the agency responsible for regulating motor vehicle safety, is edging toward decisions that could help resolve some of the big issues in the long-simmering debate—in part of its own volition and in part because of outside pressure.

The National Transportation Safety Board (*http://www.ntsb.gov*)—best known for investigating airliner crashes—recommended in 1997 that NHTSA make use of data recorders in vehicles. A NHTSA group studying the technology concluded in August 2001 that event data recorders "have the potential to greatly improve highway safety ... by improving occupant protection systems and improving the accuracy of crash reconstructions."

In addition, NHTSA received a petition two years ago from its own former administrator, Dr. Ricardo Martinez, asking that the agency adopt rules to standardize the collection, storage and retrieval of data recorder information. As it is, every manufacturer's system is different. Only those systems in vehicles built by GM and Ford Motor Co. (ford.com) are readily accessible to investigators. Vetronix Corp. of Santa Barbara, Calif. (vetronix.com), makes the devices for downloading information from the black boxes.

Finally, a year ago, NHTSA asked interested parties to provide comments on what the agency should do about data recorders. The Alliance of Automobile Manufacturers (autoalliance.org), reflecting the wariness of its members, says that development and installation of the devices should continue to be voluntary. The alliance also says answers to legal and privacy questions should "evolve through public debate and judicial precedent."

Missing the boat: Martinez says automakers are missing the boat. They have a chance to take a stand for more open and accurate information. Instead, he argues, they are hiding behind vague concerns about privacy and their proprietary products. He says the companies would benefit if more and better data confirmed that driver error, not vehicle defect, is the cause of most crashes. Martinez argues that the privacy issue is simple. First, the information belongs to the vehicle owner. Second, a legal authority should be able to get the data in the same way it seeks telephone records or computer files. Instead, Martinez says, vehicles are being designed and safety regulations are being written on the basis of inaccurate information about what is happening in crashes. "It's hard for us to say we can't do better with better information," Martinez contends.

Runge's view: Runge, the NHTSA administrator, says data recorders are needed in all vehicles to implement a widespread automatic crash notification system properly. Such a system would use wireless signals to notify police, fire and rescue services automatically when crashes occur and provide them with information about the severity and the likelihood of injuries. Runge says he's not concerned about backlash from motorists who say they don't want their vehicles to snitch on them. He said information from the devices would do as much to defend the innocent as incriminate the guilty. "I hope that's the way the American public looks at it." At the same time, Runge says: "I don't believe that it is fair to trick people. They need to know what is in their vehicle, and there should be complete disclosure about the kinds of data that are in these recorders and who may have access."

On November 3–4, 2003 the thirteenth IEEE MVEDR meeting was conducted at IEE/USA, Washington, DC.[28] The meeting was devoted to defining the data elements for a motor vehicle event data recorder.

On April 7, 2004, World Health Day was devoted to Motor Vehicle Crashes.[29] What is World Health Day? The World Health Organization (WHO) notes:

> World Health Day is an annual event held by the World Health Organization to mark the date of the establishment of the Organization. Traditionally, the day is held on 7th April every year. The World Health Organization uses World Health Day as the main tool to reach out and engage the general public in health messages. Essentially, this occasion is used to engage the international, national and local public in a health message that is known but neglected.

> The objectives of the World Health Day 2004 are: 1. To draw global attention to the growing but preventable human and economic costs of road traffic injuries. 2. To advocate for increased and sustained action in policy, programs, funding and research. 3. To place road traffic injury prevention high on the agenda of governments, international organizations, development agencies, NGOs and the private sector. 4. To launch the World Report on Road Traffic Injury Prevention. 5. To build partnerships and collaboration for road traffic injury prevention. Events will be organized around the world by governments, organizations and groups.

> The World Report on Road Traffic Injury Preventions report will be launched on 7th April 2004.[30] This is the first major and authoritative report produced and issued by the World Health Organization. The World

Report on Road Traffic Injury Prevention will be issued jointly by the World Health Organization and the World Bank. The main message of the report is that road traffic injuries are a major but neglected public health problem, requiring concerted efforts for effective and sustainable prevention. The report has five core chapters: Chapter 1 (Fundamental concepts), Chapter 2 (Global burden, intensity and impacts of road traffic injuries), Chapter 3 (Key determinants), Chapter 4 (Intervention strategies) and Chapter 5 (Conclusions and recommendations).

By late 2003, significant research and development in industry, combined with progress on global standards and privacy legislation in California, were important factors towards the implementation of EDR technologies, regardless, the NHTSA had not decided how to reply to their third petition for EDRs.

The U.S. Department of Transportation / National Highway Traffic Safety Administration conducted a Safety Performance meeting on November 11, 2003, in Romulus, Michigan.

A transcript of this meeting prepared by Neal R. Gross notes:

The next question is what are our plans and timetables for addressing event data recorders. We, as you know have been actively involved in research of event data recorders. We have been collecting event data recorder data from our crash investigation programs for a number of years.

We've worked out some common formats that allow us to read systems. So we came forward with a proposal—actually, a request for comments in October of 2002 on safety benefits, technical issues, privacy issues. and what the appropriate agency role is in continuing the development of EDRs and their installation in motor vehicles. We got 66 comments, which is quite a bit of public interest for this.

In September 2003 we participated in an ad hoc working group under the international harmonized research activities in Ottawa, Canada, to discuss with other countries what they think and what they think would be appropriate for event data recorders. And we're currently trying to discuss internally where we think we are, and what we've gotten from the comments, and we plan to reach an agency consensus on what role, if any, we should take in the standardization and implementation of EDRs by the Spring.

Following a crash there is an immediate need to clear the roadway. Eyewitnesses often disagree on what happened. The solution to this problem is to collect objective scientific crash data.

By year's end many awaited the findings of The National Academy of Sciences study NCHRP 17-24 and wondered if and when NHTSA would respond to the third petition for automotive black boxes.

CHAPTER 10 # Legal and Privacy Issues: 1999–2004

There is a crash every 5 seconds, an injury every 15 seconds and a death every 13 minutes in the United States.

$\mathbf{B}$ack on May 3–5, 1999, the National Transportation Safety Board (NTSB) conducted an International Symposium on Transportation Recorders at Arlington, Virginia. I presented a paper titled *Proactive Use of Highway Recorded Data via an Event Data Recorder (EDR) to Achieve Nationwide Seat Belt Usage in the 90th Percentile by 2002.*

In this paper I wrote about the legal implications of crash recorder data.

"The Office of Technology (OTA) assessment (1975) "Automobile Collision Data: An Assessment of Needs and Methods of Acquisition" cites the following:

> On the question of whether crash recorder data should be admitted, the main point is whether the recorder is reliable, properly read out, and provides a record of the particular event in question. The data of itself is not dispositive of liability, but merely serves as certain evidence of the event. As indicated earlier in this report, there is good correlation between crash severity a recorder might measure and the extent of crash deformation to the vehicle in which it was installed; and it would be difficult to refuse evidence on the crash severity magnitude as interpreted from vehicle deformation. Thus if the recorder provides good evidence of the event, it seems appropriate that the evidence should be admitted. It may be possible to restrict through legislation the admissibility of crash recorder evidence, particularly if the recorders are government-owned and the records are retrieved and interpreted by government employees. Consider, however, the objective of a very simple and widely used integrating accelerometer that is conveniently and readily read by any police accident investigator without special training. It would appear difficult to prevent testimony by a layman—say a tow-truck operator or an auto mechanic—as to what he saw immediately after the accident. In summary, we believe that (1) the data from a crash recorder would be admissible, if it meets necessary qualifications, in a court of law; 2) the data should be admitted if it is good evidence; (3) it will be difficult to prevent admitting crash recorder data, even by Federal law, if the record can be easily read by an untrained person...

I included a section on Respect for Ownership of the Data:

> Privacy is the most important issue regarding the success or failure of implementing the SB/EDR. In a position paper presented to the NHTSA EDR Working

Group entitled Information Privacy Principles for Event Data Recorder (EDRs) Technologies (Kowalick) 1998 noted individual motorists or others within motor vehicles have an explicit right to privacy. Although this right to privacy is not explicitly granted in the Constitution, it has been recognized that individual privacy is a basic prerequisite for the functioning of a democratic society. Indeed an individual's sense of freedom and identity depends a great deal on governmental respect for privacy. Therefore all efforts associated with introducing future EDR technologies must recognize and respect the individuals interests in privacy and information use. Thus, it is imperative to respect the individual's expectation of privacy and the opportunity to express choice. This requires disclosure and the opportunity for individuals to express choice, especially in regards to after-market products.

OEM EDR technology limits an individual's expression of both privacy and choice. After-market value added EDR products permit free market competition and sense of ownership. Several stand-alone after market technologies can easily be combined to produce an after-market EDR virtually independent of the vehicle architecture thereby readily permitting a common standard for retrofitting to a vehicle fleet. Since individuals will operate and occupy vehicles equipped with EDRs that record data elements, subsequently it follows that information is created regarding both individuals and vehicles. Individuals should have the means of discovering how the data flows. A visible means of the type of data collected, how it is collected, what its uses are, and how it will be distributed is basic to consumer acceptance. Consumers should also have a choice in making this data available for post-crash analysis.

Responsibility for disclosure should be high priority and may be achieved through methodologies via print-material formats, etc. Disclosure must be constant and consistent. Any data collected via EDR technologies should comply with state and federal laws governing privacy and information use. All data collected and stored should make use of data security technology and audit procedures appropriate to the sensitivity of the information.

EDR technology data storage should include protocols that call for the purging of individual identifier information respectful of the individual's interest in privacy. Information collected should be relevant to the purpose and mission statement associated with the EDR disclosure statement. Consumers should have the reasonable assumption that they will not be ambushed by information they are providing. Information derived from EDR technologies absent personal identifiers may be used for other purposes clearly stated in the disclosure statement. Information including personal identifiers may be permissible if individuals receive effective disclosure and have a friendly means of opting out.

Personal information should only be provided to organizations that agree to abide by the privacy principles stipulated in the disclosure statement. Should the EDR technologies be maintained in a government database Federal and State Freedom of Information Act (FOIA) obligations require disclosure. Such databases should balance the individual's interest in privacy and the public's right to know. Permanent or temporary storage of data should preclude the possibility of identifying or tracking either individual citizens or private firms and should follow the principles suggested to the EDR Working group.

A position paper presented at the Second World Congress of Intelligent Transport Systems (Yokohama, 1995) entitled *Positioning Systems and Privacy* by C.R. Drane and C.A. Scott cites:

...We put forward for discussion a stronger version of the respect for ownership principle. The stronger version holds that the driver puts time, energy, and money into moving along the road network. Accordingly, the driver has ownership of the trajectory of this movement...The idea that movement data is owned by the person who exerts effort in generating the data is a rather abstract concept.

A paper presented to the NHTSA EDR Working Group entitled Privacy Concerns for the National Highway Traffic Safety Administration by Sharon Y. Vaughn (NHTSA/OCC) concludes:

Following the same procedures that NHTSA implements with respect to operating the NASS, SCI and FARS programs, NHTSA would require a release from the owner of the vehicle in order to gain access to the data from an EDR. NHTSA would assure the owner of the vehicle that all personal information would be withheld from disclosure.

I also offered a solution to the issue of who owns the data:

There are many problems and concerns connected with the question of ownership of the EDR and the data that is generated. It has been argued that vehicles are sold to consumers without any vestigial interests retained by manufacturers and thus the vehicle owner would presumably own the data as well. If this is true, then the ability of public authorities to access the data is greatly reduced and may be impossible since the owner can withhold the data if they felt it would not serve self-interest.

Another problem results when a supplier rather than a motor vehicle manufacturer retains ownership of the data and controls access by utilizing proprietary protocols that essentially prevent anyone else from accessing the data. However, suppliers may report the result of the data extraction. It has been suggested that these problems might be overcome if the manufacturer retained ownership or if an agreement allowing access to the data could be arranged with the owner of the vehicle. The complexity of these solutions would hamper implementation of a SB/EDR system. The simple solution is to design a system that transmits the data from the vehicle to a secure archive for post-crash analysis. By transmitting data through a secure encrypted digital cell link to an archive, problems associated with permission from the owner and access to the vehicle are overcome. A simple release from the owner when the vehicle is registered is all that is legally required. Positive incentives for the owner could include reduced registration fees and a disclaimer that personal identifiers will not be collected and privacy will be preserved. An example reads:

THIS VEHICLE CONTAINS A SYSTEM TO TRANSMIT CRASH DATA ELEMENTS TO A SECURE ARCHIVE FOR POST-CRASH ANALYSIS. THE OWNER OF THIS VEHICLE MAY ACCESS THE DATA.

In 2000, the NHTSA Light Vehicle Event Data Recorder Working Group final report noted the following:

8.0 Privacy and Legal Issues

8.1 Minutes of the Breakout Session

When it comes to the collection and maintenance of data, NHTSA is obligated under the law to protect data if its release would violate the privacy rights of individuals. One of the primary sources for this obligation is the Privacy Act of 1974, 5 U.S.C. § 552a. Under the Privacy Act, Federal agencies are prohibited from disclosing any record that is contained in a system of records by any means of communication to any person, or to another agency, except pursuant to a written request by, or with the prior written consent of, the individual to whom the record pertains, unless disclosure is authorized pursuant to one of the exceptions outlined in the Act.

Under the Act, a "system of records" is a group of records under the control of an agency from which information is retrieved by the name of the individual or by some identifying number, symbol or other identifying particular assigned to the individual.

The purpose of the Privacy Act is to balance the government's need to maintain information about individuals against the right of individuals to be protected against unwarranted invasions of their privacy stemming from the collection, maintenance, use and disclosure of personal information about them.

The Act focuses on four basic policy objections: restricting disclosure of personally identifiable records maintained by agencies; granting individuals increased rights of access to agency records maintained about them; granting individuals the right to seek amendment of agency records maintained about them upon a showing that the records are not accurate, relevant, timely or complete; and establishing a code of "fair information practices" which requires agencies to comply with statutory norms for the collection, maintenance and dissemination of records.

NHTSA maintains a number of Privacy Act "systems of records" and NHTSA is restricted from releasing information from these systems under the Act. There are also other statutes that relate to NHTSA's responsibility to protect private information.

For example, NHTSA is authorized to collect statistical data on motor vehicle traffic crashes to aid in the development, implementation and evaluation of motor vehicle and highway safety countermeasures. Under this authority, the agency is not permitted to release this information in a manner that would identify individuals. In addition, the agency is required under the Freedom of Information Act (FOIA), 5 U.S.C. § 552, to make available agency records that are requested by members of the public. However, the agency is authorized to withhold any information, the release of which would constitute a clearly unwarranted invasion of personal privacy.

During this discussion, Doug Gurin of NHTSA asked whether privacy rights, under the Privacy Act, apply to actions that individuals take in public places, such as on the highway.

Sharon Vaughn responded that the Privacy Act applies to systems of records. If information is maintained in a system of records, then the agency's ability to disseminate the information will be limited. Ms. Vaughn noted, however, that names and other personal identifiers are purged from records before they are ever received by the agency and maintained in many of its databases. (Examples include FARS and NASS.)

Bob Cameron from Volkswagen asked, "What happens when an EDR is recovered from a vehicle and various people want to get access to that, whether it is for litigation, research, truck issues or the NHTSA? What are the rules regarding access to that? Are those governed by the Privacy Act?"

Sharon Vaughn explained that the recovery of an EDR would not necessarily be covered by the Privacy Act. For the Act to apply, a number of conditions would need to be met. For example, the information would need to be in the possession of a Federal agency, and maintained in a system of records (*i.e.*, a group of records under the control of the agency from which information is retrieved by the name of the individual or by some identifying number, symbol or other identifying particular assigned to the individual). If the information were maintained in a system of records maintained by NHTSA, then the agency would be unable to provide it to an OEM, unless NHTSA had the permission of the individual or met one of the other conditions under the Act, under which a disclosure can be made.

Dick Humphrey from GM said Vetronix has been developing a kit that allows the laptop to interface with the SDM. It is due out in November. Best guess for SDM installation in cars is 25%.

Volkswagen's policy regarding EDRs is to get permission from the vehicle owner to have the EDR system turned on or off.

One OEM concern is ability to access the data in a timely period to correct defects in a vehicle. Bob Cameron asked why the NHTSA can not just send the EDR data along. Sometimes third party suppliers will interpret some of the data from the components in the car. The cars already records several functions because of existing memory chips.

NHTSA does not have an investigator in-house to seize data on site.

With OEM EDRs, whoever owns the vehicle technically owns the data. For example, if the vehicle is leased, the leasing company owns the data. In leasing agreements there are clauses where the leaser retains certain rights. For example clauses that state the leaser cannot tamper with certain instruments. Collaborating with leasing companies may provide valuable information through the EDRs.

Federal Highway Administration is interested in the data to improve the defects in the highway system by having a better knowledge in how the collisions are occurring.

If the EDR data was housed with the federal government, other entities would not be able to access the EDR data without the consent of the individuals.

Alan Alminas noted State Farm is interested in research and data from a claims standpoint. There is not a question about reliability, but to what extent an expert is needed to interpret data. Could there be significant variations between plaintiff and defense between experts as to what certain data needs?

With downloading data authenticity is a critical issue. Technology must be tamper-proof. When Volkswagen downloads data they run a test program first to make sure all the circuitry is working properly so it can tell them if something is damaged or destroyed. This tells Volkswagen if they are getting accurate data. With I-Witness' DriveCam (Video EDR), data taken directly from the EDR is authentic and tamper-proof.

8.2 Event Data Recorder Privacy Concerns

The Working Group explored the issues of ownership of information collected by an event data recorder and the effect on the privacy rights of individuals involved in the recorded event.

8.2.1 Who Owns the Data?

8.2.1.1 Position of the National Highway Traffic Safety Administration

It is National Highway Traffic Safety Administration's (NHTSA) position that the owner of the subject vehicles owns the data from the EDR. In order to gain access to the data the government would have to receive a release for the data from the owner of the vehicle. In other crash investigations conducted by NHTSA, the agency assures the owner that all personal identifiable information will be held confidential because the Privacy Act and other statutory authority limit disclosure of personal information.

Some type of personal information that may be retrievable from an EDR would be name, address, age of occupant(s), location of accident (to the extent that the location of the accident would lead to personal identifiable information) and vehicle identification number. Basically, any information derived from the EDR that would lead to personal identifiable information can not be disclosed pursuant to the Privacy Act.

8.2.1.2 Position of the Federal Highway Administration

According to the Federal Highway Administration, vehicles are sold to consumers without any vestigial interests retained by the manufacturers. If the EDR is treated in this way, however, the vehicle owner would presumably own the data as well. This could hamper or stymie the ability of public authorities to access the data by requiring permission from the owner. In addition to the obvious practical difficulties of obtaining permission at the accident scene, the owner would also presumably retain the ability to withhold the data if he felt this would serve his self interest.

A further level of complexity occurs when a supplier, rather than the motor vehicle manufacturer, retains ownership of the data. In Europe, for example, the suppliers essentially control access to the data by utilizing proprietary protocols that essentially prevent anyone else from accessing the data, though they do report on the results of the data extraction.

The problems related to ownership might be resolved by some sort of retention of ownership by manufacturer, by a contractual retention of rights to access the data (perhaps similar to an easement in real property), by a provision in state motor vehicle licensing laws, or by some other federal regulation that permits public authorities to access the data regardless of ownership.

8.2.1.3 Position of Insurance Companies

Many insurance companies have not explored the legal obligation concerning the EDR. For example, one insurance company advised that they have looked into the technology, but they have not looked into any ownership issues. Another insurance company advised that they have not explored the issue of ownership extensively, but concluded summarily that if the insurance company gains ownership of the vehicle, then the data from the EDR is owned by the insurance company.

The complications develop when ownership of the vehicle does not get transferred to the insurance company. The insurance industry believes that an argument can be made that the existing policy may encompass language that would allow the insurance company access to data from the EDR. For example, in standard ISO formatted Personal Auto Insurance Policy Agreement there is language that states that the owner "authorizes us to obtain... other pertinent records." The phrase "other pertinent records" may include the data from the EDR.

8.2.1.4 Position of Volkswagen

Federal, and in many instances, state law, with certain exceptions, prohibit the disclosure of any document to any person or another agency except with the written consent of the person to whom the record pertains.

The purposes of these statutes are to protect the individual against infringing upon his or her rights to privacy as agencies embark upon data collections for multiple purposes. Certain private businesses are similarly regulated by federal and/or state law, *i.e.* the credit reporting industry.

The extent to which a vehicle owner has a right to privacy regarding EDR data depends in Volkswagen's view on whether or not the data identifies the individual person or event, or whether or not the individual person is deemed to have given his or her consent to the use of the data in the manner proposed.

It is Volkswagen's position that irrespective of how any particular data relating to the accident is proposed to be used, if it permits identification of the individual person tied to the accident, that person should be advised of its proposed collection and use regardless of whether or not the law requires it.

8.2.1.5 Position of Thomas Michael Kowalick

Mr. Kowalick advised that all data collected and stored should make use of data security technology and audit procedures appropriate to the sensitivity of the information. EDR technology data storage should include protocols that call for the purging of individual identifier information respectful of the individual's interest in privacy. Information collected should be relevant to the purpose and a mission statement associated with the EDR disclosure statement.

8.2.1.6 Position of General Motors

The risk of private citizens reacting negatively to the "monitoring" function of the EDR can be diffused through honest and open communications to customers through owners' manuals by telling them such information is recorded. The acceptance of recording this data is more likely if the "monitored" data is used to improve the product or improving the general cause of public safety.

8.2.1.7 Position of the National Transportation Safety Board

Testimony of Jim Hall, Chairman of the National Transportation Safety Board before the Committee on Commerce, Science, and Transportation in the US Senate Mr. Hall discussed as his third item of business, "Protection of Data Obtained from Event Recorders."

Mr. Hall addressed this issue focusing on the need for onboard recording devices, which has been an issue on the Board's Most Wanted List since May 1997. Mr. Hall stated that these devices can be used not only in accident investigation and reconstruction, but also by the trucking industry to identify safety trends, develop corrective actions, and can lead to operating deficiencies.

8.2.2 How Are Privacy Rights Affected When Information Is Collected by an EDR?

The fundamental issue is the need for information collected from an EDR to increase safety and the need to protect the privacy rights of individual affected by the information collected from an EDR. The National Highway Traffic Safety Administration approached these issues by applying the Privacy Act of 1974, 5 U.S.C. §552a (the "Act") and other statutory authority, (15 U.S.C. §1395, 1401 and 23 U.S.C. §403).

8.2.2.1 Privacy Act

The Privacy Act of 1974, 5 U.S.C. §552a, provides that no agency shall disclose any record which is contained in a system of records by any means of communication to any person, or to another agency, except pursuant to a written request by, or with the prior written consent of, the individual to whom the record pertains, unless disclosure of the record would be pursuant to one of the exceptions outlined in section (b) of the Act.

The purpose of the Privacy Act is to balance the government's need to maintain information about individuals with the right of individuals to be protected against unwarranted invasions of their privacy stemming from federal agencies' collection, maintenance, use, and disclosure of personal information about them. The Act focuses on four basic policy objectives:

1. To restrict disclosure of personally identifiable records maintained by agencies.

2. To grant individuals increased rights of access to agency records maintained on themselves.

3. To grant individuals the right to seek amendment of agency records maintained on themselves upon a showing that the records are not accurate, relevant, timely or complete.

4. To establish a code of "fair information practices" which requires agencies to comply with statutory norms for collection, maintenance, and dissemination of records.

8.2.2.2 Other Statutory Authority

NHTSA is authorized by Congress (15 U.S.C. §1395, 1401 and 23 U.S.C. §403) to collect statistical data on motor vehicle traffic crashes to aid in the development, implementation, and evaluation of motor vehicle and highway safety countermeasures. This also prohibits the disclosure of personal information that the agency would receive as a result of crash investigations.

Exemption 6 of the Freedom of Information Act, 5 U.S.C. §552(b)(6) prohibits disclosure of personal information received by the agency that, if disclosed, would constitute a clearly unwarranted invasion of personal privacy.

8.2.2.3 Court Decisions

Since the EDR technology is in the developmental stages, there is no case law available in this area of law. The most recent case that relates to EDR technology involves the Diagnostic and Energy Reserve Module (DERM), which was described by an engineer with General Motors as "like an airplane 'black box'." In this case, the Plaintiff sued General Motors alleging that the air bag deployed after rather than during a low-speed collision, resulting in injury to plaintiff. Although this case was decided on procedural grounds, the engineer for General Motors submitted an affidavit stating that he had downloaded data from the DERM and concluded that the DERM data from the vehicle suggests that the supplemental restraint system functioned as designed by deploying during the plaintiff's accident. *See, Harris v. General Motors Corporation*, 201 F.3d 800, 804 (6th Cir. 2000).

There are cases that mention "black boxes," but these cases describe the role of the "black box" as evidence in the case. *See, In re Korean Airlines Disaster of September 1, 1983*, 156 F.R.D. 18 (D.C. Cir. 1994)(where the release and analysis of the flight data recorder were evaluated and determined to be newly discovered evidence); *Sundstrom v. McDonnell Douglas Corp.*, 816 F.Supp. 587 (N.D. Cal. 1993)(wrongful death suit where the seat data recorder in USAF planes was alleged to be a defective design, manufacture and assembly); *In re Air Crash Disaster at Sioux City, Iowa, on July 19, 1999*, 131 F.R.D. 127 (N.D. I11. 1990)(the court held a flight simulator was not needed where sufficient evidence was available using the flight data recorder and the cockpit voice recorder).

8.2.3 Conclusion

The issues of ownership of information collected by an EDT and the effect on the privacy rights of individuals involved in the recorded events will be further explored with the development of the technology.

National Academy of Science Study

Michael Edmund O'Neill, Associate Professor at George Mason University of Law, Arlington, Virginia prepared a paper titled "Legal Issues Surrounding the Implementation and Use of Event Data Recorders" for the National Cooperative Highway Research Program, Transportation Research Board, National Research Council under NCHRP Project 17-24: Event Data Recorder (EDR) Technology for Highway Crash Data Analysis. The following unpublished and unedited excerpts (by TRB) are provided by permission granted to John Wiley & Sons, Inc. from the National Academies (11/12/03) and are copyrighted to the National Academies of Science, Transportation Research Board, 2003. The National Academies may publish a final report.

Preliminary Conclusion(s)

The initial scope of this paper was to determine whether the Fourth Amendment to the United States Constitution in any way barred the collection of data recorded by Event Data Recorders ("EDRs"). After the preliminary draft was completed, additional questions arose involving the United States Department of Transportation's ("DOT") authority to mandate the installation of EDRs in all new vehicles, as well as the admissibility of the data recorded by EDRs in court, and whether the collection of such data violates privacy rights.

The report's preliminary conclusions are as follows: First, it is clear that the DOT may require the installation of devices that demonstrably improve highway safety or advance some other significant public policy interest. The public policy interest in installing EDRs seems beyond peradventure. As a consequence, DOT likely enjoys the authority to mandate the installation of such devices on new automobiles.

Second, with respect to Fourth Amendment concerns, it appears that the police (or other governmental accident investigators) may properly seize such devices (or otherwise collect the data therefrom) without a warrant during post-accident investigations either because seizure of a required safety device does not constitute a search implicating the Fourth Amendment, or in the alternative because seizure of a safety device qualifies under the exemptions for conducting a warrantless search. Unless there are changing expectations with respect to an individual's reasonable expectation of privacy regarding EDR data, however, police may not be able to seize such data as a routine either without a warrant or express legislative authorization.

Third, although the data (and the recorder itself) may be "owned" by the automobile's owner or lessee, that data may almost certainly be used as evidence against that owner (or other driver) in either a civil or a criminal case. Certainly nothing within the federal rules of evidence or the Fifth Amendment's protection against compelled self-incrimination would exclude the use of data recorded by the EDRs. Similarly, owners might be prohibited from tampering with the data if litigation is pending.

At bottom, the issue here is not one so much of legal authority to use EDR data in court, but instead what the public will accept. While the statutory authority to require EDRs may exist, the public may not want a device installed in their automobiles that may appear to trench upon their personal privacy interests. The problem is less a legal concern than it is a battle to mold public perception. Not every life-saving device that is deployed with the best of intentions will be accepted by the public. Personal privacy and public safety must exist within the same sphere. Occasionally, respecting privacy rights will mean that harmful things may come about, but that is the cost of living in a free society.

Background
Event Data Retrieval and Analysis devices ("EDRs") act as automobile "black boxes" providing critical information about an automobile's operation and the status of its various systems in the seconds immediately preceding an accident and during the crash itself.[1] Information of this kind may assist (among others)

police agencies, accident reconstructionists, lawyers, rental car companies, safety researchers, vehicle fleet managers and insurance companies. The National Highway Traffic Safety Administration ("NHTSA") has explained that:

"The information collected by EDRs aids investigations of the causes of crashes and injury mechanisms, and makes it possible to better define safety problems. The information can ultimately be used to improve motor vehicle safety."[2]

Despite the obvious safety benefits that might accrue, however, the use of EDRs has not been without controversy.[3] Privacy concerns seem to have been a particular sore spot for those advocating the general use of EDRs.[4] Oftentimes, the concern is less about the data EDRs presently collect, but instead what future devices might be capable of recording. Presumably, the devices could be engineered to collect considerably more data and do so over even longer periods of time. On-board cameras could record driving habits, sensors could determine cell phone use, and even breathalyzers could be installed to monitor alcohol consumption. It could easily be argued that each of these innovations might improve highway safety. Such improvements, however, would plainly come at a decrease in personal privacy. Faced with such potential intrusions upon personal privacy, the public would doubtless be more willing to permit the collection of certain types of data as opposed to others.

After all, the use of EDRs to collect telemetric data is not new. General Motors Corporation ("GM") allegedly first began installing EDRs in its cars in 1990, equipping nearly 6 million cars to date.[5] Since 1990, the quantity and type of data collected by EDRs has dramatically expanded. Initially, EDRs were limited to recording data from the time between a collision and the air bag's deployment. In 1994, however, GM modified its EDRs to record and save additional information, including: the change in velocity during an accident; the change in velocity of an event even if the air bag did not deploy; whether or not the seat belt was fastened; and the time between the moment of vehicle impact and the moment of maximum change in velocity. Since the 1994 modifications, GM has enhanced the EDRs to record such information as the vehicle's overall speed, engine rpms, brake status, and its throttle position.

The data collected by currently available EDRs remains saved in the device for approximately 60 days for the average driver, or until a subsequent serious accident or other event erases the data. It cannot be erased otherwise (with the exception of intentional destruction). To make it easier to download data from the scene of an accident, GM partnered with Vetronix Corporation to produce a decoder that downloads the data from the automobile. As of this writing, at least four state trooper organizations have purchased the Vetronix decoder for highway use.[6] In order to avoid the manual, on site, method of data collection, however, Dr. Ricardo Martinez, a former NHTSA administrator, has proposed to create a "Global Safety Data Vault" through which the data from the EDRs will be downloaded automatically through telematic systems like those already in existence.[7] Although such a plan would doubtless prove controversial, the present future of EDR use appears secure. Ford and Toyota have apparently signed agreements with Renneberg-Walker and Vetronix—presently the industry's leading EDR decoders—to supply EDR decoders in their new cars.

The type of data collected by on-board sensors could readily be increased. For example, Dr. Martinez has suggested that data collection could easily be expanded to include any data that could be "gleaned from electronic sensors already installed on the vehicle."[8] Dr. Martinez was referring to the tire pressure data, telemetric data (currently used to contact emergency services), the functioning of antilock braking systems, electronic suspension information, and the routine diagnostic information used by mechanics.[9]

Although car manufacturers claim EDRs help their engineers refine on-board safety systems, privacy advocates (such as the American Civil Liberties Union) decry their use, claiming the devices unfairly erode personal privacy.[10] As a consequence of these on-going concerns, the NHTSA commissioned a panel of experts that included members of the industry, academia, and the government to study EDRs. The panel concluded in its 2001 report that EDRs would "profoundly impact highway safety" by allowing for "better design of occupant protection systems and improved accuracy of crash reconstruction." Additionally, the panel reported that studies of black boxes have shown that driver awareness of the devices can "reduce the number and severity of the crashes."[11]

The National Transportation Safety Board ("NTSB") has recommended, since 1997, that NHTSA "gather more and better real-world crash data" using EDRs. Despite the benefits EDRs seem to deliver, and the positive recommendations from the NTSB, however, NHTSA has twice rejected petitions that would require EDRs to be installed in all automobiles.[12]

Presently, the collection and use of EDR data exists in something of a legal vacuum. It has yet to be conclusively determined whether information provided by EDRs may be admitted at trial. Similarly, it is unclear whether the use of data recorded by EDRs may implicate Fourth Amendment or other privacy concerns. The federal and individual state governments are only now beginning to consider the legal implications of deploying such devices as EDRs or global positioning systems ("GPS"). For constitutional purposes, courts must address whether accessing EDR data at the scene of an accident qualifies as a search triggering the Fourth Amendment. If such access qualifies as a Fourth Amendment search (and seizure), a court must then consider whether such a search is valid without a warrant.[13] Before discussing the Fourth Amendment issues raised by EDRs, it is worth determining at the outset whether the federal government may require the installation of EDRs.

I. Regulatory Authority and Use and Collection of EDR Data

As a general matter Congress's authority to regulate interstate transportation is found within the Constitution's Commerce Clause.[14] The Constitution provides Congress the power "[t]o regulate Commerce with foreign Nations, and among the several States, and with Indian Tribes."[15] Congress's power under the Commerce Clause extends to any activities affecting commerce;[16] courts have interpreted this grant of authority broadly.[17]

In the seminal case of *Gibbons v. Ogden*, the Supreme Court described the depth and breadth of the Commerce Clause as: "complete in itself, [and] may be exercised to its utmost extent, and acknowledges no limitations, other than [those] prescribed in the Constitution."[18] The Court has expressly recognized that Congress has the authority to regulate the channels and instrumentalities of interstate commerce, which of course includes the regulation of motor vehicle safety.[19] This authority is quite broad. Nevertheless, any proposed legislation must pass rational relationship muster. Specifically, a rational relationship must exist between the activity regulated and interstate commerce.

In *Nevada v. Skinner*,[20] for example, the state of Nevada challenged the constitutionality of the national speed limit. The state of Nevada argued that Congress could not have been acting rationally in imposing a uniform national speed limit because a lower national speed limit would inhibit rather than promote the goal of rapid commercial intercourse.[21] The Supreme Court rejected that argument, holding that Congress is entitled to impose a speed limit that could create a safer highway system. The court reasoned that "commerce that proceeds safely is more efficient than commerce slowed by accident or injury."[22]

Under the rational relationship test, Congress does not have to create a perfect statute or even the best option among several. Rather, its approach need only be rationally related to an otherwise permissible, socially desired end. In the case of EDRs, this is not a particularly difficult hurdle to overcome.

A. *May the Federal Government Require Manufacturers to Install EDRs?*

Under the Motor Vehicle Safety Act, the DOT, on advice from the NHTSA, may promulgate, through informal agency rulemaking, federal highway safety standards, including manufacturer's safety component requirements. Both the Motor Vehicle Safety Act and the 1974 amendments concerning occupant crash protection standards indicate that motor vehicle safety standards are to be promulgated under the informal rulemaking procedures of the Administrative Procedure Act ("APA").[23]

For example, in 1967, based (in part) upon an understanding that seatbelts would save a substantial number of lives, DOT required manufacturers to install manual seat belts in all automobiles.[24] Similarly, after significant NHTSA testing revealed the utility of passive restraint systems, DOT required manufacturers either to install a passive restraint device, such as automatic seatbelts or air bags, or to retain manual belts and add an "ignition interlock" device that in effect forced occupants to buckle up by preventing the ignition from turning on if the seat belts were not engaged.[25]

DOT may also require manufacturers to install other "devices" in the interests of public policy. For example, in *New York v. Class*, the Supreme Court confirmed the validity of a DOT rule requiring Vehicle Identification Numbers ("VINs") in automobiles, noting that in light of the important interest served by a motor vehicle identification number, the federal and state governments were amply justified in making it a part of the web of pervasive regulation that surrounds the automobile. In addition, although it acknowledged certain privacy interests, the Court had no difficulty in upholding the regulation requiring the VIN's placement in an area ordinarily in plain view from outside the passenger compartment.[26] The regulation, of course, required the public placement of the VIN to allow police officers easily to verify ownership.[27] Effectively, this regulation compelled owners to make their automobiles identifiable to police officers (and anyone else, for that matter). As with GPS devices, use of a VIN number enables the dedicated investigator to track the location of a vehicle wherever it may be parked.

Thus, if it can be demonstrated that the installation of EDRs demonstrably improves highway safety, DOT might possess the authority under the Motor Vehicle Safety Act to require installation of these devices in all newly manufactured automobiles.[28] Even if DOT has the legal authority to do so, however, without popular support, it may be difficult to mandate the use of EDRs. In the wake of a public backlash, for example, Congress could always choose to override any DOT regulations requiring EDR use.[29]

This is precisely what happened with so-called ignition interlock devices. These devices detect the presence of alcohol on the driver's breath and, when the alcohol level is too high, prevent the car from being started. Initially, ignition interlock devices were used primarily as a means to prevent those convicted of repeated drunk driving offenses from reoccurring.

NHTSA, however, believing such devices to be a significant benefit to automobile safety, decided to require the installation of interlock devices in newly manufactured automobiles.[30] NHTSA's decision was based upon solid research demonstrating the pervasive problem of drinking and driving—a problem widely understood by the public. Despite the anticipated benefits of installing ignition interlock devices, however, public opposition was so fierce that NHTSA quickly rescinded the regulation. Thus, even though the statutory authority existed for NHTSA to require installation of the device, the public's willingness to accept it was another matter.[31] The same is doubtless true for EDRs. Merely because EDRs may seem to be a positive force for promoting highway safety and giving vital information to engineers seeking to develop safer vehicles does not mean the public will embrace them.

The similarity between EDRs and Cockpit Voice Recorders ("CVRs") has lead some to suggest that the NHTSA would be able to mandate use of EDRs in the same way the Federal Aviation Administration ("FAA") is able to require CVRs and Flight Data Recorders ("FDR") in airplanes. This is a difficult analogy to draw, however, because the FAA has substantial authority to impose requirements upon aircraft by virtue of the highly regulated status of air travel. Indeed, the FAA regulates virtually every aspect of air travel, from product design,[32] to the licensing of pilots,[33] to air traffic control.[34] As was demonstrated in the wake of the September 11, 2001 terrorist attacks, the FAA even possesses the authority to deny the right to fly over the United States. The FAA also enjoys plenary authority to investigate any problems with flight over the United States. As a consequence, the FAA can require that the FDR and CVR be turned over after an accident occurs to determine what caused the accident and to ensure the safety of future flight.

Although NHTSA has the ability to regulate certain aspects of driver behavior by requiring states to enforce certain laws, it does not have the authority to mandate that drivers reveal their driving habits via an EDR. Unlike airline pilots and even commercial vehicle drivers, no federal agency licenses the driver of a passenger vehicle. Therefore, the federal government has substantially less interest in how an individual operates her vehicle than how a pilot flies his aircraft.

Nevertheless, substantial leeway for regulation in this area does exist.

B. *What Authority Permits the NHTSA and the Various State Departments of Transportation to Include EDR Information in their own State Databases?*

Congress has authorized NHTSA to collect statistical data on motor vehicle crashes to aid in the development, implementation, and evaluation of motor vehicle and highway safety measures.[35] As a consequence, since the early 1980s, NHTSA has been obtaining crash data files derived from data recorded on PARs ("Police Accident Reports"). NHTSA refers to the collection of these computerized state crash data files obtained from 17 states as the State Data System, which is conducted by the National Center for Statistics and Analysis ("NCSA"). The crash data files are requested annually from the appropriate state agencies. In most instances, this includes the state police, the state highway safety department and the state Department of Transportation. These safety efforts, as implemented by the Secretary of Transportation, are authorized by federal statute, which provides in pertinent part:

§401. Authority of the Secretary

The Secretary [of transportation] is authorized and directed to assist and cooperate with other Federal departments and agencies, State and local governments, private industry, and other interested parties, to increase highway safety.

In addition, the Secretary of Transportation has an obligation to Congress, as detailed further in section 401, to prepare, publish and ultimately to submit a report on the highway safety performance of each State in the preceding year.[36] This report must include data on highway fatalities, injuries, motor vehicle accidents in urban as well as rural areas.[37] This data is geared to providing the Transportation Secretary with the means for comparing highway safety performance of the States in an effort to provide over all improved national safety. Increasing safety and promoting highway safety are plainly legitimate state interests. Thus, the use of EDR and other accident data is rationally related to such interests and does not violate the Equal Protection Clause of the state of federal constitution, as some opponents of collected data have argued.

At the federal level and state level, the use of EDR data is to save lives, reduce injuries and prevent property loss. This includes collecting data to assist in a better safety management system for the highway and traffic systems. The federal government, through NHTSA, utilizes this data to assess safety problems and solutions for issuing new and revised vehicle safety performance standards. At the state level, crash data is used to assist states in managing road systems and designing better roadside safety hardware, such as guardrails and crash cushions.

In the future, it is not unlikely that new statutes will permit the use of EDR data to assist in emergency medical rescues; more specifically, providing EDR data to be automatically dispatched from the crashed vehicle to the Public Safety Answering Point (PSAP) center as well as other affected parties. Furthermore, EDR data would help the local authorities assign the "right" response teams early in the event, thus fostering a more efficient emergency response system.

NHTSA's desire to make EDRs mandatory in the interest of safety thus serves a compelling national interest; however without clarification of NHTSA's authority to collect and use the data provided by EDRs, misunderstanding will continue to occur and the important public purpose of a cooperative, independent accident investigation may not be served.

Presumably, NHTSA's intention to require the installation of EDRs is motivated by a desire to protect the public by improving highway safety. Technological advances, such as the EDR, allow NHTSA to take effective actions in improving the timeliness, accuracy, completeness, uniformity, and accessibility of their highway safety data. EDRs have played major roles in NHTSA's accident investigations and will continue to do so to a greater extent as their use becomes more widespread. For this reason, it might be useful to identify methods for expanding the use and function of recorders.

Nevertheless, as the presently record stands, investigators are not empowered to halt an accident investigation and clean-up activities simply to obtain EDR data. Thus, all too often, valuable information is needlessly lost due to contradictory statutes and lukewarm mandates concerning EDR use. Consequently, it might be argued that the nation's safety could be greatly improved with a government mandate authorizing NHTSA to retrieve, preserve, copy and use EDR data as it sees fit.

Congressional and public support for any such action, however, is vital. As an illustration, one need only consider the Environmental Protection Agency's ("EPA") quest to promulgate regulations promoting a clean environment. In 1990, the EPA faced a similar predicament in its quest to protect public health and the environment through its recommendations for improving air quality. It lacked popular support, however. Alone, the EPA was severely limited in achieving its goals, by working closely with Congress, however, the EPA was able to garner support for its environmental quality recommendations. The resulting federal mandates, which regulate air emissions from area, stationary, and mobile sources, were quite controversial at the time.[38] In particular, the EPA sought the installation of catalytic converters on newly built cars.[39] The EPA prevented automobile owners from removing or otherwise interfering with the catalytic converters' function.[40] Without cultivating congressional support, it is unlikely such broad, potentially unpopular changes could ever have been made.

In similar fashion, by carefully securing legislative support of EDRs and being open with the public, NHTSA might be able to accomplish its goal of enhancing highway safety through the routine deployment of such devices. NHTSA's Strategic Execution Plan (June 1996) cites that its mission is to save lives, prevent injuries and reduce traffic related health care and other economic costs. Without clarification of NHTSA's authority to use the most advanced means available, (including the use of EDRs), misunderstandings will continue to occur and the important public purpose of a cooperative, independent accident investigation may not be served.

EDR data obviously offers a range of possibilities. EDRs could be the basis for an evolving data-recording capability that could be expanded to serve other purposes, such as emergency rescues, where their information could be combined with occupant smart keys to provide critical crash and personal data to paramedics. It is even possible that the NHTSA could prevent car owners from tampering with EDAs or otherwise interfering with their collection of data, much the same way the EPP prevented the disabling or removal of catalytic converters. The question of data ownership and data protection would have to be resolved, however; but it is entirely possible such interests can be balanced with the government's objective of ensuring consumer safety on the roads.

II. DOES THE SEARCH OF AN AUTOMOBILE TO OBTAIN THE INFORMATION CONTAINED IN AN EDR RAISE A FOURTH AMENDMENT QUESTION?

Important constitutional questions surround the use of EDRs in the field. In particular, the Fourth Amendment to the Constitution [41] protects individuals from "unreasonable" searches and seizures undertaken by the state.[42] Although, when probable cause is present, the state may conduct searches of private property and effect seizures of evidence or contraband uncovered, questions arise whenever an individual's property is searched or seized without a warrant. Of course, no legal difficulty exists if the owner consents to a search. However, even where affirmative consent is withheld (or at least not given) if an individual has no expectation of privacy in the thing to be searched, then no "search" has occurred for Fourth Amendment purposes. Accordingly, if the owner of an automobile has no expectation of privacy in the information contained in the EDR, then the acquisition of that data is not a "search," and no Fourth Amendment question exists.

Aside from the Constitution, the most pertinent federal law governing this area is likely the Privacy Act of 1974, which provides that no federal agency shall disclose any of its records that are contained in a system of records by any means of communication to any person, or to another agency, except pursuant to a written request by, or with the prior written consent of, the individual to whom the record pertains, unless disclosure of the record falls within certain specified exemptions.[43] The Privacy Act's purpose is to balance the government's need to maintain information about individuals with the right of individuals to be protected against unwarranted invasions of their privacy stemming from federal agencies' collection and use of personal information.[44] While this Act would doubtless affect the use of EDR data, plainly such novel technical innovations have yet to receive full legislative consideration. It is a near-certainty that they will in the future.

This next section addresses the extent to which the Fourth Amendment will affect the process of obtaining the data contained in EDRs, I will assume *arguendo* that individuals do have an expectation of privacy in the data contained in their EDRs.[45] Even though such a privacy right may exist, however, it is plainly not without limitation.

A. May private parties obtain the data contained in EDRs without the consent of the vehicle owner as part of discovery in preparation for trial?

First, it is worth exploring whether private parties (insurance companies, car manufacturers, litigants, etc.) may obtain EDR data. The Fourth Amendment does not govern the actions of private parties, it applies exclusively to the actions of governmental authorities. Although civil law (tort law protections and the like) may prevent a private party from obtaining EDR data without the owner's consent, those parties may retrieve the data contained in an EDR without consent as part of discovery. Rule 26(b) of the Federal Rules of Civil Procedure ("FRCP") states "that a party may obtain discovery regarding any matter, not privileged, that is relevant to the claim…" With respect to the nature of the materials that may be obtained under Rule 26, Rule 34(a)(1) allows for the discovery of "other data compilations." This language has been interpreted to include electronic data, diaries and surveillance equipment.[46] While there is no case law that authoritatively endorses the discoverability of EDR data, the data contained in an EDR has been successfully discovered in recent cases.[47]

If EDRs overcome (as they presumably will) evidentiary standards of reliability and accuracy, government agencies such as the NTSB have declared their support for the admissibility of EDR data in court.[48] By contrast, certain industry groups are concerned about the data's potential misuse in litigation and regulatory enforcement.[49] As a consequence, the American Trucking Association (ATA)—itself a target of potential regulations mandating the use of EDRs on commercial trucks—announced a policy statement:

In order to benefit from new technologies that can improve highway safety and efficiency, while providing protection against information misuse, the trucking industry supports creation of reliable data parameter standards only if: (1) they are developed and implemented for all vehicles, including passenger cars, concurrently; (2) all vehicle owners and operators are properly protected against the use of electronically-generated data in regulatory enforcement and civil litigation; (3) data are anonymous and used for safety research and trend analysis by a single lead agency or institution; (4) reasonable privacy can be assured regarding access and use of the information; (5) access to data is controlled; (6) data are recorded only for a limited period of time relative to an event; and (7) there is no burden on individual vehicle owners or operators for the reporting or collection of such data at any time.[50]

The issues of using crash data in court proceedings was addressed by in Proactive Use of Highway Recorded Data via an Event Data Recorder (EDR) to Achieve Nationwide Seat Belt Usage in the 90th Percentile by 2002, at http://www.ntsb.gov/events/symp%5Frec/proceedings/authors/kowalick.htm (last visited Jan. 26, 2003). Kowalick explains that:

"On the question of whether crash recorder data should be admitted, the main point is whether the recorder is reliable, properly read out, and provides a record of the particular event in question. The data of itself is not dispositive of liability, but merely serves as certain evidence of the event. As indicated earlier in this report, there is good correlation between crash severity a recorder might measure and the extent of crash deformation to the vehicle in which it was installed; and it would be difficult to refuse evidence on the crash severity magnitude as interpreted from vehicle deformation. Thus if the recorder provides good evidence of the event, it seems appropriate that the evidence should be admitted. It may be possible to restrict through legislation the admissibility of crash recorder evidence, particularly if the recorders are government-owned and the records are retrieved and interpreted by government employees. Consider, however, the objective of a very simple and widely used integrating accelerometer that is conveniently and readily read by any police accident investigator without special training. It would appear difficult to prevent testimony by a layman—say a tow-truck operator or an auto mechanic—as to what he saw immediately after the accident. In summary, we believe that (1) the data from a crash recorder would be admissible, if it meets necessary qualifications, in a court of law; (2) the data should be admitted if it is good evidence; (3) it will be difficult to prevent admitting crash recorder data, even by Federal law, if the record can be easily read by an untrained person

One prominent concern is the potential violation of privacy rights posed by the use of EDRs. This concern extends not only to the types of data EDRs presently record, but to the types of data EDRs might be able to record, and store, in the future. While it would be difficult to shield EDR data from civil discovery, legislation could be enacted to control the use to which such evidence could be put.

Although no federal statutory scheme directly touches upon EDR use in automobiles, there is a somewhat analogous federal statute that refers to "cockpit recordings and transcripts"[51] and "surface vehicle recordings and transcripts"[52] in the context of use, admissibility, and discovery.[53] This statute prevents the NTSB from publicly disclosing cockpit and surface vehicle voice/video recordings, while leaving disclosure of the transcripts/written depiction of those recordings at the discretion of the NTSB.[54] If the NTSB allows public disclosure, parties in a judicial proceeding are free to admit the information into evidence.[55] However, if the NTSB denies public disclosure, a party in a judicial proceeding may not use discovery to obtain the information, absent a court order.[56] Even so, statutory confidentiality safeguards exist to prevent public dissemination of the data if a court admits otherwise undisclosed data.[57] As differentiated from voice/video recordings and transcripts thereof, pure event data, collected from a recording device is generally admissible.[58] Of course, the policy considerations of prejudice and misinterpretation that may apply to graphic cockpit voice recordings from an airplane crash do not apply to basic factual EDR data. Such data would doubtless be admitted at either civil or a criminal trial unless expressly shielded by legislation.

In applying the issue of EDR data admissibility to recent case law, two cases illustrate the ways in which

EDR data has been used in civil litigation.[59] In *Harris v. General Motors Corporation*, the district court relied upon testimony from GM's expert witness regarding EDR crash data to grant GM summary judgment in a product liability suit alleging a defect in the air bag.[60] The trial judge found that the EDR data (and, presumably, the engineer's interpretation of that data) established beyond dispute that the air bag had functioned properly. Absent a factual dispute, the court felt obligated to grant GM's summary judgment motion. The court of appeals reversed the trial judge's decision to grant GM summary judgment. The appellate court determined that the trial judge erred in accepting the engineer's interpretation of the EDR data as an undisputed fact. Indeed, the appellate court raised the issue that the EDR evidence may not even have been admissible at all.[61] The Court noted:

"[While the plaintiff] did not raise the *Daubert [v. Merrell Dow Pharm., Inc.]* issue before the district court, we note that on remand, the district court must, consistent with its gate keeping role, perform a *Daubert* analysis of the proposed testimony of the defense experts, particularly [the EDR expert]. Certainly, nothing in the record as it now exists evinces either the reliability or validity of [the EDR expert's] testimony as to the [EDR]. Our own research did not reveal a single reported case addressing the *Daubert* issue as to General Motors' automotive "black box.""[62]

As will be discussed later in this paper, the appellate court was referring to the standard district courts must use in determining whether scientific or other expert testimony is admissible. The court did not opine on the reliability of EDR data, it merely held that summary judgment could not be granted on the basis of testimony regarding a novel device that was not subjected to rigorous analysis by first by the trial court in determining whether it should be admitted, and then by the plaintiffs during the course of cross examination and rebuttal evidence.

Similarly, in the product liability case of *Batiste v. General Motors Corporation*, a state trial court admitted expert EDR testimony from the same expert that testified in Harris.[63] In *Batiste*, that expert testified:

"Based on my years of experience and training and the safety aspects of automobiles, it is my opinion that the evidence in this case demonstrates that the air bag was functioning properly and should not have deployed at the time of this accident. Moreover, if there was a malfunction of the system, it would be evident from the [EDR]. The [EDR] recorded no such malfunction. Accordingly, it is my further opinion that the air bag was not defective in any respect and performed as intended and therefore, did not cause Plaintiff's injuries, if any. Furthermore, it is my opinion that the injuries, if any sustained by Plaintiff, would not have been lessened had the air bag deployed."[64]

There was little question that the EDR data, as well as the expert witness' interpretation of that data, should have been admitted. Therefore, as long as the EDR technology can pass the fairly liberal *Daubert* test, it would appear from these initial cases that EDR data and relevant expert testimony will be admissible in a civil trial.

EDR data may be analogized to the information contained in a diary or that recorded by a surveillance camera. Similarly, EDR data may be compared to personal computer files because either represents personal possessions saved in a digital format. It has long been the case that personal diaries, surveillance footage, and computer files are all discoverable. Thus, there is little doubt that (absent special statutory protection) the data contained in an EDR would likewise be discoverable.[65]

It is often overlooked that from a policy perspective, admissibility of EDR data could be a positive advancement in ensuring the integrity of litigation. Once determined to be a reliable source, EDR data appears to provide a credible and objective insight into the facts of a crash. If all relevant parties to litigation are provided initial accessibility to the data, equity will be ensured, as all parties would be able to analyze and interpret exactly the same facts. Although this may help an automobile manufacturer demonstrate that an air bag properly deployed and that the plaintiff failed to wear her seatbelt, it will also enable the plaintiff objectively to verify that he was not traveling above the posted speed limit and that the brakes failed. Objective factual determinations will greatly aid litigants and may, in fact, help reduce unnecessary litigation and impede patently fraudulent claims.

B. *May private parties, such as insurance adjusters, private attorneys, researchers, etc. obtain the data contained in the EDR at the scene of the accident without the consent of the vehicle owner?*

The simple answer is "no," private parties likely cannot obtain the information contained in the EDR without the consent of the vehicle's owner (or as part of pre-trial discovery). The owner of property has superior title to all other private parties, and can lawfully refuse possession. The NHTSA and Federal Motor Carrier Safety Administration (FMCSA) both take the position that the EDR and its data belong to the vehicle owner. Because the vehicle owner would have ownership of the EDR data superior to all others, no private party could force the individual to relinquish that data without consent.

However, private solutions may address this problem. Specifically, insurance companies, as a condition of writing an automobile insurance policy, may require that owners consent to the retrieval of EDR data. Progressive Insurance, for example, which is the nation's fifth-largest auto insurer, has placed hundreds of monitoring devices in customers' vehicles to measure how, when and where they drive.[66] According to public reports the device's patent describes a system of onboard sensors that could track whether a driver signals before turning, tailgates or stops so sharply that anti-lock brakes engage. This is in contrast with standard EDRs, which apparently only record the last seconds before a crash. Progressive has also taken a different track in that its customers *volunteered* for the test program. [67] The primary incentive is that customers can save up to 25% on insurance rates tailored to their individual driving habits, as the company expects to benefit by obtaining new business from consumers who are seeking to obtain favorable rates, or who perhaps have teenage drivers in the family whose driving habits they would like to monitor. In any event, the program is entirely voluntary.

Similarly, automobile manufacturers may include as boiler plate language within a sales agreement, a promise by the vehicle's purchaser to waive any privacy interest in EDR data in the event that he sues the manufacturer at some later date. Private law solutions such as these may provide adequate coverage to situations in which private parties may need to have access to EDR data and the owner's consent might not be likely at the time the data is needed.

C. *May Private Parties Obtain and Use EDR Data when Unrelated to Trial Discovery?*

A different sort of circumstance arises when private parties seek to obtain EDR data for purposes other than formal litigation. In a case of first impression, a car rental customer sued Connecticut-based ACME Rent-A-Car because the company fined him for exceeding the speed limit, information that was discovered by use of an on-board global positioning system (GPS).[68] The company could, in fact, pinpoint the precise location where the consumer had violated the speed limit.[69] The customer sued because he claimed he had not been afforded adequate notice that is driving would be monitored by means of a GPS device. Despite the fact he was merely in temporary control of the vehicle as a renter, he was successful.[70] Rental agencies (at least in Connecticut) may still be able to track their vehicles using GPS devices, but apparently may not issue fines unless adequate notice to the consumer is first given. Indeed, many of the problems associated with this sort of situation could be dealt with by disclosing the information to the consumer at the outset. Just as rental car companies routinely explain to prospective renters information relating to insurance coverage, they could do the same for EDRs or GPS. Leasing agencies could similarly include as standard language in lease agreements waivers related to the disclosure of recorded EDR data. Any interested party (such as a bank or a lien holder) could potentially require the car's principal driver to agree to disclose information in the event of an accident or product liability litigation. Private solutions exist that would enable EDR data to be disclosed, when necessary.

III. May Police Officers Seize EDR Data During Post-accident Investigations Without A Warrant?

A slightly more complex, but readily answerable question is whether police officers (or other government accident scene investigators) may seize data recorded by and EDR at the scene of an accident. This is perhaps the most likely scenario that would present itself; namely, an accident occurs and police officers arrive on the scene in short order to assist the injured and to investigate the crash's cause. What are their options for retrieving what may be crucial data?

The analytical framework evident in the Supreme Court's Fourth Amendment cases requires first that a reviewing court assess whether the individual claiming Fourth Amendment protection has a reasonable expectation of privacy in the object searched.[71] If not, no "search" occurs for Fourth Amendment purposes. If a privacy right is implicated, the court must next determine whether probable cause existed for the search. Finally, if a search occurred, it must have been executed subject to a valid warrant, or fit an exception to the warrant requirement.

A. *Do car owners have reasonable expectation of privacy in EDR devices as a component of their automobile?*

1) Fourth Amendment Searches

Seizure of an EDR or the data contained therein will only implicate the Fourth Amendment if it constitutes a search into a constitutionally protected area. A "search" does not occur unless the individual manifested a subjective expectation of privacy in the searched object, and society is willing to recognize that expectation as reasonable.[72]

The nature of "searches" is not quite as clear-cut in today's world as might be expected. Traditionally, a "search" has required some sort of trespass upon property. Thus, if a police officer merely happened by an open window and, in his plain view, witnessed a crime, no "search" for Fourth Amendment purposes occurred.

Automobile searches, however, present a different sort of a problem. In *United States v. McIver*,[73] the Ninth Circuit Court of Appeals upheld the warrantless placement of a global positioning system (GPS) tracking device by law enforcement on the undercarriage of a suspect's vehicle.[74] The court ruled that the officers' placement of the device was neither a search nor a seizure. The court held it was not a search because the undercarriage is part of the exterior of the vehicle, and pursuant to the Supreme Court opinion in *New York v. Class*,[75] there is no reasonable expectation of privacy in the *exterior* of a vehicle. The court determined that no seizure occurred because the device represented only a technical trespass on the automobile, and the device's placement did not deprive the defendant of dominion and control over his vehicle.[76]

EDRs may, or may not, require a specific trespass onto the owner's property. That may ultimately be a distinction without a difference, however, as the Supreme Court has recently considered a similar sort of situation in *Kyllo v. United States.*[77] In *Kyllo*, a narcotics agent used a thermal imaging scanner to determine whether the defendant was using high intensity lamps to grow marijuana in his home. The thermal imagining device did not require law enforcement officers to enter the defendant's property; rather, they simply had to point the device at the home and record the thermal image. The defendant argued that the use of a thermal-imaging device in this way, even though the officers were stationed across a public street, constituted a "search" within the meaning of the Fourth Amendment. The Court agreed, holding that when the government uses a device that is not in general public use to explore details of a private home that would previously have been unknowable without physical intrusion, the surveillance is a Fourth Amendment "search," and is presumptively unreasonable without a warrant.

Thus, while there was no physical "trespass," the Court nevertheless determined that a search had occurred. To compare this to the situation involving EDRs, it might be possible to conclude that data retrievable through external means (like vehicular speed obtained through the use of a radar gun) may not constitute a search, while data such as the functioning of air bags, seat belts, or braking, which could only previously have been done during a physical examination of the vehicle, would implicate a Fourth Amendment search. This analogy is difficult to draw, however, because homes have traditionally received far greater constitutional protections than have automobiles. The *Kyllo* Court also suggested that should the use of such devices become routine, the privacy interest thereby affected might diminish. In other words, if such devices became so common that the general public knew of their routine use and existence, the scope of the privacy right itself might change.

Presently, of course, anyone who operates a motor vehicle knows that both the license plate and the VIN numbers are readily accessible to the police (or other on-lookers). Drivers accept the fact that the police can run those identification numbers and find out a great deal about the car's owner. Physical descriptions, police records, driving records, home addresses, telephone numbers, and a good deal of other personal information is available to law enforcement authorities just by having access to VIN and license plate numbers. Yet, no one demands that the police obtain a warrant before obtaining such otherwise private information. If the general public is willing to accept such warrantless intrusions into their lives, it is possible that should wireless access to EDR data become customary, the public will also become cavalier about permitting authorities routine access to that data as well.

2) Privacy and the Fourth Amendment

Car owners doubtless have a reasonable expectation of privacy in their car. [78] That zone of privacy generally applies to the passenger compartment, but, depending upon the circumstances, may not extend to cover containers located within the automobile.[79] Thus, analogizing EDRs to "containers" which may "hold" evidence, officers investigating car accidents may seize EDRs knowing the device "holds" critical crash data which will assist in the investigation. As the car owner has no privacy interest in the "container," [80] such a seizure does not implicate the Fourth Amendment.[81] However, this is not a particularly satisfactory analogy. A better comparison might be made to other data that is retrievable from the accident scene, such as skid marks or roadway conditions. Essentially, one must consider whether engineering data of this sort is really "owned." After all, a radar gun deployed by a police officer records data about auto speed. The car's owner, however, has no privacy interest in the data recorded by that radar gun. This analogy is also tricky, however, because the car's owner controls the EDR as well, so the fit is a bit odd.

In addition, whether a car owner maintains a reasonable expectation of privacy in a car and its component parts after an accident may depend on the owner's actions. A property owner must manifest some intention to maintain the privacy of the property the police intend to search.[82] Analogizing to cases involving so-called "fire searches," where property owners take affirmative steps to protect their damaged property, they retain a valid expectation of privacy.[83] Similarly, by voluntarily abandoning property, an individual forfeits any reasonable expectation of privacy in the property.[84]

Thus, if the vehicle is only slightly damaged and can be driven away, and the owner demonstrates an intention to drive the car away, the owner likely retains his or her privacy interest in the automobile and its component parts. However, if the car is damaged beyond reasonable repair, then the owner must make some further effort to secure the car to maintain a privacy interest. If the owner takes no affirmative steps to secure the car, then police officers may conduct a valid search for causal evidence without a warrant.[85] Similarly, if the driver (and any occupants) is injured or unconscious, the police may be able to retrieve the EDR and any data contained therein during the process of assisting the injured.

Finally, if an EDR is regarded as a safety or "other" device required under the Motor Vehicle Safety Act, and the data thereby recorded is considered important in advancing a significant public policy interest, a car owner may not necessarily possess a privacy interest in the data superior to that of the public. If one analogizes to the privacy interest that is abrogated by the use of mandatory VIN numbers, it could similarly be argued that a car owner may have no reasonable expectation of privacy in EDR-recorded data. In *New York v. Class*, for example, the Supreme Court stated that because of the important role the VIN plays in the pervasive government regulation of the automobile, and the efforts by the federal government to ensure that the VIN is placed in plain view, there is no reasonable expectation of privacy in the VIN, for Fourth Amendment purposes.[86] Thus, if the EDRs are required because they play an important role in government regulation of the automobile industry, there may be no reasonable expectation of privacy in EDR data for Fourth Amendment purposes.

B. *Does a car owner have a reasonable expectation of privacy in the telemetry data provided by EDR devices?*

Under *Katz*, an individual must manifest a subjective expectation of privacy in the searched object that society is willing to recognize as reasonable to protect property from unreasonable search.[87] In pursuing the second inquiry, the test of legitimacy is not whether the individual chooses to conceal assertedly "private activity," but whether the government's intrusion infringes upon the personal and societal values protected by the Fourth Amendment. Thus, regardless of car owner's subjective expectations, it is unlikely the courts will validate an objective expectation of privacy in vehicle safety data.

In *Smith v. Maryland*, the Supreme Court held that the installation, at the request of the police, of a pen register[88] at the telephone company's offices to record the telephone numbers dialed on the petitioner's telephone did not violate the Fourth Amendment.[89] The petitioner had no legitimate expectation of privacy in the telephone numbers since he voluntarily conveyed them to the telephone company when he used his telephone. "This Court consistently has held that a person has no legitimate expectation of privacy in information he voluntarily turns over to third parties."[90] Telephone numbers dialed from one's home arguably raise more significant privacy concerns than the speed of one's vehicle immediately prior to a crash—information that would be obtainable if, for example, a police officer equipped with a radar gun were standing by immediately prior to the accident. Yet, telephone numbers dialed from one's home do not enlist a reasonable expectation of privacy. In light of the fact that the same information may be ascertained by other means, it seems unlikely that vehicle safety data will itself be the subject of an inviolate privacy interest. Of course, some information, such as the engine's RPM's, may not be externally measurable and thus may not fit readily within this determination.

In *Oliver v. United States*, the Supreme Court offered further that the privacy issue turns on "whether the government's intrusion infringes upon the personal and societal values protected by the Fourth Amendment."[91] While no single consideration has been regarded as dispositive, "the Court has given weight to such factors as the intention of the Framers of the Fourth Amendment, … the uses to which the individual has put a location, … and our societal understanding that certain areas deserve the most scrupulous protection from government invasion."[92]

It seems unlikely under *Oliver* that vehicle safety data warrants Fourth Amendment protection. Individuals make no use of EDRs. In fact, very few individuals are even aware their automobiles contain such a device. EDRs exist to provide information to third parties, much the same way that a license plate or a registration certificate provides information to other parties. Thus the "uses to which the individual" puts on the EDR weighs against a reasonable expectation of privacy. Similarly, it is unlikely that our "societal understanding" supports protection of vehicle safety data. Many forms of data provided by EDRs (speed, brake application, seat belt use, air bag deployment) are readily and regularly gathered as part of accident reconstruction and investigation. This data is gathered without objection because it is not regarded as a search into a constitutionally protected area. Simply changing the method of collecting this now unprotected data does not create a privacy interest where none before existed.

The current understanding of both NHTSA and the Federal Highway Administration appears to be that the vehicle's owner also owns any data recorded by the EDR.[93] The NHTSA takes the position that:

The owner of the subject vehicle owns the data from the EDR. In order to gain access to the data NHTSA must obtain a release for the data from the owner of the vehicle. In crash investigations conducted by NHTSA, the agency assures the owner that all of NHTSA's personal identifiable information will be held confidential pursuant to the Privacy Act *(5 U.S.C. 552a)* and other statutory authorities which limit disclosure of personal information. Any information derived from the crash investigation, including an EDR, that would lead to personal identifiable information may not be disclosed pursuant to the Privacy Act.[94]

Similarly, the Federal Highway Administration's ("FHWA") Office of Chief Counsel observes that:

Vehicles are sold to consumers without any vestigial interests retained by the manufacturers. The problems related to ownership might be resolved by some sort of retention of ownership by manufacturer, by a contractual retention of rights to access the data (perhaps similar to an easement in real property), by a provision in the state motor vehicle licensing laws, or by some other federal regulation that permits public authorities to access the data regardless of ownership.[95]

In addition to the NHTSA and FHWA, the Federal Motor Carrier Safety Administration ("FMCSA") branch of the Department of Transportation and American Trucking Associations, Inc. ("ATA") has asserted that the owner of a vehicle with an EDR is the exclusive owner of that EDR's data.[96]

A potential implication of the ownership issue is that a vehicle owner, if found to be the owner of the EDR data, may preserve her right to withhold or erase the data if that decision is in her self-interest. The FMCSA notes that only the vehicle owner, or a party having the owner's consent, can access the EDR data, unless a law enforcement official has obtained a warrant to investigate a crash. As will be discussed, infra, I do not think this is necessarily correct. [97]

However, the data ownership issue is not quite as obvious as it might appear at first blush, however. Similarly, it may be the case that no one really "owns" the data recorded by the EDR. To borrow an example from copyright law, it is clear that:

"Original works . . . fixed in any tangible medium of expression, . . . from which they can be . . . communicated, either directly or with the aid of a machine or device are subject to copyright protection. Ideas or facts, however, are not protected by copyright."

Thus, while the data recorder itself is owned by the person who controls the vehicle's title, the data, which is not original work-product but rather merely factual information, may not be owned. While I may own a piece of real estate, for example, I do not "own" the information contained in the county plat recording its physical boundaries. A person may own an automobile, but she doesn't own the information relating to that vehicle's speed as it hurtles down the freeway. The issue may therefore be more one of access to the data, as opposed to ownership of the data itself.[98] It therefore seems unlikely that information gathered by EDRs warrants protection under the Fourth Amendment. As such, it may be subject to search without the Fourth Amendment's strictures.

C. *Wireless Communications and Electronically Stored Data*

An important issue meriting further consideration is the increasing use of wireless means to access electronically stored data. This is true not only in the context of data stored in EDRs, but access to wireless network connections that lead to data contained in hard drives or servers, or wireless connections to the internet where information may be maintained. With respect to EDRs, the issue is whether police need a warrant to access data stored on the devices if they can do so wirelessly—in other words, without directly trespassing on private property. In an early effort to deal with the nascent field of wireless communications, Congress comprehensively overhauled federal wiretap law with the enactment of the 1986 Electronic Privacy Act ("ECPA").[99] The ECPA, which was designed to protect and secure the privacy of wire and oral communications between individuals, extended the wiretap provisions to include wireless voice communications and electronic communications such as e-mail or other computer-to-computer transmissions.[100] As previously noted, such communications were not given protection by most courts. Among other things, the ECPA prohibits the willful interception or willful use of "wire" or "oral" communications.[101] EDR data does not fall within these definitions, but it is useful to see the way in which Congress prohibited the interception of communications where no physical trespass was necessary. Although the interception of all wire communications is prohibited by the ECPA, interception of oral communications is only prohibited if made under circumstances justifying a reasonable expectation of privacy. The expectancy of privacy is a necessary precondition to obtaining the Act's protection.

It is important to understand that EPCA only applies to "communications," fairly narrowly defined, and does not pertain to the wireless downloading of mere data. Arguably, the interception of communications between individuals merits even greater protection than the mere wireless downloading of data. Nevertheless, the majority of federal courts held that individuals had *no* reasonable expectation of privacy in such wireless communications, whether one or both of the parties communicated by a cordless LAN telephone or cellular telephone.[102] If one has no reasonable expectation of privacy in a telephone call, then should one have a reasonable expectation of privacy in factual information recorded by an EDR? This is certainly not a clear-cut situation. Nevertheless, Congress and most states intervened to ensure the privacy of such communications in light of these federal court decisions. Similarly, Congress (and the individual states) certainly could choose to protect EDR recorded data as well. However, it is difficult to conclude that EDR data would receive *greater* privacy protections than those afforded to actual person-to-person communications.

Interestingly, the law treats electronically stored data, such as EDR data, quite differently either from data intercepted in real time, or data stored in more traditional ways. The ECPA provides protection for e-mail and other forms of "electronic communication" held in "electronic storage," which could arguably include data stored in an EDR. In order for the government to seize any "electronic communications"[103] in "electronic storage"[104] for 180 days or less requires only an ordinary warrant, and seizure of electronic communications in storage for more than 180 days[105] on an "electronic communications service,"[106] requires only a subpoena or an order issued pursuant to an offering of "specific and articulable" facts showing reasonable grounds to believe that the contents of an "electronic communication" are relevant to an ongoing criminal investigation is required.[107] Presumably, the government could create different requirements for EDR-stored data as well.

III. May Police Officers Obtain The Data Without The Owner's Consent after Obtaining A Warrant for Both Criminal And Non-criminal Investigations?

Clearly, if police were to acquire a warrant they would be able to obtain the information contained in an EDR. Police have been able to obtain computer files, diaries, and video surveillance footage with a warrant.[108] As described *infra*, EDRs are comparable to diaries, surveillance data, and computer files. Personal diaries, surveillance information, and computer files, are all obtainable with a warrant, or in a civil case as a routine part of pre-trial discovery. Thus, Police would be able to obtain the data contained in an EDR.

A. *May police officers seize EDR information without a warrant?*

Perhaps the more difficult question is whether the police may obtain the data recorded by the EDRs without a warrant. Assuming *arguendo* that car owners do have a reasonable expectation of privacy in EDRs or EDR data, police officers may nonetheless seize EDRs without a warrant based on exigent circumstances or "special needs." If an individual maintains a reasonable expectation of privacy in the object to be searched, then seizure is within scope of Fourth Amendment, and is prohibited without a warrant or a valid exception to the warrant requirement. Exceptions to the warrant requirement, being founded on public interest requirement for some flexibility in application of the general rule, arise in those cases where the societal costs of obtaining a warrant, such as danger to law officers or risk of loss or destruction of evidence, outweigh reasons for recourse to a neutral magistrate.[109]

Furthermore, the Fourth Amendment guarantees "the right of the people to be secure in their persons, houses, papers, and effects, against unreasonable searches and seizures," and further provides that "no Warrants shall issue, but upon probable cause."[110] However, the government has consistently recognized exceptions to the warrant requirement even the "the First, Second, and Fourth Congresses ... authorized federal officers to conduct warrantless searches" of ships and vessels, to find property that owed duty.[111]

Courts and other commentators often point to *Carrol v. U.S* as the case that created the so-called "automobile exception" to the Fourth Amendment.[112] The "automobile exception" allows police officers to stop and search automobiles without a warrant, as long as the police officers have probable cause to believe that there is evidence of criminal activity within the automobile. The rationale behind the decision in *Carrol*, was that the mobile nature of cars creates an exigent circumstance making a warrant impractical and counter productive to law enforcement. *Carrol* follows earlier cases the enabled authorities to search carriages and maritime vessels as part of a system of inspections without needing to resort to a warrant.

1) Exigent Circumstances

Circumstances that justify warrantless searches include those in which officers reasonably fear for their safety, where firearms are present, where there is risk of criminal suspect's escaping, or fear of destruction of evidence.[113] The exigent circumstances doctrine bears special application where the object of the search is damaged property, as in the inspection of burnt homes or businesses. Analysis under the Supreme Court's line of cases involving "fire searches" is framed by two Supreme Court decisions: *Michigan v. Tyler*,[114] and *Michigan v. Clifford*.[115] In *Tyler*, a furniture store fire was reduced to "smoldering embers" by the time the local Fire Chief reported to the scene.[116] The Chief concluded that the fire was possibly the result of arson, and called a police detective, who took some photographs, but "abandoned his efforts because of the smoke and steam."[117] Four hours later, the Chief returned with the Assistant Chief, whose task it was to determine the origin of all fires in the township. The fire was effectively out when they returned, and the building was empty. The investigators quickly left, returning with the police detective around 9:00 a.m.

They found suspicious burn marks, not visible earlier, and took samples of carpet and stairs. Rejecting the premise that "the exigency justifying a warrantless entry to fight ends, and the need to get a warrant begins, with the dousing of the last flame, " the Court found the two searches conducted on the morning after the fire were constitutionally permitted.[118] After noting that the investigation on the night of the fire was hindered by the darkness as well as the steam and smoke, the Court found that the fire officials' morning-after entries were no more than an actual continuation of the first valid search.[119] The *Tyler* Court promulgated a yet-undisturbed rule: "a warrantless entry by criminal law enforcement officials may be legal when there is compelling need for official action and no time to secure a warrant."[120]

In *Clifford*, a fire department reported to a residential fire about 5:42 in the morning. The fire was extinguished, and the fire officials and police left the premises at 7:04 a.m. At about 1:00 p.m. that afternoon a fire investigator arrived at the scene, having been informed that the fire department suspected arson.

Despite the fact that the house was being boarded up on behalf of the out-of-town owners, the Cliffords, and despite their knowledge that the Cliffords did not plan to return that day, the fire investigator and his partner searched the house. After determining that the fire had been set in the basement, and how, the investigators searched the entire house, taking photographs.[121] In finding that the challenged search by the fire investigator was not a continuation of an earlier search, as in *Tyler*, and in distinguishing between the two cases, the Court explained:

Between the time the firefighters had extinguished the blaze and left the scene and the arson investigators first arrived . . . the Cliffords had taken steps to secure the privacy interests that remained in their residence against further intrusion. These efforts separate the entry made to extinguish the blaze by that made later by different officers to investigate its origin. Second, the privacy interests in the residence—particularly after the Cliffords had acted—were significantly greater than those in the fire-damaged furniture store [in *Tyler*], making the delay between the fire and the mid-day search unreasonable absent a warrant, consent, or exigent circumstances.[122]

Thus, the *Clifford* Court laid out three factors for analyzing the constitutionality of warrantless searches of fire-damaged premises: 1) whether there are legitimate privacy interests in the fire-damaged property that are protected by the Fourth Amendment; 2) whether exigent circumstances justify the government intrusion regardless of any reasonable expectation of privacy; and 3) whether the object of the search is to determine the cause of the fire or to gather evidence of criminal activity.[123] Fire officials need no warrant to enter and remain in a building for reasonable time to investigate the cause of a blaze after it has been extinguished.[124] Should police officers charged with investigating car accidents be afforded the same authority? Perhaps so. Fire officials are charged not only with extinguishing fires, but with finding their causes. Prompt determination of the fire's origin may be necessary to prevent its recurrence...."[125] For this reason, fire officials need no warrant to enter and remain in a building for a reasonable time to investigate the cause of a blaze after it has been extinguished.[126]

Accident investigation officials are similarly charged with finding an accident's cause. EDRs provide data critical to that inquiry. In *Commonwealth of Massachusetts vs. Mamacos*, a defendant, charged with two counts of homicide by negligent operation of motor vehicle, moved to suppress results of the testing of his truck and all items removed from his truck on the ground that such evidence was obtained without search warrant.[127] However, after the case was transferred from the Appeals Court, the Supreme Judicial Court, Essex County, O'Connor, J., held that: (1) the police department had a right to remove any truck involved in a fatal accident from the scene of the accident and to hold such truck in storage for a reasonable time, and (2) even if the owner of the truck involved in the fatal accident had subjective expectation of privacy with respect to truck's brakes, society would not recognize such an expectation of privacy as reasonable when the truck came into possession of the police following the death of motorists, and accordingly, police officer's examination and testing of the brakes conducted after the owner requested that the truck be returned to him was not a "search" within the meaning of the Fourth Amendment.[128]

While prompt discovery of an accident's origin is not necessary to prevent its reoccurrence, a prompt seizure of the EDR may be required to prevent loss of the EDRs critical data. This may be especially true where vehicles are only slightly damaged, and may be driven from the scene by their owners. "[A] warrantless entry by criminal law enforcement officials may be legal when there is a compelling need for official action and no time to secure a warrant."[129] Where a driver may remove a vehicle from the accident scene there exists the possibility that critical evidence may be lost, thus creating a "compelling need for official action."[130]

Alternatively, police officers may seize EDRs without a warrant during accident investigations because the EDR contains critical evidence of the accident's potential causes, and may furnish other evidence used to prosecute drivers from criminal offenses. It is well settled that warrantless searches of automobiles are permitted by the Fourth Amendment if the officers have probable cause to believe that the vehicle contains contraband or other evidence of a crime.[131] Whether an officer has probable cause to search a vehicle depends on the totality of the circumstances viewed "in light of the observations, knowledge, and training of the law enforcement officers involved in the warrantless search."[132] As the Supreme Court stated in *Ross*, "If probable cause justifies the search of a lawfully stopped vehicle, it justifies the search of every part of the vehicle and its contents that may conceal the object of the search."[133]

Accident investigators, because of their training and experience, may have reason to suspect that drivers involved in accidents have committed a criminal offense or are not answering the officer's questions regarding potential crimes truthfully. In those instances, police officers may be justified in seizing EDRs because they know or have probable cause to believe that the EDR contains evidence of a crime. A police officer in that situation may seize the EDR without violating the Fourth Amendment. However, the exigent circumstance rational has been supplemented by subsequent cases such as *South Dakota v. Opperman*, which held that in addition to the mobile nature of cars "less rigorous warrant requirements govern because the expectation of privacy with respect to one's automobile is significantly less than that relating to one's home or office."[134] *Cady v. Dombrowski*, explained that the reduced privacy interest derived "from the pervasive regulation of vehicles capable of traveling on the public highways."[135] *Opperman*, elaborated on the "pervasive regulation" rationale stating that automobiles are subjected to "continuing governmental regulation and controls including periodic inspection and licensing requirements."[136] *Opperman*, pointed out that cars are subjected to inspections for expired license plates, inspection stickers and for other violations "such as exhaust fumes or excessive noise" and "[defective] headlights or other safety equipment." [137]

Explaining the boundaries of a search conducted without a warrant, *U.S. v. Ross* held "if probable cause justifies the search of a lawfully stopped vehicle, it justifies the search of every part of the vehicle and its contents that may conceal the object of the search."[138] *Ross*, further held that the search was determined not by the nature of the containers within the car, but by the nature of the evidence searched for.[139]

When a police officer arrives at the scene of an automobile accident, he is investigating and inspecting both safety and criminal concerns. The police officer will need to assess at the scene if there are any safety issues that need to be resolved immediately such as a gas leak, or an obstructed lane of traffic. Additionally, the police officer will need to determine whether the cars at the scene are safe to be driven from the scene. As a separate empowering interest, the police officer will have to determine if any of the drivers were committing criminal offenses prior to the accident. These criminal offenses could be as minor as driving without a license, expired tags, defective tail light, improper or unsafe suspension (such as, but not limited to, an unsafe low-rider, or a truck that is improperly raised), speeding, failing to yield, improper lane change etc. On the other hand, the criminal offenses could also be quite severe such as driving while intoxicated, driving under the influence, reckless driving, and reckless homicide etc.

The individuals involved in the car accident already have a reduced privacy interest.[140] The rationale for this reduced sense of privacy is that society objectively recognizes that cars are already subjected to multiple inspections, safety requirements, and licenses.[141] Finally, if a police officer has probable cause to believe that one of the cars involved in the accident was violating criminal guidelines, or in the alternative that the act of driving away from the scene would violate minimal safety statutes, the police officers can search and inspect the car to the extent necessary to resolve those issues. [142] *Ross* gives police officers the authority to search in any "container" where they have probable cause to believe they will find evidence of the crime or violation they are investigating. By downloading the EDR data, a police officer could quickly evaluate if the brakes are working properly, if the brakes were used at all, if the driver was speeding, if the driver was speeding to the point of being reckless, if the driver was in an accident of the magnitude that would likely damage a car to the point that it would be unsafe to drive. Depending on the particular EDR, the police may be able to obtain even more information.

2) "Special Needs" Exception

Moreover, authority derived from legislative action mandating seizure of EDRs during accident investigations may arise through the "special needs" exception to the Fourth Amendment's proscription against warrantless searches. A search unsupported by either warrant or probable cause can be constitutional when special needs other than the normal need for law enforcement provide sufficient justification.[143] Under the "special needs doctrine," a court identifies a special need which makes impracticable adherence to warrant and probable cause requirements, then balances the government's interest in conducting a particular search against an individual's privacy interests upon which the search intrudes.[144] The special needs doctrine allows the state to dispense with the normal warrant and probable cause requirements when two conditions are satisfied. First, the state must show that it has some "special need" or governmental interest beyond the normal law enforcement activities that make the search or seizure necessary. Second, the state must show that its interest cannot be achieved or would be frustrated if a court imposed normal warrant and probable cause requirements. If the state satisfies these two conditions, the court engages in an independent analysis, balancing the state's interest against individual privacy interests. Only if the court is satisfied that the state's interest in the search or seizure outweighs the individual's privacy interest will it uphold the search and dispense with the warrant and probable cause requirements.

In addition, the Motor Vehicle Safety Act makes clear that the state has a vested interest in highway safety. If EDRs are required in automobiles, it must be that NHTSA found EDRs instrumental in promoting highway safety. Thus it seems likely that the state, in mandating seizure of EDRs during accident investigation, has a special interest beyond normal law enforcement, such as the state promotion of safety on highways and the efficiency of the civil tort system.[145] Thus, it seems likely that significant state interests cannot be achieved if accident investigators are subject to normal warrant and probable cause requirements.

Finally, even assuming citizens can claim some privacy interest in an EDR or its data, that interest is likely to be relatively small considering the state's interest in promoting highway safety and public welfare. Whatever intrusion that results from the seizure of an EDR is therefore minimal. Seizure of EDRs thus fulfills the "special needs" exception, and accident investigators can be authorized by statute[146] to seize EDRs without a warrant.

B. *Additional Considerations Regarding the Use of EDR Data*

Most of the forgoing discussion dealt with the seizure of information at the scene of an accident. Law enforcement authorities have fairly broad authority to secure information from the scene of a crash. Different rules, however, come into play when police are simply seeking information related to the prosecution of a crime. There have been a number of cases in which EDR data was used to prosecute a crime.[147] It is clear that the information recorded by EDRs could prove crucial in future criminal prosecutions—particularly if the devices are altered so as to record a greater variety of data. A recent Florida case highlights the uses to which EDR data may be put. In that case, the defendant was charged with four counts of DUI manslaughter and two counts of vehicular homicide killing two teenagers in a an accident.[148] Although a blood test showed that the defendant was intoxicated at the time of the accident, the trial court found that test inadmissible because the defendant had not given consent to take it voluntarily. Absent data from the EDR, which measured the defendant's speed at 114 mph five seconds before the crash and detected that he was pressing the gas pedal very hard—at 99 percent of its maximum capacity—it might have been difficult to prosecute the case.

Prosecutors are not the only ones potentially to benefit from the use of EDR data. Just like plaintiffs may use the data to bring lawsuits against manufacturers, defendants may also be able to use EDR data to defend themselves successfully. In *Colorado v. Cain*, [149] the defendant in a vehicular homicide prosecution used EDR evidence to show that he was not speeding when the accident occurred. His defense was successful, as the jury acquitted him.

Presumably, police officers obtained the data in these cases at the accident scenes. If police officers seek to obtain data from an EDR *after* the event has occurred and the vehicle has been released into the owner's custody, then it is clear that they must obtain a search warrant. Of course, the practical difficulty is whether the necessary data would still exist. Unless so ordered, or if involved in litigation, the defendant vehicle owner would be under no obligation to preserve the data unless such a duty were statutorily created. Although no cases have yet addressed the issue of EDR tampering, court rulings in cases involving similar devices in trains and trucks indicate that deliberate erasure or tampering with EDR data will move courts to invoke so-called evidence spoilation remedies.[150]

Perhaps the more interesting question is whether the police could simply download the vehicle's data via a wireless connection. Under those circumstances, police would not be committing a physical trespass. It might also be argued that to obtain data through those means, police are in effect doing nothing more than that which they do when they use a radar gun to record speed.

Kyllo v. United States[151] provides some instruction here. There, the Supreme Court held that when the government uses a device that is not in general public use to explore details of a private home that would previously have been unknowable absent physical intrusion, a Fourth Amendment "search" has occurred and it is presumptively unreasonable without a warrant. In that case, the officers were standing on the other side of the street from the targeted home and used a thermal imaging device to record the home's heat signature. The Court's decision turned largely upon the novelty of the device used and the fact that it involved the invasion of one's privacy at home. Traditionally, the of protection from the prying eyes of law enforcement is considerable less than that afforded to homes. Similarly, if EDRs become routinely used by manufacturers, the expectation of privacy that vehicles owners may enjoy might become diminished over time.

IV. The Fifth Amendment and EDRs
The Constitution's Fifth Amendment provides that no person "shall be compelled in any criminal case to be a witness against himself."[152] It has been suggested that the Fifth Amendment's protection against compelled self-incrimination could be invoked as a means of preventing the admission of EDR data at trial. This simply represents a misunderstanding of what the amendment was designed to protect. The Fifth Amendment grants an evidentiary privilege that

permits an individual to refuse to give testimony that may tend to incriminate him. Thus, the privilege applies only to testimony or statements, it does not apply to other forms of evidence.[153] Thus, in Schmerber v. California,[154] the Supreme Court had little difficulty in determining that the state could compel the defendant to produce a blood sample without violating the Fifth Amendment. There, the Court clarified the circumstances in which the self-incrimination protection applies:

"[T]he privilege protects an accused only from being compelled to testify against himself, or other wise provide the state with evidence of a testimonial or communicative nature, and that the withdrawal of blood and use of the analysis in question in this case did not involve compulsion to these ends."[155]

Thus, the protection only exists to shield *statements* from legal compulsion.[156] In other words, the government may not, in a criminal proceeding, move into evidence statements obtained from the defendant in violation of the Fifth Amendment. While it could certainly be argued (and was by counsel in *Schmerber*) that the withdrawal of the defendant's blood over his objection amounted to compelled testimony, the Court refused to so hold. Even diaries, business records and journals, all of which may be written by the defendant, are admissible at trial pursuant to the Fifth Amendment. The only evidence that is shielded by the privilege against self-incrimination is testimonial evidence made by the defendant to authorities or in the context of a judicial or administrative proceeding of some sort. The admission of EDR data simply does not run afoul of the Fifth Amendment.

Moreover, it is well-established that the protection against self-incrimination applies only to criminal, not civil cases.[157] Thus, whether in pre-trial discovery or cross-examination at trial, the Fifth Amendment can not used to shield testimony or evidence to be used in a civil proceeding. The only caveat is that the defendant may not be forced to make statements in a civil trial that might incriminate him in a future criminal proceeding. In the celebrated O.J. Simpson civil trial, for example, the plaintiffs were able to put Simpson on the stand and cross examine him because he had already been acquitted of murder and thus could no longer be subject to criminal prosecution. In any event, EDR data would not constitute "testimony" for Fifth Amendment purposes.

V. The Federal Rules of Evidence And The Use of EDR Data at Trial

The Federal Rules of Evidence ("FRE") are the rules by which courts control information presented to the fact-finder (normally, the jury).[158] These rules determine, among other things, what information is relevant for the jury to hear and to consider. Courts would almost certainly treat information recorded by EDRs no differently than that recorded by any other reliable means for purposes of discovery and admission as evidence. Although it is true that discovery criteria and admissibility thresholds vary from jurisdiction to jurisdiction, there appears to be a general consensus that electronically recorded data should be treated no differently from so-called hard copy documents.[159]

The Federal Rules of Civil Procedure ("FRCP"), which govern the admissibility of evidence in federal court, and serve as a model for many of the states are quite liberal in permitting the introduction of evidence at trial. Pursuant to those rules, "[p]arties may obtain discovery regarding any matter, not privileged, which is relevant to the subject matter involved in the pending action."[160] Ordinarily, a Rule 34 Request for Production of Documents is the way in which electronically recorded evidence is discovered. It provides:

[a]ny party may serve on any other party a request (1) to produce and permit the party making the request ... to inspect and copy, any designated documents, including writings ... and other data compilations from which information can be obtained, translated, if necessary, by the respondent through detection devices into reasonably usable form.[161]

As part of routine trial preparation, courts will impose a duty on litigants to produce electronic data that is otherwise subject to discovery.[162] Indeed, Federal Rule of Evidence (FRE) 1001, expressly observes that written documents may be electronically stored.[163] Whether EDR data would be discoverable and ultimately admissible in a civil proceeding would be governed by the boundaries of what is discoverable and the parameters of admissibility in a given jurisdiction. However, as a general rule, anything that is "not privileged [and] relevant to the subject matter involved in the pending action" may be discovered in a civil proceeding. [164]

It is obvious that such data is not privileged, nor would it be reasonable to argue that the data would be irrelevant. Standards for relevancy are broad. Relevant evidence is simply that which tends to prove or disprove any fact at issue in a case.[165] EDR data, generated immediately prior to and contemporaneous with an accident, would plainly be probative of any facts at issue in a trial of this sort.

Nevertheless, there are several issues that must be considered. Special rules apply to testimony presented by so-called experts—individuals who were not themselves fact-witnesses, but because of special training and/or experience may be able to assist the jury in its deliberations.

A defining feature of *expert* testimony is that the personal knowledge requirement is suspended. A qualified expert may testify to matters not within the expert's personal sensory experience, and to opinions not ultimately based on personal knowledge. This means not only that experts may testify based on hearsay reports of sensory observations made by others, but also, in principle, that experts may testify to propositions not based on anyone in particular's sensory observations.

The theories that experts may bring to bear are not confined to their particular expertise. Indeed, they may testify *as experts* only if they can claim *scientific*, *technical*, or other *specialized* knowledge that will help the jury to understand the evidence or determine a fact in issue. Under *Fed. R. Evid.* 702, this specialized knowledge may be derived from experience, training, or education. The issue whether the expert possesses specialized knowledge thus derived—*i.e.*, whether the expert is *qualified*—is decided, in the first instance, by the trial court in its sound discretion. Once qualified, Rule 702 says that experts may offer testimony concerning their expert knowledge "in the form of an opinion or otherwise."[166] The rules permit an expert to testify to general scientific or technical principles, leaving their application to the jury.

A. *The* Daubert *Test*

For many years, the admissibility of expert scientific evidence was governed by a common law rule of thumb known as the *Frye* test, after a 1923 decision by the United States Court of Appeals for the District of Columbia in which it was first articulated.[167] Under the *Frye* test, expert scientific evidence was admissible only if the principles on which it was based had gained "general acceptance" in the scientific community.[168]

Despite its widespread adoption by the courts, this "general acceptance" standard was viewed by many as unduly restrictive, because it sometimes operated to bar testimony based on intellectually credible but somewhat novel scientific approaches. In *Daubert*, the Supreme Court was asked to decide whether the *Frye* test had been superseded by the adoption, in 1973, of the *Federal Rules of Evidence.*[169] In particular, *Fed. R. Evid.* 702 broadly governs the admissibility of expert testimony and it simply provides that: "If scientific, technical, or other specialized knowledge will assist the trier of fact to understand the evidence or to determine a fact in issue, a witness qualified as an expert by knowledge, skill, experience, training, or education, may testify thereto in the form of an opinion or otherwise."

The majority opinion in *Daubert*, authored by Justice Blackmun, held that Rule 702 did supersede *Frye.*[170] This did not mean, however, that all expert testimony purporting to be scientific was now admissible. Rather, Rule 702 required that the testimony must be founded on "scientific knowledge." This implied, according to the Court, that the testimony must be grounded in the methods and procedures of science and would possess the requisite scientific validity to establish evidentiary reliability.[171] Furthermore, the Court held that the testimony must be sufficiently tied to the facts of the case.[172]

B. *EDRs and the* Daubert *Test*
Although the data itself recorded by EDRs would be treated no differently than any other document for purposes of discovery, there might be an issue with respect to the data's accuracy and reliability. In assessing the reliability of scientific expert testimony, the Daubert Court gave future trial courts a number of factors to consider.[173] Those factors included (1) whether

the expert's technique or theory can be or has been tested-that is, whether the expert's theory can be challenged in some objective sense, or whether it is instead simply a subjective, conclusory approach that cannot reasonably be assessed for reliability; (2) whether the technique or theory has been subject to peer review and publication; (3) the known or potential rate of error of the technique or theory when applied; (4) the existence and maintenance of standards and controls; and (5) whether the technique or theory has been generally accepted in the scientific community.[174] This final factor is essentially a restatement of the *Frye* rule; although Daubert makes clear that it is only *one* factor to be subjectively considered by the judge.[175]

An important element trial judges must use to determine the admissibility of expert testimony is whether the expert's theory can be challenged in some objective sense.[176] Thus, it will be important for EDR manufacturers to make available information concerning the reliability of their devices. If the device may itself be tested by third parties, and if the relevant scientific community accepts the accuracy and reliability of such devices, there should be no legal constraints on the admissibility of EDR data.[177] In fact, Rule 702 is sufficiently broad that it may allow contradictory expert testimony that is the product of competing principles or methods in the same field of expertise.[178] Thus, an automobile manufacturer might call an expert to testify to the precision of its data, opposing counsel can attempt to impeach this testimony by offering its own analysis. Rule 702 allows all of this evidence to be adduced at trial, and then permits the trier of fact to determine its legitimacy.

In *Harris v. General Motors Corporation*,[179] for example, the United States Court of Appeals for the Sixth Circuit reversed the decision of summary judgment for the defendant because it had been based solely on a GM engineer's interpretation of crash data collected from an EDR. GM filed a motion for summary judgment relying on the engineer's testimony that, based upon his interpretation of the EDR data, the air bag had properly deployed. The engineer's interpretation of the data contradicted both the driver's and the passenger's testimony. They claimed that the driver was injured because the air bag deployed late. The trial court, not only admitted the testimony based on the EDR data, but granted GM summary judgment on that basis. The lower court applied the "physical facts" rule[180] to conclude that the plaintiff's *could not* have been correct based upon the indisputable physical facts of the case. Explaining that it had failed to unearth any cases dealing with the interpretation of EDR data, the court of appeals held that "the [EDR] data suggests that the air bag deployed properly; it does not establish beyond factual dispute that the air bag could not have deployed belatedly in the manner" the plaintiff contended.[181] Because the court found the record devoid of any demonstration that GM had proved the reliability of its engineer's testimony beyond factual doubt, the court remanded the case to the trial court for a hearing on admissibility.

The admissibility of information gathered from EDRs in the courtroom would almost certainly be admissible under the *Daubert* test. As the Court held in *Daubert*, any information grounded in the methods and procedures of science that possesses the requisite scientific validity to establish evidentiary reliability pertaining to the case at hand was to be admissible.[182] By way of offering further guidance, the Court noted the availability of other mechanisms of judicial control, including summary judgment and the ability to exclude confusing or prejudicial evidence under *Fed. R. Evid.* 403.

VII. Conclusion

In my opinion, the DOT may have the authority to require the installation of EDRs if demonstrated to improve highway safety or advance other significant public policy interests. Although the legal authority may exist, however, that does not mean that the public will necessarily accept the implementation of such devices—particularly if it is felt that EDRs trench upon legitimate privacy interests.

Although the privacy concerns are doubtless serious, they are not insurmountable. First, police may properly seize EDRs without a warrant during post-accident investigations either because seizure of a required safety device does not constitute a "search" implicating the Fourth Amendment, or in the alternative because seizure of a safety device qualifies under the one of the exemptions for conducting warrantless searches. In other, non-emergency situations, however, the police may need to obtain a warrant before seizing the data—at least insofar as the data is intended to be used in a criminal investigation and a privacy interest in the data continues to be recognized. It is possible, as the *Kyllo* Court appears to suggest, that if the downloading and use of EDR data becomes sufficiently widespread and routine, individuals may no longer have a reasonable privacy interest in the data.

Second, although the data (and the recorder itself) is in all likelihood "owned" by the automobile's owner, that data may almost certainly be used as evidence against that owner (or other driver) in either a civil or a criminal case. The Fifth Amendment's privilege against compelled self-incrimination will not shield the data from being introduced at trial. Nor will it be possible to prevent adverse litigants from obtaining the data as a matter of routine pre-trial discovery. Of course, depending upon the nature of the data collected, Congress (or the individual states) could choose to privilege use of the data, but that seems quite unlikely. In reality, the only real obstacle to the use of EDR data in court would be either a demonstration that the devices themselves, or the decoders used to download the data, were consistently inaccurate and unreliable, or if automobile owners were able to tamper with the devices thereby rendering the data unusable. Courts have a long history of creating remedies for the spoiling of data, however; similarly, laws could be enacted criminalizing the willful destruction or alteration of EDR data.

Finally, it is worth noting that some concerns may be mollified if insurance companies or the car manufacturers themselves take active steps to provide individual purchasers/insurers with incentives to waive any interests they may have in the EDR data. For example, insurers could premise policy decisions on whether a consumer agreed to allow the company unimpeded access to the data. Similarly, insurers could provide discounts to such consumers in exchange for their waivers. Automobile manufacturers could require purchasers to agree to permit them to review data in the event of an accident either for safety engineering purposes, or to protect the company from suit. Private solutions to these potential problems no doubt exist; these are only a few of the possible avenues that could be taken to ensure that EDR collected data is easily recovered."

Privacy Issues?

Besides the National Academies Study cited above, others have written about legal issues and privacy. David M. Katz noted in an article in *Rutgers Computer & Technology Law Journal*, Spring 2003, titled *"Privacy in the private sector: use of the automotive industry's "event data recorder" and cable industry's "interactive television" in collecting personal data"*, that, while the mention of the term "Big Brother" typically arouses suspicion of the government, many privacy advocates have been expressing concern over the private sector's access to and use of valuable personal information. One example of this possible encroachment upon privacy rights occurs within the automotive industry, specifically the use of "Event Data Recorders" ("EDR") in vehicles. These devices, akin to the "black boxes" of the aviation industry, record the vital statistics of an automobile's operation in five-second loops, thereby saving the final five seconds prior to impact. Some of the notable data recorded include air bag deployment, seatbelt use, speed prior to impact, acceleration or deceleration, and braking.

Common Law Invasion of Privacy Claim as Applied to the EDR

As mentioned above, the common law tort of invasion of privacy, based on intrusion upon seclusion, serves as a plaintiff's most viable claim against a manufacturer who installs EDR devices. This very cause of action was asserted against GM in the recent class action case of *Valan v. General Motors Corp.* in the New Jersey state courts. In Valan, the plaintiff, as a class representative, alleged that personal information was being gathered from within the vehicle without the owner's knowledge or consent. However, the Valan court held that the complaint fails on its face, reasoning:

The data collected by the EDR is not private information, but rather is reasonably observable to anyone watching the vehicle on the road. Certainly, if someone doing a safety study stood on the side of the road with a device that could measure this information it would not be considered an invasion of privacy. Police routinely use radar to detect the speed of vehicles for enforcement purposes and the law does not recognize this as an invasion of privacy.

In reaching its conclusion on the invasion of privacy cause of action, the Valan court relied on the well-established principle: "[W]hatever the public may see from a public place cannot be private." The Valan court also cited a United States Supreme Court doctrine that states that "an expectation of privacy as to automobiles is further diminished by the obviously public nature of automobile travel."

Aside from the facial failure of the intrusion upon seclusion claim, the Valan court's grant of summary judgment to GM also noted that even if the EDR was tortiously intrusive, the plaintiff failed to allege that an intrusion had actually occurred. In other words, the permanent data collection does not occur without a crash, as the EDR re-records continuously in five-second loops. Finally, the Valan court makes a significant note in dicta: "[A]bsent impoundment or court order, control over the [EDR] and its data rests with the owner of the vehicle and, thus, remains private." This ownership principle, while not essential to the holding, endorses the popular belief of agencies such as the NHTSA. The Valan court also noted that the plaintiffs' true concerns lie in the idea that data was being gathered without their knowledge. Therefore, "[r]ather than distort the law of privacy, the issues [of consumer expectations] in this case are more properly resolved under the consumer fraud laws." This point is valid, as the success and legality of the EDR may hinge on effective communication and disclosure by automobile manufacturers.

One final postscript on the Valan case is a cause of action not asserted by the plaintiffs: a Constitutional invasion of privacy claim. This claim will necessarily fail if the common law tort claim fails. On the other hand, if the plaintiff is able to prevail on his/her tort claim, he/she may have a viable Constitutional claim. The obvious hurdle here is making the argument that the automobile manufacturer engaged in state action. The Supreme Court has recently liberalized the state action standard in Brentwood Academy v. Tennessee Secondary School Athletic Ass'n. The Brentwood Academy court held that a not-for-profit, otherwise private, athletic association's regulatory activity was state action due to the "entwinement" of state school officials in the association's structure. Therefore, a plaintiff could seek to illustrate automobile manufacturers' "entwinement" with arms of the federal government in the areas of research and policy-making. This rather tenuous argument may fail simply because of the fact that no government agency has currently mandated the use of the EDR in automobiles. A hearty endorsement of the technology may not suffice."

In summary, it is true that current and emerging technology may tell all about your driving habits and who may have caused the crash. Just imagine a situation where a crash occurs and both drivers are suffering from brain trauma combined with other injuries that leave both unable to recall what happened. Even if law enforcement suspects that one party is guilty it will be difficult is not impossible to prove without piecing together "evidence" from others. But what if they have conflicting opinions? What if there are no witnesses? Then perhaps the most compelling testimony will come from the onboard silent witnesses—the event data recorders.

Everyone knows that automotive crashes are usually notoriously described as subjective events because of a lack of objective data. Subjective evidence and victims stories vary widely. Ironically, the fact is that virtually every new light vehicle (car or truck) sold today contains some means of collecting and storing objective crash data. The significance of this fact is that the automakers are impeding progress in safety by encrypting the data and making it inaccessible to the public. The automakers argument that by encrypting the data they are preserving individual privacy is woefully lacking and will be challenged by those who advocate public safety.

However, it is also true that certain automakers have allowed the data on their vehicles to be retrieved by mechanics or crash investigators with simple handheld devices. Eventually other automakers may follow this route, but the average motorist is still restricted from the crash data unless they pay in one form or other for data retrieval.

Voluntary Consensus Standards vs. Federal Rulemaking

Sincere involvement in a voluntary standards initiative is the ideal, specially if the technology has the possibility of greatly enhancing vehicle and highway safety. There is also a 'negative impact' that such action has on enhancing vehicle and highway safety via voluntary consensus standards. If federal regulators abdicate their authority, then voluntary pacts are hard if not impossible to enforce and make recalls more difficult to impose. Therefore, a compromise solution would be for a regulatory body (NHTSA, FMCSA, *etc.*) to incorporate important parts of voluntary consensus standards (by reference) in rulemaking. This is a win-win for all parties.

California Privacy Law

On September 22, 2003, California Senate Bill AB 213: EDR Disclosure was signed by Governor Gray Davis. The *New York Times* reported the following:

> *Privacy Law in California Shields Drivers* by Matthew L. Wald.
>
> WASHINGTON, Sept. 22—California today adopted the nation's first law meant to protect the privacy of drivers whose cars are equipped with "black boxes," or data recorders that can be used to gather vital information on how a vehicle is being driven in the last seconds before a crash. Gov. Gray Davis signed the law, which takes effect on July 1, requiring carmakers to disclose the existence of such devices and forbidding access to the data without either a court order or the owner's permission, unless it is for a safety study in which the information cannot be traced back to the car. More than 25 million cars and trucks have the boxes that measure speed, air-bag deployment and the use of brakes, seat belts and turn signals. But California's privacy law is the first of its kind, says Thomas M. Kowalick, co-chairman of a committee convened by the Institute of Electrical and Electronics Engineers to set standards for the boxes. Most of the recorders are on General Motors vehicles, but Ford and others have deployed some. Other manufacturers have plans to do the same. The police in South Dakota sought information from such a recorder to determine whether Bill Janklow, a member of the House of Representatives and a former governor of the state, had run a stop sign and was speeding on Aug. 16 when he hit and killed a motorcycle rider near Flandreau. But Maj. James Carpenter of the South Dakota Highway Patrol said that because the car was not a recent model—it was a 1995 Cadillac—it had limited information. The

California bill was introduced by Tim Leslie, a Republican assemblyman, who contended that the devices were installed without the owner's knowledge or consent and that the information they gathered should be subject to the same legal protections as provided by the Fourth Amendment for other kinds of private information. He compared it to the process for getting permission to tap a telephone. Mr. Leslie's legislative director, Kevin O'Neill, said in a telephone interview that in the case of a crash that resulted in civil litigation or criminal prosecution, the data would be obtainable by court order. But the information should be protected by a process, Mr. O'Neill said.

The bill that Governor Davis signed reads as follows:

JANUARY 29, 2003—An act to add Section 9951 to the Vehicle Code, relating to vehicles. LEGISLATIVE COUNSEL'S DIGEST: AB 213, Leslie. Vehicles: manufacturers: disclosure. Existing law sets forth various provisions governing vehicle manufacturers. Those provisions include the requirement that manufacturers disclose in the owner's manual, or other written material, as specified, of a new motor vehicle sold in this state, the fact that the vehicle, as equipped, may not be operated with tire chains. This bill would require a manufacturer of a new motor vehicle sold or leased in this state that is equipped with one or more recording devices, commonly referred to as "event data recorders (EDR)" or "sensing and diagnostic modules (SDM)," to disclose that fact in the owner's manual for the vehicle. The bill would prohibit specified data that is recorded on a recording device from being downloaded or otherwise retrieved by a person other than the registered owner of the motor vehicle, except under specified circumstances. The bill would also require a subscription service agreement to disclose that specified information may be recorded or transmitted as part of the subscription service. The bill would provide that it applies to all motor vehicles manufactured on or after July 1, 2004. Because a violation of the Vehicle Code is an infraction, the bill would create new infractions, thereby imposing a state-mandated local program. The California Constitution requires the state to reimburse local agencies and school districts for certain costs mandated by the state. Statutory provisions establish procedures for making that reimbursement. This bill would provide that no reimbursement is required by this act for a specified reason.

THE PEOPLE OF THE STATE OF CALIFORNIA DO ENACT AS FOLLOWS: SECTION 1. Section 9951 is added to the Vehicle Code, to read: 9951. (a) A manufacturer of a new motor vehicle sold or leased in this state, which is equipped with one or more recording devices commonly referred to as "event data recorders (EDR)" or "sensing and diagnostic modules (SDM)," shall disclose that fact in the owner's manual for the vehicle. (b) As used in this section, "recording device" means a device that is installed by the manufacturer of the vehicle and does one or more of the following, for the purpose of retrieving data after an accident: (1) Records how fast and in which direction the motor vehicle is traveling. (2) Records a history of where the motor vehicle travels. (3) Records steering performance. (4) Records brake performance, including, but not limited to, whether brakes were applied before an accident. (5) Records the driver's seatbelt status. (6) Has the ability to transmit information concerning an accident in which the motor vehicle has been involved to a central communications system when an accident occurs. (c) Data described in subdivision (b) that is recorded on a recording device may not be downloaded or otherwise retrieved by a person other than the registered owner of the motor vehicle, except under one of the following circumstances: (1) The registered owner of the motor vehicle consents to the retrieval of the information. (2) In response to an order of a court having jurisdiction to issue the order. (3) For the purpose of improving motor vehicle safety, including for medical research of the human body's reaction to motor vehicle accidents, and the identity of the registered owner or driver is not disclosed in connection with that retrieved data. The disclosure of the vehicle identification number (VIN) for the purpose of improving vehicle safety, including for medical research of the human body's reaction to motor vehicle accidents, does not constitute the disclosure of the identity of the registered owner or driver. (4) The data is retrieved by a licensed new motor vehicle dealer, or by an automotive technician as defined in Section 9880.1 of the Business and Professions Code, for the purpose of diagnosing, servicing, or repairing the motor vehicle. (d) A person authorized to download or otherwise retrieve data from a recording device pursuant to paragraph (3) of subdivision (c), may not release that data, except to share the data among the motor vehicle safety and medical research communities, to advance motor vehicle safety, and only if the identity of the registered owner or driver is not disclosed. e) (1) If a motor vehicle is equipped with a recording device that is capable of recording or transmitting information as described in paragraph (2) or (6) of subdivision (b) and that capability is part of a subscription service, the fact that the information may be recorded or transmitted shall be disclosed in the subscription service agreement. (2) Subdivision (c) does not apply to subscription services meeting the requirements of paragraph (1). (f) This section applies to all motor vehicles manufactured on or after July 1, 2004. SEC. 2. No reimbursement is required by this act pursuant to Section 6 of Article XIII B of the California Constitution because the only costs that may be incurred by a local agency or school district will be incurred because this act creates a new crime or infraction, eliminates a crime or infraction, or changes the penalty for a crime or infraction, within the meaning of Section 17556 of the Government Code, or changes the definition of a crime within the meaning of Section 6 of Article XIII B of the California Constitution.

A press release from The Society of Automotive Analysts dated November 11, 2003 announces a Black Box Telematics Debate:

Who Owns the Data? scheduled for November 19, 2003 in Detroit, Michigan.

ANN ARBOR, Mich., Nov. 11 /PRNewswire/—Information superhighway or not, the automotive industry is on a clear collision course with civil liberties.

Already, onboard data about how a vehicle is driven can be and are being downloaded without the knowledge or consent of the driver. That information is being studied by fleet owners, inspected by law enforcement, and even being subpoenaed into courts of law.

Ironically, these data are being made available courtesy of the vehicle's manufacturer and paid for by an unknowing consumer.

Industry and legal experts will discuss this "black box" technology and the issues of privacy and intellectual property in "The Censure of Sensors—A Telematics Debate" on Wednesday, November 19, 2003, at the Westin Hotel in Southfield, Michigan.

In cooperation with SAA, the Society of Automotive Engineers (SAE) International will issue a public statement relating to the standardization of data transfer, the first step in expanding the application of telematics in automobiles.

Csaba Csere, editor of *Car and Driver* magazine, in Ann Arbor, will moderate. Panelists confirmed to date are: Philip Haseltine, president of the Automotive Coalition for Traffic Safety; Arlington, VA., Michael Khoury, head of the technology practice at the law firm of Raymond & Prokop P.C.; Southfield, Michigan; Kathleen Konicki, Director of Safety for Nationwide Insurance; Columbus, OH

Reporting on the event in a *Detroit News* article titled *"Auto 'Black Boxes' defended"*, Ed Garsten wrote:

> The federal government should resist regulating the use of so-called "black boxes" in cars and light trucks that record driver behavior, experts said Wednesday.

I spoke with Garsten on the telephone to ask about the event and his story. He related the sense that the event would have been better if a wider range of speakers representing alternative views were present. *An abstract of the article notes:*

> The federal government should resist regulating the use of so-called "black boxes" in cars and light trucks that record driver behavior, experts said Wednesday. There are now 40 million vehicles equipped with devices and sensors that record the speed, direction, location and other data at the time of a crash or accident, providing police agencies and insurance companies with valuable information. "These technologies are going to advance faster than the three years it takes to write federal regulations," Michael Haseltine, president of the Automotive Coalition for Traffic Safety, told a panel sponsored by the Society of Automotive Analysts. While the boxes, called event data recorders, are not yet in widespread use, the computer sensors that "decide" whether or not to deploy an air bag already record some very basic information, such as changes in speed, that police and insurance investigators are depending on to help prove fraud or other criminal activity among motorists. "The insurance industry

> literally loses billions of dollars every year in fraudu-
> lent claims," said Kathy Konicki, director of safety at
> Nationwide Insurance in Columbus, Ohio. "If we have
> sufficient data … then we could take some pretty sig-
> nificant steps forward to start reducing fraud." But
> there are concerns the information may be used to
> harass some drivers or encroach on consumer privacy
> rights. "It won't be long before someone says you can't
> have access to my event data recorder because it is a
> violation of my Fifth Amendment rights," said Michael
> Khoury, a lawyer with Southfield law firm Raymond &
> Prokop.

The quote from Mr. Haseltine, president of the Automotive Coalition for
Traffic Safety, cited that "The technologies are going to advance faster than the
three years it takes to write federal regulations," leaves the impression that perhaps it would be better to permit the automakers to create their own voluntary
standards.

On this point, safety advocated note that the problem is that historically
the automakers have excluded other end-users of the data from participating.
Automakers support standards that protect their interests. This policy is a dead
end for vehicle and highway safety in America.

Shortly afterwards, an article titled "Court limits in-car FBI spying" by
Kevin Poulsen in the November 19, 2003 edition of *Security Focus* detailing the
findings of an appeals court ruling on privacy:

> An appeals court this week put the brakes on an FBI
> surveillance technique that turns an automobile
> driver's on-board vehicle navigation system into a
> covert eavesdropping device, after finding that the
> spying effectively disables the system's emergency and
> roadside assistance features.

> The case arose from a 2001 FBI surveillance operation
> in Las Vegas, in which agents obtained a court order
> compelling a telematics company to secretly activate
> the stolen vehicle recovery feature in a customer's car.
> The feature, designed to listen-in on car thieves as
> they cruise around in a stolen auto, turns on a dash-
> board microphone and pipes conversations out over a
> cell phone connection—normally to the company's
> response center, but in this case to an FBI listening
> post.

> After initially complying for 30 days, the company
> asked a federal judge to block the order. It lost, and
> filed the appeal with 9th U.S. Circuit Court of Appeals
> while complying with the order. The proceeding were
> handled in strict secrecy, and the text of the final ruling

omits the name of the company. Geri Lama, a spokesperson for General Motors subsidiary OnStar®, says it wasn't them.

Court records strongly point to OnStar's Texas-based competitor ATX Technologies, which makes the "Tele Aid" systems used in Mercedes vehicles: the description fits the Tele Aid systems, and the Dallas-based attorney listed as arguing the appeal is also representing ATX in unrelated civil litigation in Texas. ATX spokesman Gary Wallace said he couldn't immediately comment.

Emergency Services Blocked

Under federal law, the FBI can obtain court orders compelling telecommunications companies, ISPs, landlords and others to assist the Bureau in spying on customers. But the law requires that surveillance in such cases be conducted "unobtrusively and with a minimum of interference with the services" provided by the company. With the navigation system's cellular link dedicated full time to eavesdropping, the system had no way to communicate with the company's response center if the roadside assistance or emergency reporting features were activated, according to the court's split 2–1 decision.

"Pressing the emergency button and activation of the car's air bags, instead of automatically contacting the Company, would simply emit a tone over the already open phone line," the majority wrote. "[T]he FBI, however well-intentioned, is not in the business of providing emergency road services, and might well have better things to do when listening in than respond with such services. The result was that the Company could no longer supply any of the various services it had promised its customer, including assurance of response in an emergency."

The decision, released Tuesday, is only binding in the 9th Circuit, which covers eight western U.S. states and Hawaii. Other federal circuits have not addressed the issue.

Despite the reversal, David Sobel, an attorney with the Electronic Privacy Information Center, says the ruling is not a victory for privacy. "Although the bottom line is that the surveillance order was rejected, the real effect of it is that this kind of monitoring is permissible as long it does not interfere with the service," says Sobel. "It underscores the fact that it's becoming increasingly difficult to escape the reach of surveillance capabilities."

Canadian Case

In a April 15, 2004 news story titled "*Car's 'black box' helps convict Quebec man*" Miro Cernetig, Quebec Bureau Chief of the *Toronto Star* reported that a Quebec judge used the data from an event data recorder to hand out an 18-month prison term to 26-year old Eric Gauthier for speeding through a downtown intersection and slamming into another car that killed Yacine Zinet, whom he accused of running a red light. The significance of this case is that, according to Crown prosecutor Jeannot Decarie, this is the first Canadian jailed as a result of information captured by a car's on-board computer.

In response to this court case the Toronto Star staff solicited feedback via a website. Readers were notified that "Many cars now come with event data recorder devices—so-called black boxes—installed. These devices track location, speed and time and purchasers may not know that they are there. Do these devices raise privacy issues."

The following comments are derived from this public website. Although these comments were placed in the public domain, names and addresses of the respondents have been deleted in order to preserve their individual privacy. By reading these remarks you will have a better understanding of public views.

- The fact that there are very few police cars on the road is proven by the number of reckless drivers [on the road]. If technology can provide the 'baby sitting' that many of us seem to need and improve driving habits and save lives, then I am all for it.

- Anyone that hates the idea does not want to be accountable for their irresponsible driving.

- CBC had a program a few months back about these black boxes. What was funny was that car dealers didn't know [they were there] either. Apparently the owner's manual was so vague you couldn't tell what it was making reference to.

- This man killed someone while driving at 150 kph. Thank goodness there was a black box in his car.

- I have never been convicted of a major driving violation. I never drink and drive and I generally drive the speed limit. Yet, I am totally against these 'black boxes'. I have nothing to hide, [but] I do not want to live, nor want my child, to live in a Big Brother world.

- Event recorders are typical of an emerging technology in that it has the potential of being good or bad. The information in the recorder is private and should require a Court Order by the Police to allow access to the information.

- Public safety is much more important than the privacy.

- As a privacy issue I see little difference between electronic recording devices (ERD) and surveillance cameras. Cars are driven on public highways. Vehicle accidents are the third leading cause of death in Canada. In a motor vehicle accident an ERD is an unbiased witness.

- Just who is entitled to the data that it is collecting?

- Just like DNA evidence can place someone at the scene of a crime, it can't tell the whole story. Unless the data from the car can be synced with the traffic light computer data, the only thing that we can be certain of is the car's speed. There is no proof that the other driver did not run a red light. Imagine having a glass of wine with a friend that's later murdered. Is the DNA evidence from the glass enough to convict?

- As long as they're only used to determine fault in a car collision, it's alright with me. Going beyond that, with satellite surveillance to let police/the government know the car's whereabouts, is going too far.

- It is strange to think that in an established and civilized society such as ours, the use of valuable technology is still scrutinized because of people reluctant to admit to their offensive driving practices. Responsible drivers would willingly accept any advances that work to catch dangerous drivers on the road.

- We are being monitored more and more. There are security cameras, red light cameras, 407 ETR cameras, GPS(Onstar), music download-ing, going on any international flight, and using your computer hard drive. 1984 was 20 years too soon, but it is arriving with blinding speed.

- The people who are raising privacy issues against these black boxes may not realize that if ever they are a part of a horrible car accident, or their loved one is, it may help convict whoever is responsible. As long as people obey the law, then there should be no real problem with these boxes.

- What ever happened to full disclosure? Shouldn't we be advised when we are being recorded—the same way "this call is being recorded for quality assurance" advises us that other parties other than the intended recipient will be listening in?

- I'm glad that the person who caused the accident was caught. But I also think that you should be told if there is a black box in your new car—and have the option to disconnect it if you want to.

- It seems to me that the issue of black boxes in cars is related to that of photo radar. I wonder how many people who are in favour of one oppose the other? I, for one, do not agree with either.

- I think black boxes do raise privacy issues, especially when drivers have no idea such equipment is in their cars. As the article sug-gested, people often don't know that the black box can even "provide real-time details of a driver's ... destinations." People have a right to know when there's something in their car that records their speed, the number of times they press the brake, and other information that could possibly incriminate them, just as it did with Eric Gauthier.

- Driving a vehicle on a public highway is in no way a private act. The boxes make no judgements, they merely record facts, which could equally aid the innocent as help convict the unsafe. It is not a game, drivers do not have a right to drive unsafely unless they are caught. Traffic laws are not guidelines, the catch-me-if-you-can attitude costs far too much in human life and suffering.

- I think it's a great idea, if these devices were installed in all cars and if all drivers were notified. However, it's quite ridiculous when the driver has no idea that a device like that is installed in their car or that it could be used against them in court.

- I feel that devices like these should not raise privacy issues. I do not feel that it is an invasion of privacy because the devices are not really recording anything like a voice or taking pictures of any sort. They are just recording speeds, brake times, heat emanations, etc. This information will help police solve crimes at a faster rate.

- Using a black box from the car that you used while involved in an accident is no different than forcing the accused to testify against himself. This is a sacrifice of one more of our freedoms that is used solely to make the police's and Crown prosecutors' jobs easy. The state has no business knowing where I am because the Charter allows us to move freely within our country. These black boxes are a potential for Big Brother to watch us every second of our lives. Do we really want this kind of society?

- Lets get off the invasion-of-rights bandwagon. At least the black boxes won't manipulate the truth and perhaps justice can finally have real meaning again. Once again the only people calling for a ban on such items are those that don't follow the rules and those that make money defending them.

- What is next? Putting CPU chips in the brains of each child born so they will do the bidding of the rich, the powerful and their political masters? It is time our leaders started doing things for the people instead of to the people.

- Responsible drivers will applaud this useful tool, and victims of irre- sponsible drivers will benefit by the information disclosed. They should be mandatory in all motor vehicles using public roads.

- There is nothing private, ever, about the use of a car on a public highway—especially not something as obvious as its speed and when the brakes were last used. If dangerous drivers want to protect their so-called right to privacy then they should drive up and down their own driveways and stay off public highways.

- The idea of a so-called "black box" should be no concern for any honest, responsible driver. Who knows—maybe the bad drivers of the world might think a little when they know the box is recording their actions.

- I am very afraid of verdicts like this. Apparently you can be blamed for any accident you get into based on driving too fast rather than the other guy maybe did go through that red light. That's what bothers me about this. How can a black box tell the police investigators if this person really went through a red light?

- The driver of a car is granted a license. This is not a right. A licensee has rules that are to be followed. Making the holder aware that a monitoring of the use of the license to ensure compliance is not an invasion of privacy. An invasion of privacy would be monitoring what goes on in the car when it is stopped and the license is not being used.

- I think this is a great idea! We have them for planes: why not cars? Maybe now that people know they are being watched they'll slow down and be a little more cautious on the roads.

- Privacy is a treasured right in our society and is not to be taken If such black boxes were mandatory, and therefore made known at purchase, then you have a choice—drive responsibly, or don't drive at all.

- Driving is a privilege, not a right. If you break the laws and it is recorded then you should pay the penalty for your actions. I don't think this is going to curb stupid drivers at all, but can help after the accident has occurred to assist in figuring out what happened.

- Considering how often a story of a fatal car crash appears in the news, the black boxes would do more good than harm. If it is not used for surveillance, but only to reveal what happen at a crash scene, people's privacy is not at risk. In any investigation, privacy is never an issue; it is set aside for the truth. As not all people are honest, the black boxes seem essential.

- The fact that the black box in Eric Gauthier's car proved he was driving dangerously shows the device can have an important role in highway safety. However, most people are not even aware that their car may carry such a device. I wonder that if more people knew about these devices then they might drive more carefully in the first place?

- I have a BMW 528i and the speedometer is off by 10 km/hr. I asked a BMW dealer to correct it and was told that it is within tolerance. They also told me that they are unable to change it. I think the government should mandate a tighter control on this if it is going to be used as evidence. 10 km/hr could make a big difference in a court of law.

- While I believe in the right to privacy—a city street, a piece of high-speed metal on wheels and a speed limit that is ignored all add up to "in the public interest". If you speed, you break the law. If you break the law, you are punished. There is no privacy on the street.

- I really don't think the "black box" is a privacy issue. I say it's high time that drivers get what they deserve when they break the law. And if the data recorded by the black box will help, more power to it. The public doesn't seem to have a problem with the black box on commercial airliners. It's the same thing.

Motor Vehicle Event Data Recorder Case Law

The following citations point you toward legal cases. In some instances the decisions have not yet been published.

United States of America

- *Wright v. CSX Transportation*, 5:01-cv-324-4 (M.D. Ga. Oct. 1, 2002)

- *Harris v. General Motors Corp.*, Electronic Citation: 2000 FED App. 0039P(6th Cir.), File Name: 00a0039p.06

- *Bachman, et al v. General Motors Corp., Uftring-Chevrolet-Oldsmobile, Delphi Automotive Systems and Delco Electronics Systems*, Illinois App. Ct., 4th Dist., No. 4-01-0237, Appeal from Circuit Court of Woodford County, Case No. 98L21 (2002)

- *Illinois v. Barham*, Illinois App. Ct, 5th Dist., No. 5-02-0047, Appeal from the Circuit Court of Johnson County, Case No. 00-CF-90 (2003)

- *Anderson-Baratona v. General Motors Corp.*, No. 99A19714, GA, Cobb County Cir. Ct., Apr. 7, 2000

- *Colorado v. Cain*, 1st Judicial District Court, Division 3, Jefferson County, Case No. 01 CR 967 (2002)

- *Florida v. Walker*, 20th Judicial Circuit, Lee County, Case No. 00-002866CF RTC (2003)

- *Pennsylvania v. Rhoads*, Montgomery County, Court of Common Pleas, Criminal Division, Docket No. 746701 (2002)

- *Wisconsin v. Furst*, Outagamie County Circuit Court, Case No. 00CF667 (2001)

- *California v. Beeler*, San Diego Superior Court, Case No. SCD158974 (2002)

- *Florida v. Matos*, 17th Judicial Circuit, Broward County, Case No. 02015762 CF 10A (2003)

- *Wisconsin v. Martinez*, Brown County Circuit Court, Case No. 01CF766 (2002)

- *Wisconsin v. Jahner*, Outagamie County Circuit Court, Case No. 02CM649 (2002)

- *South Carolina v. Cassels*, Beaufort County, General Session Indictment No. 2002 GF 070372 (2003)

- *Florida v. Ubals*, 17th Judicial Circuit, Broward County, Case No. 1017144 CF 10A (2003)

- *California v Sanchez*, Ventura County, Case No. 2001 9000 34 (2003)

- *Michigan v. Wood*, Charlotte, Eaton County, Case No. 02 283 FH (2003)

- *South Dakota v. Janklow*, 3rd Cir., Moody County, Case No. 03-147 (2003)

- *New York v. Christmann*, Newark Village Court, Case No. 03110007 (2004)

- *Arizona v. O'Brien*, Superior Court of Arizona, Maricopa County, Case No. CR 2003-016197-001 DT (2004)

- *Pennsylvania v. Weaver*, Bucks County Juvenile Court (2004)

Canada

- *Canada R. v. Daley*, 2003 NBQB 20, S/CR/7/02 (2002)

- *Canada R. v. Gauthier*, 2003 QCCQ, Case No. 500-01-013375-016 (2003)

- *Canada R. v. Gratton*, 2003 ABQB 728 , Case No. 016051344Q (2003)

- *Canada R. v. Brander*, 2003 ABQB 756, Case No. 017156316Q1 (2003)

Safety Culture?

On April 26, 2004, former National Transportation Safety Board (NTSB) Chairman James Hall called for the surface transportation industry to work together to create a "safety culture" during his keynote speech at the opening session of the 2004 Intelligent Transportation System (ITS) America annual meeting in San Antonio, Texas. Hall noted:

> I've always believed in the importance of accident recorders, or so-called 'black boxes.' The evolution of recorders is why aviation is so safe and why it is so important that this technology be integrated into our surface transportation modes. The truth is that even in the most sophisticated systems, things go wrong. The best way to be sure we stay one step ahead is with recorders, so we're able to catch incidents before they become accidents that cause fatalities. We now have the ability to collect so much data, the question really becomes, not whether we should have data recorders or even what kinds, but how we use the information that the recorders provide. How do we share the data among all the groups who can benefit from it? How do we get public safety officials and private business and

government to work together to make the most of the information? How do we protect privacy rights? Forums like this are where these questions get asked, and where the answers begin to emerge. You can't expect any one element of the transportation system to go off in a corner and address issues as complex as standards, policy, and privacy, because indeed you are dealing with an inter-connected system of needs. You can't look to one another and point the finger expecting the person sitting next to you to solve it alone. It can't be just a private-sector problem, and I don't think you want it to become solely a government problem either. These are issues that can be solved only in a public-private-academic setting much like you've established right here.

Pennsylvania Studies Vehicle Event Data Recorders

On July 13, 2004, the Pennsylvania House Democratic Policy Committee held a public hearing at the Mon Valley Community Health Center in Monessen, Pennsylvania to investigate consumer privacy concerns caused by motor vehicle event data recorders.

Hosted by state Rep. Ted Harhai, D-Westmoreland/Fayette, the committee received testimony from Cpl. Michael Schmidt, collision analysis and reconstruction specialist, Troop A, Indiana, Pennsylvania State Police; the American Civil Liberties Union; the Pennsylvania Automotive Association; the Pennsylvania AAA Foundation; the Alliance of Automobile Manufacturers, and myself. I presented a high-level overview titled, "Motor Vehicle Event Data Recorder Initiatives in the U.S.A: 1997–2004."

Representative Harhai had introduced a bill (H.B. 2106) that would require written disclosure by dealers of any new or used vehicle that contains an event-data recorder or similar device and the data it can collect.

Lawmakers in more than 20 states have considered bills that would require dealers to notify consumers of EDRs. However, California is the only state to pass such a measure. The California bill went into affect during the first week of July 2004.

Under Harhai's bill, a dealer who did not notify consumers of the devices would face fines of $1,000 per violation, or $3,000 per violation in cases of victims who are 60 or older. The violations would fall under the unfair trade practices and consumer protection law.

Several local residents attended the hearing to watch five members of the Democratic Policy Committee query EDR experts. Two newspaper journalists covered the event and articles appeared in the *Monessen Valley Independent* (*Big Brother Taking a Ride?* by Stacey Wolford), the *Pittsburgh Tribune-Review* (*Access to Auto Data Subject of Bill* by Chris Foreman) and the *Connellsville Daily Courier* (*In Pennsylvania: Access to Auto Data is Subject* of Bill by Chris Foreman).

The participants were very interested in the topic.

Representative Harhai noted, "I just think it's unbelievable. You create this data but you don't have any access to it. What begun as a research tool for car manufacturers has expanded to become a "Big Brother" mechanism for accident investigation and even court cases. This is not necessarily a bad thing since the recorded data can be helpful, but it is wrong when consumers don't know these EDRs are installed in their vehicles and can be used against them. I requested this hearing to learn more about the different perspectives surrounding this complex issue and we certainly did learn a lot today."

The Alliance of Automobile Manufacturers representing BMW Group, DaimlerChrysler, Ford Motor Company, General Motors, Mazda, Mitsubishi Motors, Porsche, Toyota and Volkswagen provided the following written statement to the public hearing dated July 12, 2004.

Alliance OF AUTOMOBILE MANUFACTURERS

July 12, 2004

RE: House Bill 2106 – Disclosure of Event Data Recorders on Motor Vehicles

Dear Representative Stetler and Members of the Policy Committee,

The Alliance appreciates the opportunity to submit comments on HB 2106 for your consideration. Further, we support and look forward to assisting the sponsors' efforts to fully evaluate and understand the use of these recording devices in motor vehicles and the complexities associated with the proposed legislation.

The Alliance supports disclosing the presence of event data recording devices ("EDRs") in the owner's guides and in subscription service agreements provided for new motor vehicles. The Alliance is prepared to work with the sponsors, to help provide information on EDRs, and to study the issue so that appropriate legislation, if any, can be drafted. To that end, the Alliance has provided model legislative language on EDRs. The Alliance model language does two things. One, it specifies that the existence of an EDR should be disclosed in owner's manuals and subscription agreements and (2) further states that EDR data recorded can only be accessed with permission of the vehicle's owner and in other limited circumstances, e.g., with a court order, etc.

However, it should be noted that not all vehicles have EDRs, and there is no standard definition of an EDR. HB 2106 would apply to new or used motor vehicles, not just those produced after the effective date. Retroactive application of a disclosure requirement would create a significant administrative and financial burden on manufacturers to amend their owner's manuals. The bill would also require "separate written disclosure" which would seem to be a document entirely separate from the owner's manual. This is a departure from the approach both NHTSA and other states have taken when evaluating this issue.

A patchwork of individually enacted State EDR laws raises substantial questions about extraterritorial effect, e.g., cross-border sales, and administration of conflicting requirements. Recognizing these concerns, the federal regulatory agency responsible for motor vehicle safety has begun to evaluate how to regulate the devices. NHTSA recently issued a notice of proposed rulemaking, NPRM Docket No. NHTSA-2004-18029, that would require a specific statement in the owner's manual, so that disclosure of EDRs soon will be addressed on a national level. Therefore, the Alliance believes it would be better to wait and see the rule NHTSA promulgates before proceeding with any state legislation.

Please feel free to contact our local representative, Mary Keenan, at (717) 236-0443, if you have any questions or would like further information.

Sincerely,

Kris Kiser
Vice President, State Affairs

**BMW Group • DaimlerChrysler • Ford Motor Company • General Motors
Mazda • Mitsubishi Motors • Porsche • Toyota • Volkswagen**

When the Alliance of Automobile Manufacturers testimony was discussed several issues were raised. Legislators wanted to know if it was true that after years of research and development there "was no standard definition of a event data recorder." They were informed that although several terms are used to describe the same functionality, in fact a global standard for motor vehicle event data recorders would be finalized within months which included a definition commonly accepted by industry, government and the public.

Another issue that raised concern was the statement that, "Therefore, the Alliance believes it would be better to wait and see the rule NHTSA promulgates before proceeding with any state legislation." The legislators were concerned with the time line for the federal rulemaking which baring any delays might be issued by September 1, 2008. To them, the question was what will happen between now and then. They were informed that government estimates forecast 200,000 fatalities and 10 million injuries suffered by victims of motor vehicle crashes before September 1, 2008. Hearing this the legislators expressed there intent to proceed with their legislation.

On July 13, 2004, the Alliance of Automobile Manufacturers requested a two month extension from NHTSA to reply to the federal Notice of Proposed Rulemaking. Their request was partly based on the fact that General Motors Corporation was on vacation for two weeks in July.

July 13, 2004

The Honorable Jeffrey W. Runge, M.D.
Administrator
National Highway Traffic Safety Administration
400 Seventh Street, S.W.
Washington, D.C.   20590

Dear Dr. Runge:

**RE: Event Data Recorders; Notice of Proposed Rulemaking (NPRM)
(69 Fed. Reg. 32932, June 14, 2004); NHTSA Docket No. NHTSA-2004-18029
Petition to Extend Comment Due Date**

The Alliance of Automobile Manufacturers is writing to petition for an extension of sixty days in which to comment on the agency's recent proposal addressing event data recorders (EDR's). The proposal was published on June 14, 2004, 69 Fed. Reg. 32932. If the agency grants this petition, the new due date for comments would be October 12, 2004.

The Alliance is seeking this extension for three reasons.

First, General Motors is traditionally closed for the first two weeks in July for a company-wide shutdown. General Motors has been an active participant in various working groups on the use of EDR's in motor vehicles. The Alliance expects General Motors to be involved significantly in the preparation of the Alliance comments on the EDR proposal but will be unable to receive substantial input from General Motors until after General Motors reopens in the middle of July. Although the Alliance members who are not closed will continue to work to develop information supportive of Alliance comments in the meantime, the absence of General Motors' contribution will, as a practical matter, defer the development of a comprehensive Alliance position until nearly the close of the scheduled comment period.

Second, the proposed definition of EDR covers a wide variety of vehicle systems, not all of which are currently integrated or in communication with one another. In order to provide useful comments on the proposed definition of "EDR," Alliance members must conduct extensive reviews of current and planned vehicle platforms to understand the scope of what would be required, should the definition of EDR be adopted as proposed. This effort will require substantially more time than available within the 60

**BMW Group • DaimlerChrysler • Ford Motor Company • General Motors
Mazda • Mitsubishi Motors • Porsche • Toyota • Volkswagen**

Dr. Jeffrey W. Runge, M.D.   2   July 13, 2004

day comment period. If the requested extension is not granted, Alliance members may not have enough time to collect the information needed to understand the cost and leadtime implications of the proposed definition, and may not be able to provide useful input to the agency on those important issues..

Finally, the Alliance understands that NHTSA does not wish its proposed EDR rule to hinder or deter the continued voluntary introduction of EDRs in new motor vehicles. The Alliance shares that goal, and it is the Alliance's intention to provide constructive comments that could form the basis for a consensus standard on a standardized definition of EDR and on standardized minimum data elements for EDRs. Particularly in light of the factors discussed above (General Motors two-week shut down and the need to inventory current and future platforms to understand the scope of the proposed definition of EDR), the Alliance does not expect that it can complete the task of developing constructive suggestions by the close of the original comment period. The Alliance believes that it can do so if NHTSA agrees to extend the comment period to October 12, 2004.

For all of these reasons, the Alliance requests that the agency grant this petition and extend the comment period to October 12, 2004.

Sincerely,

Robert Strassburger
Vice President
Safety and Harmonization
Alliance of Automobile Manufacturers

cc: Mr. Stephen R. Kratzke, Esq.
       Mr. J. Edward Glancy, Esq.
       Dr. William Fan

Written testimony representing the American Civil Liberties Union (ACLU) was submitted at the Pennsylvania EDR Policy Hearing by Sara Restauri of the Greater Pittsburgh Chapter of the ACLU:

> Tuesday, July 13, 2004
>
> House Democratic Party Committee Meeting
>
> Testimony of Sara Restauri, ACLU Intern
>
> Good afternoon. My name is Sara Restauri, and I am a legal intern at the Greater Pittsburgh Chapter of the American Civil Liberties Union. First, I would like to thank you for inviting me here this afternoon to address the issue of consumer privacy and the information collected by Event Data Recorders. Protecting the Constitutional right to privacy is a matter of great importance, and I thank you for you interest in considering the ACLU's views on this issue.
>
> New technology is often a double-edged sword that facilitates daily activities but also has the potential to invade our right to privacy. The Event Data Recorders (also known as EDRs) and similar devices found in newer automobiles are a prime example of this duality. EDRs are a necessary component of an airbag system, and in that respect EDRs help to make our daily activities safer. However, EDRs also record information, such as rate of speed and application of brakes, which may be downloaded by police officers and used as evidence in court.
>
> Under the Fourth Amendment to the United States Constitution and Article I, Section 8 of the Pennsylvania Constitution, persons have a right to privacy in relationship to the state in their bodies and personal effects, homes, and other areas or items maintained as private, unless the person has affirmatively waived that right by freely, knowingly, and voluntarily either consenting to state intrusion or abandoning the area or item. If the person does not consent to the intrusion or abandon the area or item, then the state may only intrude if government agents first obtain a search warrant. Searches may not be conducted without a search warrant unless there are exigent or emergency circumstances and there is probable cause.
>
> It is likely that a police officer would obtain EEDR data while investigating a car accident of some sort. Such searches usually do not involve exigent circumstance;

therefore, an investigating officer must either get consent from the vehicle owner to conduct the search, or obtain a search warrant prior to conducting the search. If the vehicle owner consents to the search but is not aware of the presence of an EDR and does not know that the data recorded can be retrieved by the police and used as evidence, then the police seizure of EDR data may be an invasion of the driver's reasonable expectation of privacy.

In the interest of protecting the right to privacy, the public should be informed of the presence of electronic devices that could be used to invade areas where people reasonably expect privacy. House Bill 2106 protects consumers from electronic invasions of privacy by informing them about EDRs before they decide to purchase a vehicle or consent to a police search of their vehicle.

There are obvious benefits to the use of EDRs; not only are they a necessary component of life-saving airbag systems, but they also assist police officers in investigating the causes of motor vehicle accidents. The data can be used to make cars and roadways safer for everyone. Other forms of technology, such as video cameras, cellular phones, Global Positioning System, biometrics, the Internet and e-mail are often used in a way that benefits society and complies with the Constitution. But, the government has also abused these devices to monitor and collect information about private citizens. Video cameras installed in private places for the purpose of discouraging crime and terrorism are missued for other purposes, such as racial profiling and stalking. Likewise, EDRs could be misued as a means of tracking and snooping on individuals. These invasions of privacy are unconstitutional, and should be stopped.

The misuse of electronic devices will have a chilling effect on public life; knowing that a camera could be watching you, or that your car's EDR system could be tattling on you, will inevitably result in self-consciousness and modified behavior. While these issues are beyond the scope of this particular Bill, I encourage this committee to consider the problems and address these issues in future legislation. As technology continues to develop at a rapid pace, so too will the abuse of electronic devices; the result is a growing threat to the right to privacy that we all enjoy and depend on.

Thank you for your time and consideration.

In concluding the hearing, Representative Harhai summarized by noting, "My bill would simply ensure full disclosure and ultimately better sales practice. As a member of the House Consumer Affairs Committee, I will continue to fight to protect consumer's rights."

There may not be a short term solution to the complex privacy and legal issues. Its worth repeating that despite the obvious safety benefits that might accrue, the use of MVEDRs will not be without controversy. Individual motorists or others within a motor vehicle expect a right to privacy. Although the right to privacy is not explicitly granted in the Constitution, it has been recognized that individual privacy is a basic prerequisite for the functioning of a democratic society. Indeed an individual's sense of freedom and identity depends a great deal on government respect for privacy. Efforts associated with introducing MVEDRs must recognize and respect the individual's interest in privacy and information use. The real issue is striking a balance between privacy concerns and the quest for the freedom to travel safely. Research of privacy concerns indicate that oftentimes, the concern is less about the data MVEDRs presently collect, but instead what future devices might be capable of recording. Presumably, the devices could be engineered to collect considerably more data and do so over even longer periods of time. Privacy can be used as a smokescreen by those who already have access to the data. Whose privacy is being protected is a good question? Everyone should beware of those that claim they are protecting individual or consumer privacy while at the same time excluding the creator of the data from disclosure or access. Look at it this way: you purchase the vehicle, create the data during a crash and may be excluded from access unless you purchase a sophisticated electronic decoder or hire an accident reconstructionist. The very idea that crash data is proprietary data and not public data is a major unresolved issue. Undoubtedly as vehicles become more electronic than mechanical this notion will be challenged. Motorists and occupants may be forced to choose between privacy and safety until someone offers an "on" or "off" switch. The road ahead will be difficult in this area as the public gains more knowledge and understanding.

CHAPTER 11

Pros and Cons and Customers of Safety Data and NHTSA Rulemaking

Rational decision-making in public policy is dependent on impartial research and information. Developing research capacity nationally is a central feature of the new model of road safety. Without research capacity, there exist few means to overcome misconceptions and prejudices about road crash injuries.

In this chapter I note the real world perceptions about MVEDRs and users of automotive black box technologies. I also provide the NHTSA rulemaking proposal on EDRs.

The following research focuses on what college-age motorists perceive to be the positive and negative aspects of implementing on-board Motor Vehicle Event Data Recorders (MVEDRs) in vehicles. A number of key issues ranging from perceived safety benefits versus fear or privacy invasion are included. This research was conducted at Sandhills Community College (SCC), Pinehurst, North Carolina. [1]

Positive Aspects

The data may aid in:

- regulatory initiatives;
- alleged defect investigations;
- litigation cases;
- initiatives to improve driver behavior;
- law enforcement efforts;
- designs for safer vehicles;
- gathering accurate statistics;
- decreasing vehicle prices;
- decreasing insurance premiums;
- identifying hazardous conditions;
- identifying needs for safety devices;
- providing information as to why some crashes are fatal;

- providing actual crash velocity in real time conditions;
- reducing the amount of crash testing in labs;
- providing quicker emergency response time to crashes;
- providing a better understanding as to how driver's respond to a crash;
- providing a better understanding as to how occupants in various positions respond to a crash;
- providing a better picture of overall crash behavior;
- may help catch people who intentionally crash to collect insurance;
- determine the number of people within a vehicle and help reduce insurance fraud;
- providing critical information that will determine causes of injuries and fatalities;
- increasing the safety of future vehicles;
- increasing seat belt usage;
- encouraging safer driving habits;
- determining defective vehicle systems and parts;
- leading to improved airbag safety;
- determining if children were in-position or out-of-position;
- providing an accurate number of daily, weekly, monthly and yearly crashes in specific locations;
- helping to reduce road rage behavior;
- aiding in school bus safety;
- signaling and timing emergency response.

Negative Aspects

The data may

- reduce informational and personal privacy;
- be misused by government, law enforcement, and insurance companies;
- be missued by vehicle manufacturer's in product warranty disputes;
- create a "big brother" syndrome.

The Use of Safety Data

Automotive black box technologies can serve as a catalyst towards using emerging transportation safety technologies. MVEDRs will accelerate deployment of driver-assisted technologies, collision avoidance systems, vehicle diagnostic systems and advanced medical response capabilities.

MVEDR technologies include retrieving, gathering, and storing scientific objective data which may improve highway efficiency, mobility, productivity and environmental quality by providing compelling evidence of the types of crashes, the role of human error, systems engineering and systems integration issues.

MVEDR safety data will enhance highway transportation. Transportation impacts every facet of American life, providing people access to work, school, loved ones, and recreation.

Manufacturers

Vehicle manufacturers are typically installing MVEDRs to collect data to improve the design of motor vehicles and diagnose vehicle systems. The automotive industry utilizes data-driven design of vehicles. They benefit from knowledge of larger number of crashes across a continuum of severity and early evaluation of system and vehicle design performance. The trend is toward international harmonization of safety standards.

Government

Government users fall into several levels—the federal level, state level, and other local users. The federal role includes uses of MVEDR that to carry out its mission: to save lives, reduce injury and proper loss. This includes collecting data to assist in a better safety management for highway and traffic systems. The federal government could also utilize data to assess safety problems a provide solutions for issuing new and revised vehicle safety performance standards. The federal government would enhance safety by promulgating and evaluating standards, identifying problem injuries and mechanisms, stipulating injury criteria and investigating defects.

At the state level, crash data could be used to assist states in managing road systems and designing better roadside safety hardware, such as guardrails and crash cushions. These groups are very interested in collecting crash location information that would vastly improve their ability to improve roadside safety. State officials require crash information to identify problem intersections and road lengths, to determine hazardous countermeasures and to evaluate the effectiveness of safety interventions.

At the local level, MVEDR data could be used to assist emergency medical services (EMS) control, especially if MVEDR data could be automatically dispatched from the crashed vehicle to the Public Safety Answering Point (PSAP) as well as other affected parties.

Law Enforcement

The users benefit greatly from obtaining quick and impartial information regarding the crash. They are often changed with determining the facts associated with a crash, and these data provide additional tools to validate field collision data, determine crash causation and fraud.

Insurance Companies

Insurance companies often analyze a collision for validity prior to paying the claim. MVEDR data provide more accurate related to a crash. Fraudulent claims cost more than $20 billion annually. Insurers require accurate data for subrogation of claims and recovery of expenses. Data from MVEDRs improve risk management, expedite claims and decrease administrative costs.

Plaintiffs, Defense Attorneys, Judges, Juries, Courts and Prosecutors
This group of users often obtain costly experts in the field of crash reconstruction to assist them in providing their position. The use of MVEDR data will put more "science" on the table during these actions and could lead to shorter actions or no action altogether. Juries would get objective information, too. Courts could require vehicles be equipped with recording devices.

Human Factors Research
Human factor researchers are continuously looking for more data to understand the human's involvement associated with crash causation. Pre-crash MVEDR data could be used by these researchers to understand driver performance and conduct further analysis of this complicated issue. Research such as man—machine interface, crash causation, the effects of aging and harmonized crash dummy development and injury causation would enhance safety.

State Insurance Companies
Insurance officials could use MVEDR data to support decisions regarding insurance rates, such as, approving discounts for owners who pre-agree to release MVEDR data should a crash occur.

Parent Groups
The Customers, such as Mothers Against Drunk Drivers (MADD) and other parent groups, could use MVEDR data to support trends in crashes.

Fleets and Drivers
MVEDRs could be used by drivers/fleet owners in many ways, including: improving driver safety, educating drivers about technology on vehicles, auto-downloading data for driver use, providing information vehicle safety characteristics (data element related), and providing information regarding the general performance of the vehicle. Another primary use of MVEDR data by the driver/owner could be the use of the data to demonstrate their proper vehicle operation during a collision.

Medical Injury Guideline Data Usage
Hospital officials, EMS providers, and other EMS decision makers could use MVEDR data to improve field triage decisions. These data could be used to trigger a series of events which would ensure that the "right" help got to the crash and Emergency Room (ER) staff to look for non-visible injuries. While more related to Automatic Crash Notification (ACN), these new methodologies could save lives.

Vehicle Owner
What's in it for you? The vehicle owner could review MVEDR data to determine if the vehicle had been in a previous crash. These data would indicate the severity of the crash, which may relate to the level of repairs the vehicle had undergone

during its lifetime. The driving public would benefit from better policies, vehicle design, emergency response, roadway design and driving habits, lowered insurance costs, decreased possibility for fraud, fewer crashes, and more efficient systems.

Transportation Researchers and Academia

Transportation researchers could use MVEDR data to conduct research related to vehicles, highway, medical treatments, and so forth.

Breaking the Stalemate in Safety

By widespread usage of MVEDR safety data it is possible to break the stalemate in safety. This will also require political stamina and courage from an administrator of NHTSA who will stand firm for an interest that has no organized consistuency—the public. Since the public bears the impact of auto industry and government safety programs, the decision of who defines safety should not be left in the hands of a self-promoting industry or a timid bureaucracy. We must debunk the notion that if you disturb or restrict the automakers then you jeopardize the entire economy. The American public should understand the difference between marketing safety and making safer vehicles. The bottom line is that everyone can do better.

NTSB Symposia

On June 3–4, 2004 a Highway Vehicle Event Data Recorder Symposium was conducted at the National Transportation Safety Board (NTSB) Academy at George Washington University, Virginia Campus, Ashburn, Virginia.

The Society of Automotive Engineers (SAE), the National Transportation Safety Board (NTSB), the Federal Motor Carrier Safety Administration (FMCSA), Forensic Accident Investigations, and KEVA Engineering coordinated to sponsor the event.

Approximately 110 professionals, including accident reconstructionists, police, insurance industry professionals, members of the legal community, research institutions, and government agencies, exchanged ideas and gave technology updates and case studies regarding motor vehicle event data recorder technology.

I was invited to attend and presented a high-level overview of IEEE standard activities.

A number of automakers, including Honda R&D Americas, Inc., Nissan Technical Center North America, BMW of North America LLC, Mitsubishi Motors R & D of America, inc., Nissan North America, Inc., DaimlerChrysler Corporation, and the Association of International Automobile Manufacturers (AIAM), were listed in the participant list and had representatives at the symposia.

The welcome and introduction were extended by the Chairman of the NTSB, Ellen Engleman Conners, and the Administrator of the Federal Motor Carrier Safety Administration (FMCSA), Annette M. Sandberg. Conners' message was that "out of tragedy, good will come." She eloquently stated how the NTSB's mission is to use science, data, and facts to find good in tragedy. Sandberg followed-up noting the great need for more data to reduce injury and fatalities. Both were strong proponents of using motor vehicle event data recorders in the highway mode of transportation.

Ford Motor Company executives provided two presentations:

Bill Ballard, a design analysis engineer employed since 1985, provided a discussion of the current status of EDRs within vehicles sold by Ford in North America. Richard R. Ruth, employed by Ford Motor Company since 1973 in a wide variety of product engineering projects, and currently manager of the design analysis vehicle dynamics and interior department, provided a perspective of why manufacturers record data. He also explained what a manufacturer considers when designing an EDR function into existing electronic modules, listed some examples of what is recorded in specific systems and why, and detailed a recap of EDR-related issues as seen by a manufacturer.

Ford Motor Company is generally considered to be the second largest original equipment manufacturer in the world. Company information about Ford Motor Company highlights the following points:

- Our Vision: To become the world's leading consumer company for automotive products and services.

- Our Mission: We are a global family with a proud heritage passionately committed to providing personal mobility for people around the world. We anticipate consumer need and deliver outstanding products and services that improve people's lives.

- Our Values: Our business is driven by our consumer focus, creativity, resourcefulness, and entrepreneurial spirit. We are an inspired, diverse team. We respect and value everyone's contribution. The health and safety of our people are paramount. We are a leader in environmental responsibility. Our integrity is never compromised and we make a positive contribution to society.

I took detailed notes during this presentation to capture the essence of the remarks. Mr. Ballard went into great detail describing the Restraints Control Module (RCM) technology. The RCM has three main functions:

1. It monitors system inputs and deploys various devices, such as air bags, when it is appropriate to do so.

2. It monitors system input and output signals to insure certain devices are operating properly, and if a monitored device is not operating properly, then warns the driver of the concern.

3. It records inputs and outputs to document that the system operated properly.

Once the general functions of the RCM were explained, Mr. Ballard then noted how an RCM records those items that go into the decision making process within the algorithm for a very short duration of approximately 0.1 second. Furthermore, an integration of acceleration trace provides a change in velocity, but not absolute speed. He also emphasized that a backup power supply is intended to support the real time internal analysis of an impact and to provide deployment of restraint devices as required, not to ensure the recording of data.

At this point, Mr. Ballard noted, "An RCM is not an EDR." He then attempted to distinguish between the two. His noted that an Event Rata Recorder (EDR) is intended to record an event for the purposes of accident reconstruction and used the example of an airplane EDR or "black box" that records events but does not affect the functioning of the airplane. Then he indicated that an RCM is intended to deploy the various devices when needed to provide occupant protection. Finally, he noted that recording is tertiary and is limited to ensuring proper function and repair of the system.

Following his presentation there were several questions about the usage of the terms RCM and EDR. People in the room wanted clarification. Mr. Ballard explained that a follow-up presentation by Mr. Ruth would address this issue.

Mr. Ruth began his presentation titled, "Event Data Recorders: Ford Motor Company," by saying, not without irony, "Absolutely, Ford Motor Company does not have any EDRs, nor do we record the speed of a vehicle."

My sense was that this opening statement was extremely confusing to the assembled group of crash reconstructionists.

In an attempt to clarify the matter, he explained the Ford Motor Company EDR philosophy as such: certain information recorded by vehicle systems can enhance automotive safety. It is important to obtain consent or other sufficient legal authority prior to retrieving electronic data from an individual vehicle relating to a crash or a near-crash event, for either internal purposes or for release to third parties. And finally, Ford Motor Company understands the value of a customer's trust and recognizes a customer's privacy.

Then Mr. Rush explained why manufacturers record any data to begin with. He stated that auto manufacturers need the ability to ascertain if their safety-related components or systems performed properly in field events to decide if a product improvement is necessary. Looking to the past, he reminded everyone that earlier automobiles included mostly mechanical systems, including early restraint systems, and allowed a mechanical reconstruction of system performance. Back then, no EDR was necessary, nor was there any electronic data available to record. He acknowledged that electronic systems may need to record critical system inputs and outputs to know if systems performed as intended in the field.

To the specific question, "Does Ford have EDRs?" Mr. Rush answered that Ford has Restraint Control Modules (RCMs) which primarily make decisions to deploy restraint devices and run diagnostics. He continued: "They also record some information in the event of a crash, but this is not the primary purpose of the RCM. Furthermore, Ford has Power Control Modules (PCMs) which primarily determine fuel, spark, and fire for the engine and run diagnostics but may also record some information in the event of a crash."

When pressed on the terminology, he again emphasized that neither is called an Event Data Recorder (EDR), and once again noted (with emphasis) that Ford did not design either module for the express purpose of helping accident reconstructionists. However, he did acknowledge that some of the recorded data may help them.

The audience was very interested in hearing a seasoned engineer from one of the major automakers explain this point: "How can EDRs enhance automotive safety?" Mr. Rush said there are basically two different philosophies: 1) component/subsystem performance facilitating manufacturer product enhancements, and 2) total vehicle performance and operator behavior, which would help accident reconstructionists and improve the quality of the data in NASS and FARS for long term study of vehicle safety, facilitating future industry or government policy programs or regulation.

His next point was "The EDR value question," which he explained in this manner:

1. Manufacturers record information to validate component/subsystem performance.

2. Accident reconstructionists and safety researchers want EDRs to make the job easier, less costly, and more accurate.

3. Retail customers may be concerned about privacy.

Therefore, he concluded that the question of how much an EDR is worth to society is still open to discussion.

On the issue of data extraction, he firmly stated the Ford Motor Company viewpoint (even though he claimed Ford Motor Company did not have EDRs), that:

1. EDR memory contents are proprietary to the auto manufacturer or EDR module supplier.

2. Data download and interpretation is resource intensive and Ford Motor Company does not have the resources to do this.

3. Manufacturers may choose third party companies like Vetronix to help respond to accident reconstructionist requests for assistance.

He noted that Ford Motor Company had little interest in providing these services.

His closing thoughts were:

EDRs may be a simple concept, but there are many complex technical issues in implementing them.

Demands by data-users for increasing kinds and amounts of data must be balanced by consideration of costs, the interests and protection of customers, and priorities of data use.

As a safety community, we have to communicate to the public the potential safety benefits of EDR's better to counteract negative press touting that Big Brother is watching.

A healthy dialogue between industry and government is needed before any potential regulation is finalized to maximize value.

If regulation is inevitable, give enough lead time for efficient design and phase-in.

Overall, I found Mr. Ruth's presentation educational and insightful. However, in the follow-up questions, many in the room were still confused with his statement that the Ford Motor Company did not have any Event Data Recorders (EDRs), nor did they ever record speed.

After the symposia concluded for the day, and on my return to my motel, I decided to stop in at the local Ford Motor Company dealership and ask if any of the employees were familiar with the technology Mr. Rush described in detail: the so-called "RCM."

A Ford Motor Company salesman told me that he was only familiar with what was disclosed in a vehicle owner's manual.

That made sense. So I took a look in an online manual. Page 4 of the Ford F-350 SD manual said, "Congratulations on acquiring your new Ford. Please take the time to get well acquainted with your vehicle by reading the handbook. The more you know and understand about your vehicle, the greater the safety and pleasure you derive from driving it."

Page 5 included the following (verbatim):

- Event Data Recorder
- The computer in your vehicle is capable of recording detailed data potentially including but not limited to information such as:
- The use of restraint systems including seat belts by the driver and passengers,
- Information about the performance of various systems and modules, and
- Information related to engine, throttle, steering, brake or other system status potentially including how the driver operates the vehicle including but not limited to vehicle speed.

Page 6 continued (verbatim):

- This information may be stored during regular operation or in a crash or near crash event. This stored information may be read out and used by:
- Ford Motor Company
- Service and repair facilities
- Law enforcement or government agencies
- Others who may assert a right or obtain your consent to know such information

After reading this, I was totally confused. The next day I shared my concern with a professional accident reconstructionist who happened to be sitting next to me. He just shook his head, then simply commented that perhaps the automakers were positioning themselves for the upcoming federal rulemaking.

NHTSA Rulemaking

Ironically, on the same day that the Ford Motor Company explained they did not have any event data recorders, and less than thirty miles away, at the Office of Budget and Management located next to the White House in Washington, the Federal government began the rulemaking process by approving a Notice of Proposed Rulemaking (NPRM). The Rulemaking Analyses and Notices noted the following:

Executive Order 12866 and DOT Regulatory Policies and Procedures

NHTSA has considered the potential impacts of this proposed rule under Executive Order 12866 and the Department of Transportation's regulatory policies and procedures. This document was reviewed by the Office of Management and Budget under E.O. 12866, "Regulatory Planning and Review." This document has been determined to be significant under the Department's regulatory policies and procedures. While the potential cost impacts of the proposed rule are far below the level that would make this a significant rulemaking, the rulemaking addresses a topic of substantial public interest.

The agency has prepared a separate document addressing the benefits and costs for the proposed rule. A copy is being placed in the docket.

As discussed in that document and in the preceding sections of this NPRM, the crash data that would be collected by EDRs under the proposed rule would be extremely valuable for the improvement of vehicle safety by improving and facilitating crash investigations, the evaluation of safety countermeasures, advanced restraint and safety countermeasure research and development, and advanced ACN. However, the improvement in vehicle safety would not occur directly from the collection of crash data by EDRs, but instead from the ways in which the data are used by researchers, vehicle manufacturers, ACN and EMS providers, government agencies, and other members of the safety community. Therefore, it is not presently practical to quantify the safety benefits.

We estimate that about 67 to 90 percent of new light vehicles are already equipped with EDRs. As discussed earlier, vehicle manufacturers have provided EDRs in their vehicles by adding EDR capability to their vehicles' air bag control systems. The costs of EDRs have been minimized, because they involve the capture into memory of data that is already being processed by the vehicle, and not the much higher costs of sensing much of that data in the first place.

The costs of the proposed rule would be the incremental costs for vehicles equipped with EDRs to comply with the proposed requirements. As discussed in the agency's separate document on benefits and costs, we estimate the total annual costs of the proposed rule to range from $5.7 to $8.6 million. While the potential costs include technology costs, paperwork maintenance costs, and compliance costs, the paperwork maintenance and compliance costs are estimated to be negligible. The proposal would not require additional sensors to be installed in vehicles, and the major technology cost would result from a need to upgrade EDR memory chips. The total cost for the estimated 11.2 to 15.2 million vehicles that already have an EDR function to comply with the proposed regulation is estimated to be $5.7 to $7.7 million. If manufacturers were to provide EDRs in all 16.8 million light vehicles, the estimated total cost is $8.6 million. A complete discussion of how NHTSA arrived at these costs may be found in the separate document on benefits and costs.

Regulatory Flexibility Act

NHTSA has considered the impacts of this rulemaking action under the Regulatory Flexibility Act (5 U.S.C. 601 et seq.). I certify that the proposed amendment would not have a significant economic impact on a substantial number of small entities.

The following is the agency's statement providing the factual basis for the certification (5 U.S.C. 605(b)). If adopted, the proposal would directly affect motor vehicle manufacturers, second stage or final manufacturers, and alterers. SIC code number 3711, Motor Vehicles and Passenger Car Bodies, prescribes a small business size standard of 1,000 or fewer employees. SIC code No. 3714, Motor Vehicle Part and Accessories, prescribes a small business size standard of 750 or fewer employees.

Only four of the 18 motor vehicle manufacturers affected by this proposal would qualify as a small business. Most of the intermediate and final stage manufacturers of vehicles built in two or more stages and alterers have 1,000 or fewer employees. However, these small businesses adhere to original equipment manufacturers' instructions in manufacturing modified and altered vehicles. Based on our knowledge, original equipment manufacturers do not permit a final stage manufacturer or alterer to modify or alter sophisticated devices such as air bags or EDRs. Therefore, multistage manufacturers and alterers would be able to rely on the certification and information provided by the original equipment manufacturer. Accordingly, there would be no significant impact on small businesses, small organizations, or small governmental units by these amendments. For these reasons, the agency has not prepared a preliminary regulatory flexibility analysis.

Summary of the Collection of Information

To improve the availability and usability of data collected by motor vehicle sensors during a crash event, the proposed regulation would require manufacturers that voluntarily equip vehicles with an EDR to record specified data elements and to standardize the format of the resulting data.

Motor vehicle manufacturers voluntarily equipping vehicles with an EDR would also be required to submit information to the agency on accessing and retrieving the stored data. The technical specifications would be required to be of sufficient detail to permit an individual to design and build a tool for accessing and downloading the data in the specified format. This information would be required to be submitted not later than 90 days before the beginning of the production year in which the EDR equipped vehicles are to be offered for sale.

Description of the Need for the Information and Proposed Use of the Information

The information sought by NHTSA in this collection would be used by the agency and crash investigators (*e.g.*, other government agencies, police investigators, motor vehicle crash researchers, etc.) to access and retrieve standardized crash data from voluntarily installed EDRs. Improving the availability of crash event data would permit the agency to improve analysis of a restraint system's crash protection performance and the determination of crash-avoidance system effectiveness. Improving the data elements and data available to the agency would allow NHTSA to make more targeted rulemaking decisions, thus improving overall vehicle safety in the future.

Description of the Likely Respondents (Including Estimated Number, and Proposed Frequency of response to the Collection of Information)

NHTSA estimates that a maximum of 18 vehicle manufacturers would submit the required information. The manufacturers are makers of passenger cars, multipurpose passenger vehicles, trucks and buses that have a GVWR of 3,855 kg (8,500 pounds) or less and an unloaded vehicle weight of 2,495 kg (5,500 pounds). For each report, a manufacturer would provide, in addition to its identity: (1) non-proprietary technical information of sufficient detail to permit an individual to design and build a tool to download the EDR data in the specified format and (2) information of sufficient detail to permit access to the data in each vehicle make and model produced by the manufacturer that is equipped with an EDR.

Manufacturers would be required to submit the above information once per year.

Estimate of the Total Annual Reporting and Record-keeping Burden Resulting from the Collection of Information

NHTSA estimates the total annual burden hours to be $18,900. (30 burden hours × 18 manufacturers × $35/burden hour) NHTSA estimates that each manufacturer would incur a total of 30 burden hours per year under this collection. The agency estimates that each manufacturer would incur 20 burden hours per year to comply with the information collection and 10 burden hours per year for data standardization. The estimate for the hour burden arising from the information submission is based on the fact that manufacturers would be submitting existing information from its vehicle production data and equipment specification data. As the industry voluntarily standardizes EDR output, the agency anticipates this burden would decrease because manufacturers will be able to cite voluntary industry standards in place of technical specifications. The burden arising from the record-keeping portion of this request would be a result of manufacturers reprogramming existing sensor systems to meet the data standardization requirements of this program. Given the lead time of the proposed regulation, this reprogramming could be accomplished during a scheduled upgrade of a motor vehicle's sensor systems. This one time reprogramming cost is estimated between $100,000 and $180,000, for the entire industry. Once a manufacturer has standardized all of the existing sensors, we would anticipate this burden to be reduced to a minimal number.

If a manufacturer needed to increase the electronic storage capability of the existing sensors to comply with the proposal, this would result in an additional cost of $0.50 per vehicle. As discussed above and in the separate document on costs and benefits, the estimated cost for the entire industry from the increased memory and software reprogramming is $5.7 to $8.6 million.

Persons desiring to submit comments on the information collection requirements should direct them to the Office of Information and Regulatory Affairs, OMB, Room 10235, New Executive Office Building, Washington, DC, 20503; Attention: Desk Officer for U.S. Department of Transportation.

The agency will consider comments by the public on this proposed collection of information in:

Evaluating whether the proposed collection of information is necessary for the proper performance of the functions of NHTSA, including whether the information will have a practical use;

Evaluating the accuracy of the agency's estimate of the burden of the proposed collection of information, including the validity of the methodology and assumptions used;

Enhancing the quality, usefulness, and clarity of the information to be collected; and

Minimizing the burden of collection of information on those who are to respond, including collection techniques or other forms of information technology; e.g., permitting electronic submission of responses.

OMB is required to make a decision concerning the collection of information contained in the proposed regulation between 30 and 60 days after publication of this document in the Federal Register. Therefore, a comment to OMB is best assured of having its full effect if OMB receives it within 30 days of publication. This does not affect the deadline for the public to comment to NHTSA on the proposed regulation.

NHTSA requests comments on its estimates of the total annual hour and cost burdens resulting from this collection of information. Please submit comments according to the instructions under the Comments heading of this notice. Comments are due by [insert date that is 60 days after the date of publication in the Federal Register].

Executive Order 13132 (Federalism)

Executive Order 13132 requires NHTSA to develop an accountable process to ensure "meaningful and timely input by State and local officials in the development of regulatory policies that have federalism implications." "Policies that have federalism implications" is defined in the Executive Order to include regulations that have "substantial direct effects on the States, on the relationship between the national government and the States, or on the distribution of power and responsibilities among the various levels of government." Under Executive Order 13132, the agency may not issue a regulation with Federalism implications, that imposes substantial direct costs, and that is not required by statute, unless the Federal government provides the funds necessary to pay the direct compliance costs incurred by State and local governments, or the agency consults with State and local officials early in the process of developing the proposed regulation. NHTSA may also not issue a regulation with Federalism implications and that preempts State law unless the agency consults with State and local officials early in the process of developing the proposed regulation.

The agency has analyzed this rulemaking action in accordance with the principles and criteria contained in Executive Order 13132 and has determined that, although the proposed regulation would preempt conflicting State law, it does not have sufficient federalism implications to warrant consultation with State and local officials or the preparation of a federalism summary impact statement. The proposed rule would have no substantial effects on the States, or on the current Federal-State relationship, or on the current distribution of power and responsibilities among the various local officials.

Executive Order 12778 (Civil Justice Reform)

This proposed rule would not have any retroactive effect. Under section 49 U.S.C. 30103, whenever a Federal motor vehicle safety standard is in effect, a state may not adopt or maintain a safety standard applicable to the same aspect of performance which is not identical to the Federal standard, except to the extent that the state requirement imposes a higher level of performance and applies only to vehicles procured for the state's use. This section would not apply to the proposed rule, because it would not be a Federal motor vehicle safety standard. General principles of preemption law would apply, however, to displace any conflicting state law or regulations. If the proposed rule were made final, there would be no requirement for submission of a petition for reconsideration or other administrative proceedings before parties could file suit in court.

National Technology Transfer and Advancement Act

Section 12(d) of the National Technology Transfer and Advancement Act of 1995 (NTTAA), Public Law 104-113, section 12(d) (15 U.S.C. 272) directs us to use voluntary consensus standards in regulatory activities unless doing so would be inconsistent with applicable law or otherwise impractical. Voluntary consensus standards are technical standards (e.g., materials specifications, test methods, sampling procedures, and business practices) that are developed or adopted by voluntary consensus standards bodies, such as the Society of Automotive Engineers (SAE). The NTTAA directs us to provide Congress, through OMB, explanations when we decide not to use available and applicable voluntary consensus standards.

As discussed above, both the SAE Vehicle Event Data Interface (J1698-1) Committee and the IEEE Motor Vehicle Event Data Recorder (MVDER) working group (P1616) are developing standards specific to EDRs. While there are currently no voluntary consensus standards for EDR data elements or data format, the agency will consider such standards when they are available. Where appropriate, the agency has incorporated by reference SAE J211, Class 60 for the specified data filtering requirements.

Unfunded Mandates Reform Act

Section 202 of the Unfunded Mandates Reform Act of 1995 (UMRA) requires Federal agencies to prepare a written assessment of the costs, benefits, and other effects of proposed or final rules that include a Federal mandate likely to result in the expenditure by State, local, or tribal governments, in the aggregate, or by the private sector, of more than $ 100 million in any one year (adjusted for inflation with base year of 1995). Before promulgating a rule for which a written statement is needed, section 205 of the UMRA generally requires NHTSA to identify and consider a reasonable number of regulatory alternatives and adopt the least costly, most cost-effective, or least burdensome alternative that achieves the objectives of the rule. The provisions of section 205 do not apply when they are inconsistent with applicable law. Moreover, section 205 allows NHTSA to adopt an alternative other than the least costly, most cost-effective, or least burdensome alternative if the agency publishes with the final rule an explanation why that alternative was not adopted. If adopted, this proposed rule would not impose any unfunded mandates under the Unfunded Mandates Reform Act of 1995. This proposed rule would not result in costs of $100 million or more to either State, local, or tribal governments, in the aggregate, or to the private sector. Thus, this proposed rule is not subject to the requirements of sections 202 and 205 of the UMRA.

NHTSA Proposal and Response to Petition

As discussed earlier, in the late-1990s, NHTSA denied two petitions for rulemaking requesting the agency to require the installation of EDRs in new motor vehicles, because the motor vehicle industry was already voluntarily moving in the direction recommended by the petitioners, and because the agency believed "this area presents some issues that are, at least for the present time, best addressed in a non-regulatory context."

Today, after the completion of the NHTSA-sponsored EDR Working Group's tasks and after considering the public comments and the petition from Dr. Martinez, we have tentatively concluded that motor vehicle safety can be advanced by a limited regulatory approach. In order to promote safety, we are particularly interested in ensuring that when an EDR is provided in a vehicle, the EDR will record the data necessary for effective crash investigations, analysis of the performance of advanced restraint systems, and ACN systems, and that these data can be easily accessed and used by crash investigators and researchers.

Given what the motor vehicle industry is already doing voluntarily in this area, we are not at this time proposing to require the installation of EDRs in all motor vehicles. As indicated earlier, we estimate that 65 to 90 percent of model year 2004 passenger cars and other light vehicles have some recording capability, and that more than half record such things as crash pulse data.

We are proposing a regulation that would specify requirements for light vehicles that are equipped with EDRs, i.e., vehicles that record information about crashes. The proposed regulation would (1) require the EDRs in these vehicles to record a minimum set of specified data elements; (2) specify requirements for data format; (3) require that the EDRs function during and after the front, side and rear vehicle crash tests specified in several Federal motor vehicle safety standards; (4) require vehicle manufacturers to make publicly available information that would enable crash investigators to retrieve data from the EDR; and (5) require vehicle manufacturers to include a brief standardized statement in the owner's manual indicating that the vehicle is equipped with an EDR and discussing the purposes of EDRs. A discussion of each of these items is provided in the sections that follow.

The proposed regulation would apply to the same vehicles that are required by statute and by Standard No. 208 to be equipped with frontal air bags, i.e., passenger cars and trucks, buses and multipurpose passenger vehicles with a GVWR of 3,855 kg (8500 pounds) or less and an unloaded vehicle weight of 2,495 kg (5500 pounds) or less, except for walk-in van-type trucks or vehicles designed to be used exclusively by the U.S. Postal Service. This covers the vast majority of light vehicles. Moreover, these are the vehicles that will generally have advanced restraint systems, since they are the ones subject to the advanced air bag requirements now being phased in under Standard No. 208.

We are not addressing in this document what future role the agency may take related to the continued development and installation of EDRs in heavy vehicles. We will consider that topic separately. Any action we might take in that area would be done in consultation with the Federal Motor Carrier Safety Administration.

Similar to our approach in the area of vehicle identification numbers, we are proposing a general regulation rather than a Federal motor vehicle safety standard. Thus, while a failure to meet EDR requirements would be subject to an enforcement action, it would not trigger the recall and remedy provisions of the National Traffic and Motor Vehicle Safety Act, currently codified at 49 U.S.C. Chapter 301.

A. *Data Elements to be Recorded*

As indicated above, we are proposing to require light vehicles that are equipped with EDRs to meet a number of requirements, including one for recording specified data elements.

Before discussing the proposed set of specified data elements, we will briefly address the issue of the crash recording capability that would trigger application of the regulation's requirements. We are proposing to apply the regulation to vehicles that record any one or more of the following elements just prior to or during a crash, such that the information can be retrieved after the crash: the vehicle's longitudinal acceleration, the vehicle's change in velocity (delta-V), the vehicle's indicated travel speed, the vehicle's engine RPM, the vehicle's engine throttle position, service brake status, ignition cycle, safety belt status, status of the vehicle's frontal air bag warning lamp, the driver's frontal air bag deployment level, the right front passenger's frontal air bag deployment level, the elapsed time to deployment of the first stage of the driver's frontal air bag, and the elapsed time to deployment of the first stage of the right front passenger's frontal air bag. Thus, if a vehicle has a device that records any of the basic items of information typically recorded by EDRs, the proposed regulation would apply to that vehicle.

In analyzing what minimum set of specified data elements to propose, we focused on the elements that would be most useful for effective crash investigations, analysis of the performance of safety equipment, e.g., advanced restraint systems, and ACN systems. We believe these are the areas where information provided by EDRs can lead to the greatest safety benefits.

EDRs can improve crash investigations by measuring and recording actual crash parameters. They can also measure and record the operation of vehicle devices whose operation cannot readily be determined using traditional post-crash investigative procedures. For example, EDRs could determine whether the ABS system functioned during the crash.

EDRs can also directly measure crash severity. Currently, NHTSA estimates crash severity using crash reconstruction tools. One product of these tools is an estimate of the vehicle's delta-V. With an EDR, delta-V could be directly measured. Another assessment made by the crash investigators is the principal direction of force (PDOF). This is currently estimated based on physical damage. With x-axis and y-axis accelerometers, this could be measured or post-processed for planar (non-rollover) crashes, providing PDOF as a function of time.

EDRs can be particularly helpful in analyzing the performance of advanced restraint systems. They can record important information that is not measurable by post-crash investigations such as time of deployment of pre-tensioners and the various stages of multi-level air bags, the position of a seat during the crash (a seat is often moved by EMS personnel during their extrication efforts), and whether seat belts were latched.

Improved data from crash investigations will enable the agency and others to better understand the causes of crashes and injury mechanisms, and make it possible to better define and address safety problems. This information can be used to develop improved safety countermeasures and test procedures, and enhance motor vehicle safety.

EDRs can also make ACN systems more effective. An important challenge of EMS is to find, treat, and transport to hospitals occupants seriously injured in motor vehicle crashes in time to save lives and prevent disabilities. ACN systems, such as the GM On-Star system, can automatically and almost instantly provide information about serious crashes and their location to EMS personnel, based on air bag deployment or other factors. GM has announced that it will begin equipping vehicles with advanced ACN systems that provide measurements of crash forces for improved EMS decision-making. Data from EDRs can be used as inputs for advanced ACN systems.

As discussed earlier, vehicle manufacturers have made EDR capability an additional function of vehicles' air bag control systems. The air bag control systems necessarily process a great deal of vehicle information. EDR capability can be added to a vehicle by designing the air bag control system to capture, in the event of a crash, the relevant data in memory. The costs of EDR capability have thus been minimized, because it involves the capture into memory of data that is already being processed by the vehicle, and not the much higher costs of sensing much of that data in the first place.

In developing our proposed regulation for EDRs, we have followed a similar approach. That is, we have focused on the recording of the most important crash-related data that -are already being processed by vehicles, and not using this rulemaking to require such things as additional accelerometers. (The addition of an accelerometer to a vehicle could add costs on the order of $20 per vehicle.)

For a variety of reasons, including the fact that the light vehicles covered by this proposal are subject to Standard No. 208's requirements for air bags, some of the most important crash-related data we have identified are already being processed (or will soon be processed) by all of these vehicles. Under our proposal, these data elements would be required to be recorded for all vehicles subject to the regulation.

Other important crash-related data are currently processed by some, but not all vehicles. This reflects the fact that some advanced safety systems are provided on some but not all vehicles. Under our proposal, these data elements would be required to be recorded only if the vehicle is equipped with the relevant advanced safety system or sensing capability.

The following table identifies the data elements that would be required to be recorded under our proposal. We note that the vast majority of the elements in the table are being considered by SAE and/or IEEE in their ongoing efforts to develop standards for EDRs.

Data Elements that Must Be Recorded (R=Required; IE=If Equipped)			
Data Element	R/IE	Recording Interval/Time	Condition for Requirement (IE)
Longitudinal acceleration	R	−0.1 to 0.5 sec	n.a.
Maximum delta-V	R	Computed after event	n.a.
Speed, vehicle indicated	R	−8.0 to 0 sec	n.a.
Engine RPM	R	−8.0 to 0 sec	n.a.
Engine throttle, % full	R	−8.0 to 0 sec	n.a.
Service brake, on/off	R	−8.0 to 0 sec	n.a.
Ignition cycle, crash	R	−1.0 sec	n.a.
Ignition cycle, download	R	At time of download	n.a.
Safety belt status, driver	R	−1.0 sec	n.a.
Frontal air bag warning lamp, on/off	R	−1.0 sec	n.a.
Frontal air bag deployment level, driver	R	Event	n.a.
Frontal air bag deployment level, right front passenger	R	Event	n.a.
Frontal air bag deployment, time to deploy, in the case of a single stage air bag, or time to first stage deployment, in the case of a multi-stage air bag, driver	R	Event	n.a.
Frontal air bag deployment, time to deploy, in the case of a single stage air bag, or time to first stage deployment, in the case of a multi-stage air bag, right front passenger	R	Event	n.a.
Multi-event, number of events (1,2,3)	R	Event	n.a.
Time from event 1 to 2	R	As needed	n.a.
Time from event 1 to 3	R	As needed	n.a.
Complete file recorded (yes, no)	R	Following other data	n.a.
Lateral acceleration	IE	−0.1 to 0.5 sec	If vehicle is equipped to measure acceleration in the vehicle's lateral (y) direction

Data Elements that Must Be Recorded *(Continued)* (R=Required; IE=If Equipped)			
Data Element	R/IE	Recording Interval/Time	Condition for Requirement (IE)
Normal acceleration	IE	−0.1 to 0.5 sec	If vehicle is equipped to measure acceleration in the vehicle's normal (z) direction
Vehicle roll angle	IE	−1.0 to 6.0 sec	If vehicle is equipped to measure or compute vehicle roll angle
ABS activity (engaged, non-engaged)	IE	−8.0 to 0 sec	If vehicle is equipped with ABS
Stability control status, on, off, engaged	IE	−8.0 to 0 sec	If vehicle is equipped with stability control, ESP, or other yaw control system
Steering input (steering wheel angle)	IE	−8.0 to 0 sec	If vehicle equipped to measure steering wheel steer angle
Safety belt status, right front passenger (buckled, not buckled)	IE	−1.0 sec	If vehicle equipped to measure safety belt buckle latch status for the right front passenger
Frontal air bag suppression switch status, right front passenger (on, off, or auto)	IE	−1.0 sec	If vehicle equipped with a manual switch to suppress the frontal air bag for the right front passenger
Frontal air bag deployment, time to N^{th} stage, driver *	IE	Event	If vehicle equipped with a driver's frontal air bag with a second stage inflator
Frontal air bag deployment, time to N^{th} stage, right front passenger *	IE	Event	If vehicle equipped with a right front passenger's frontal air bag with a second stage inflator
Frontal air bag deployment, N^{th} stage disposal, Driver, Y/N (whether the Nth stage deployment was for occupant restraint or propellant disposal purposes) *	IE	Event	If vehicle equipped with a driver's frontal air bag with a second stage that can be ignited for the sole purpose of disposing of the propellant

Data Elements that Must Be Recorded *(Continued)* (R=Required; IE=If Equipped)			
Data Element	**R/IE**	**Recording Interval/Time**	**Condition for Requirement (IE)**
Frontal air bag deployment, N^{th} stage disposal, right front passenger, Y/N (whether the Nth stage deployment was for occupant restraint or propellant disposal purposes) *		Event	If vehicle equipped with a right front passenger's frontal air bag with a second stage that can be ignited for the sole purpose of disposing of the propellant
Side air bag deployment, time to deploy, driver		Event	If the vehicle is equipped with a side air bag for the driver
Side air bag deployment, time to deploy, right front passenger		Event	If the vehicle is equipped with a side air bag for the right front passenger
Side curtain/tube air bag deployment, time to deploy, driver side		Event	If the vehicle is equipped with a side curtain or tube air bag for the driver
Side curtain/tube air bag deployment, time to deploy, right side		Event	If the vehicle is equipped with a side curtain or tube air bag for the right front passenger
Pretensioner deployment, time to fire, driver		Event	If the vehicle is equipped with a pretensioner for the driver safety belt system
Pretensioner deployment, time to fire, right front passenger		Event	If the vehicle is equipped with a pretensioner for the right front passenger safety belt system
Seat position, driver (whether or not the seat is forward of a certain position along the seat track)		−1.0	If the vehicle is equipped to measure the position of the driver's seat
Seat position, passenger (whether or not the right front passenger seat is forward of a certain position along the seat track)		−1.0	If the vehicle is equipped to measure the position of the right front passenger's seat

Data Elements that Must Be Recorded *(Continued)* (R=Required; IE=If Equipped)			
Data Element	R/IE	Recording Interval/Time	Condition for Requirement (IE)
Occupant size classification, driver		–1.0	If the vehicle is equipped to determine the size classification of the driver
Occupant size classification, right front passenger		–1.0	If the vehicle is equipped to determine the size classification of the right front passenger
Occupant position classification, driver		–1.0	If the vehicle is equipped to dynamically determine position of the driver.
Occupant position classification, right front passenger		–1.0	If the vehicle is equipped to dynamically determine position of the right front occupant.

* List this element n-1 times, once for each stage of a multi-stage air bag system.

As indicated above, in developing this list, we focused on the elements that would be most useful for effective crash investigations, analysis of the performance of safety equipment, e.g., advanced restraint systems, and ACN systems. Some of the data elements will be useful for all three of these purposes; others, for only one or two. The following table shows NHTSA's assessment of the application for each element.

Data Elements and Application			
Data Element Name	Crash Investigation	Advanced Restraints Operation	ACN
Longitudinal acceleration	X	X	X
Maximum delta-V	X	X	X
Speed, vehicle indicated	X		
Engine RPM	X		
Engine throttle, % full	X		
Service brake, on/off	X		
Ignition cycle, crash	X		
Ignition cycle, download	X		

Data Elements and Application *(Continued)*			
Data Element Name	Crash Investigation	Advanced Restraints Operation	ACN
Safety belt status, driver	X	X	X
Frontal air bag warning lamp, on/off	X	X	
Frontal air bag deployment level, driver	X	X	
Frontal air bag deployment level, right front passenger	X	X	
Frontal air bag deployment, time to first stage, driver	X	X	
Frontal air bag deployment, time to first stage, right front passenger	X	X	
Frontal air bag deployment, time to second stage, driver	X	X	
Frontal air bag deployment, time to second stage, right front passenger	X	X	
Frontal air bag deployment, second stage disposal, driver, Y/N	X	X	
Frontal air bag deployment, second stage disposal, right front passenger, Y/N	X	X	
Multi-event, number of events	X	X	
Time from event 1 to 2	X		
Time from event 1 to 3	X		
Complete file recorded	X	X	X
Lateral acceleration	X	X	X
Normal acceleration	X		
Vehicle roll angle	X		X
ABS activity	X		
Stability control, on, off, engaged	X		
Steering input	X		
Safety belt status, right front passenger	X	X	X
Frontal air bag suppression switch status, right front passenger	X	X	
Side air bag deployment, time to deploy, driver	X	X	
Side air bag deployment, time to deploy, right front passenger	X	X	
Side curtain/tube air bag deployment, time to deploy, driver side	X	X	
Side curtain/tube air bag deployment, time to deploy, right side	X	X	
Pretensioner deployment, time to fire, driver	X	X	
Pretensioner deployment, time to fire, right front passenger	X	X	
Seat position, driver	X	X	

Data Elements and Application *(Continued)*			
Data Element Name	Crash Investigation	Advanced Restraints Operation	ACN
Seat position, right front passenger	X	X	
Occupant size classification, driver	X	X	
Occupant size classification, right front passenger	X	X	
Occupant position classification, driver	X	X	
Occupant position classification, right front passenger	X	X	

Several of the elements are associated with crash severity. These include longitudinal acceleration, lateral acceleration, normal acceleration, delta-V, and vehicle roll angle. The longitudinal, lateral, and normal accelerations are vehicle crash signatures in the x, y, and z directions. Delta-V represents the overall crash severity. These are important elements used in determining vehicle crash severity. Vehicle roll angle is important to determining crash severity in non-planar (rollover) crashes and useful for advanced ACN systems.

The service brake on/off and steering input elements are important to understanding the human response to avoiding a pending crash. Several elements cover pre-crash vehicle dynamics and system status: vehicle speed, engine RPM, engine throttle (% full), ABS activity, and stability control (on, off, or engaged). These elements are helpful in determining crash causation.

The elements concerning ignition cycle provide data on how many times the ignition has been switched on since its first use. The difference in the two measurements provides the number cycles between the time when the data were captured and when they were downloaded. GM, in its EDRs, currently records these data. They aid investigators in determining the interval between the recorded event and the time when it occurred. Small differences between these data indicate that the event in the EDR was generated recently, while large differences indicate that they are from an earlier event that may not be associated with a current crash.

Many of the data elements relate to the usage and operation of restraint systems. These elements are important in analyzing advanced restraint operations. For example, without an EDR, it may not be possible after a crash to determine whether a multi-stage air bag deployed at a low or high level.

As discussed above, we are proposing to require some of the data elements to be recorded only if the vehicle is equipped with the relevant safety system or sensing capability. We note that as manufacturers equip greater numbers of their vehicles with advanced safety systems, a number of these data elements would be required to be recorded on an increasing number of vehicles, or even all vehicles. Of particular note, as manufacturers upgrade the side impact performance of their vehicles it is expected that all light vehicles will measure lateral acceleration.

We request comments on the data elements listed in Table I, including whether the list sufficiently covers technology that is likely to be on vehicles in the next five to 10 years. NHTSA encourages manufacturers to develop, to the extent possible, additional data elements for inclusion in the EDR as these new technologies emerge.

B. *Data Standardization*

As discussed earlier, one of our goals in this rulemaking is to ensure that data are recorded and can be accessed in a manner that enables crash investigators and researchers to use them easily. One aspect of this is the format of the recorded data. To increase the value of these data in assessing motor vehicle safety, the proposed regulation would require that the data be recorded in a standardized format.

We believe that data standardization would enable crash investigators and researchers to more easily identify, interpret, and compare data retrieved from vehicles involved in a crash. Currently, the data format of an EDR is established by individual manufacturers and is based on that manufacturer's specific technical specifications. In the absence of any standardization, there is presently a wide variation among vehicle manufacturers as to the format of data recorded by an EDR. Comparisons between data recorded by different manufacturers are less precise when differences exist between the parameters recorded and the precision and accuracy specified. Such comparisons become even less useful if manufacturers do not rely on a common definition of a given data element.

To address this issue, the Society of Automotive Engineers (SAE) established a committee to establish a common format for the display and presentation of the data recorded by an EDR. The SAE Vehicle Event Data Interface Committee (J1698-1), which held its first meeting in late February 2003, has been considering common data definitions for specific data elements, as well as other aspects of EDR standardization.

The Institute of Electrical and Electronics Engineers (IEEE) is also addressing the standardization of EDR data formats. The IEEE Motor Vehicle Event Data Recorder (MVEDR) working group (P1616) is drafting a data dictionary and standards document for EDRs. P1616 is considering specifying the data format with a set of attributes for each defined data element. IEEE stated that it expected to complete a standard to standardize data output and retrieval protocols by March 2004.

In light of the current lack of adopted industry standards, we are proposing a standardized format that would ensure the usability of EDR data, while still providing manufacturers flexibility in design. The proposed regulation would define each data element and specify the corresponding recording interval/time, unit of measurement, sample rate, data range, data accuracy, data precision, and where appropriate, filter class. The proposed data format would require EDRs to capture crash data of sufficient detail and time duration to ensure the usefulness of the data in crash reconstruction without threatening its integrity. NHTSA crash testing has shown that the typical offset frontal crash may last as long as 250 milliseconds. We are also aware that underride and override crashes may last even longer. Furthermore, rollover crashes can last several seconds, depending on the number of rolls.

The proposed time periods (set forth in Table I above) would establish a recording duration of 8 seconds prior to beginning of the event to capture relevant pre-crash and event data. Acceleration data would be required to be captured during the event. Finally, only rollover data would be required to be recorded for several seconds after the event. To the extent possible, the specified recording duration is limited to reduce the likelihood of data being corrupted by failure in the vehicle's electric system resulting from the crash

The proposed format would not mandate storage or output parameters.

C. *Data Retrieval*

A second aspect of accessibility is the necessity for crash investigators and researchers to have the capability of downloading crash data from the EDR. To ensure the availability of these data, we are proposing to require vehicle manufacturers to submit to the NHTSA docket specifications for accessing and retrieving the recorded EDR data that would be required by this regulation. We are also seeking comment on alternative approaches.

At the present time, investigators and researchers can access crash data stored by EDRs for only a limited number of vehicles. Prior to 2000, only vehicle manufacturers could access the EDR data for their vehicles. In 2000, Vetronix released its Crash Data Retrieval (CDR) tool for sale to the public. The CDR tool is a software and hardware device that allows someone with a computer to communicate directly with certain EDRs and download the stored data. It is estimated that about 40 million vehicles on the road have EDRs that can be read using the CDR tool, including many late model GM vehicles and some new Ford vehicles.

However, Vetronix is licensed by only a limited number of vehicle manufacturers to build these devices. Vetronix must presently use proprietary vehicle manufacturer information to develop and configure the hardware and software needed to allow the CDR tool to retrieve data from a vehicle's EDR. If a vehicle manufacturer declines to license or otherwise provide any proprietary information needed to build a device, tool companies will not be able to produce them.

Both the SAE Vehicle Event Data Interface Committee (J1698-1) and the IEEE Motor Vehicle Event Data Recorder working group (P1616) discussed above have considered the downloading of EDR data by means of the On Board Diagnostic (OBD) connector developed in conjunction with the Environmental Protection Agency (EPA). EPA has established requirements for onboard diagnostic technologies, which manage and monitor a vehicle's engine, transmission, and emissions. The EPA regulations include a new standardized communications protocol for the next generation of onboard diagnostic technology that allows a single common interface between the OBD connector and diagnostic tools used to read and interpret vehicle data and convert them into engineering units.

The EPA communications protocol utilizes a Controller Area Network (CAN) to provide a standardized interface between the OBD connector and the tools used by service technicians and vehicle emission inspections stations. CAN employs a serial bus for networking computer modules as well as sensors. The standardized interface allows technicians to use a single communications protocol to download data to pinpoint problems and potential problems related to a vehicle's emissions.

Full implementation of the CAN protocol is required by 2008. Because it is a universal system, the use of the OBD connector and the CAN serial bus could assure uniform access to EDR data and alleviate concerns that the data would only be accessible through the use of multiple interfaces and different kinds of software, if at all.

While standardizing the means of downloading EDR data, possibly using the OBD connector, offers potential benefits, we are at this time proposing only to require vehicle manufacturers to submit to the agency docket specifications necessary for building a device for accessing and retrieving recorded EDR data. This approach will help ensure that EDR data can be accessed in a manner readily usable by crash investigators and researchers. It will also allow motor vehicle manufacturers the flexibility to standardize protocols for data extraction.

We note that the context of NHTSA's proposal is quite different from the context of EPA's requirements for collecting, storing, and downloading emissions-related data. The EPA approach is very structured. It needed to be appropriate for facilitating the routine monitoring and servicing of mandated emission control systems on motor vehicles, thus helping to ensure that those systems perform properly over the useful life of those vehicles. Establishing that approach has required many years of effort and the development of numerous industry standards.

On the other hand, we are proposing a standard for voluntarily installed EDRs, and need to ensure that it is appropriate for the much more limited purpose of crash investigations. We are interested in a simple, flexible approach, while maintaining the ability to extract data efficiently from a motor vehicle's voluntarily installed EDR. To obtain the desired outcome, NHTSA believes that it need not and should not become involved in managing the interface between the auto industry and the companies that may manufacture EDR download tools. But it is evident that some interface is needed, and to that extent we are proposing that certain information be provided.

We are proposing to require that each manufacturer of vehicles equipped with EDRs provide information of sufficient detail to permit companies that manufacture diagnostic tools to develop and build devices for accessing and retrieving the data stored in the EDRs. The vehicle manufacturer would be required to specify which makes and models (by model year) of its vehicles utilize the corresponding EDR system and to specify the interface locations. The leadtime we are providing for implementing this proposed regulation (discussed below) would enable vehicle manufacturers to design their EDRs so that the data may be accessed by use of a standardized interface and communications protocol. In the event that SAE, IEEE, or other voluntary standard organization establishes a standard for a protocol to be used in downloading EDR data, manufacturers would be able to reference the industry protocol in their submissions.

Manufacturers would be required to submit this information in a timely manner to ensure that the specifications were received by NHTSA's docket not less than 90 days before the start of production of makes and models utilizing EDR systems. This would give tool companies time to develop a tool before an EDR-equipped vehicle is used on public roads.

We are also seeking comment on alternative approaches to providing access to EDR crash data, such as permitting the vehicle manufacturer to demonstrate that a reasonably priced tool is publicly available for a particular make/model or to offer to licence at a reasonable price any proprietary information needed to build such tools. We note that EPA permits manufacturers to request a reasonable price for provided OBD-related information. See EPA final rule at 68 FR 38427, June 27, 2003. Comments are requested on the similarities and differences between OBD and EDR related information, the uses of that information, and relevant statutory authorities, and on whether this type of approach would be appropriate for EDR information. We note that one difference is that OBD tools are used as part of commercial activity, i.e., routine servicing and repair of motor vehicles, while EDR tools as used in crash investigations. The market for EDR tools would likely be much smaller. If we were to adopt an approach along these lines, what factors should be used for determining a "reasonable price?"

Commenters supporting any of these or other alternative approaches are encouraged to suggest specific regulatory text and to explain how the recommended approach would ensure that crash investigators and researchers have the capability of downloading data from EDRs. Depending on the comments, we may adopt an alternative approach in the final rule.

D. *Functioning of Event Data Recorders and Crash Survivability*

If an EDR is to provide useful information, it must function properly during a crash, and the data must survive the crash. We are proposing several requirements related to the functioning of the EDR and survivability.

Performance of EDRs in crash tests. First, we are proposing to require EDRs to meet the requirements for applicable data elements and format in the crash tests specified in Standards No. 208, 214, and 301. These tests are (some have been issued as final rules, but not yet taken effect) a frontal barrier crash test conducted at speeds up to 35 mph, a frontal offset test conducted at 25 mph, a rear-impact crash test conducted at 50 mph, and a side impact test conducted at 33.5 mph. Data would be required to be retrievable by the method specified by the vehicle manufacturer (discussed above) after the crash test.

This requirement would provide both a check on EDR performance and also ensure a basic level of survivability. Manufacturers are familiar with these crash tests since they are specified in the Federal motor vehicle safety standards.

As to the issue of survivability, the EDRs of light vehicles are currently part of the air bag module. These modules are located in the occupant compartment of vehicles, providing protection against crush in all but the most severe crashes. Moreover, because EDRs are part of the air bag module, their electronics are designed to operate in a shock environment. However, current EDRs lack protection from fire and immersion in water and motor vehicle fluids.

While requiring EDRs to function properly during and after the crash tests specified in Standards No. 208, 214, and 301 would ensure a basic level of survivability, it would not ensure that EDR data survive extremely severe crashes or ones involving fire or fluid immersion. While EDR data would be useful to crash investigators and researchers analyzing such crashes, we do not have sufficient information to propose survivability requirements that would address such crashes. Research is needed to develop such requirements, and information on the costs of countermeasures to meet these additional requirements would need to be developed. Countermeasures that would ensure the survivability of EDR data in fires may be costly. For all of these reasons, we are not including such requirements in this proposal.

Trigger threshold. We are also proposing requirements concerning the level of crashes for which EDRs must capture data. These requirements would ensure that EDRs capture information about crashes of interest to crash investigators and researchers.

The EDR operates in two modes. One is the steady state monitoring of pre-crash data. EDRs operate continuously in this mode whenever the vehicle is operating. This process allows momentary recording of the pre-crash data. EDRs operate in the second mode when the vehicle is involved in a crash. In this mode, two decisions are made. The first is the determination of the occurrence of a crash and is accomplished by use of a trigger threshold. The second is the decision to capture the recorded data and accomplished using a comparative process. Based on the outcome of this process, the recorded data associated with a crash are captured or deleted.

In current light-duty vehicle applications, the trigger threshold is associated with the air bag crash severity analyzer. The circumstances that cause the threshold to be met are called an "event." The beginning of the event that causes current EDRs to start capturing data in its permanent memory is sometimes defined as the vehicle's exceeding a specified deceleration threshold, typically around 2 g's. After the event is over, and the air bags are deployed, the data are stored in the EDR, if appropriate.

For determination of the beginning of an event, we are proposing to require the EDR to start recording data when the vehicle's change in velocity during any 20 millisecond (ms) time interval equals or exceeds 0.8 km/h. That is equivalent to slightly more than 1 g of steady-state deceleration.

The vehicle's change in velocity is determined in one of two ways, depending on the data collected by the EDR. In the case of a vehicle that does not record and capture lateral acceleration, the delta-V is based on the longitudinal acceleration only. In the more complex case of a vehicle whose EDR records and captures both longitudinal and lateral acceleration, the delta-V is calculated based on both sets of data, or, simply stated, change in velocity of the vehicle in the horizontal plane.

Timing of the unique, non-recurrent actions like the deployment of an air bag in an event is very important. The trigger threshold is used to define time zero. Time zero is used to determine many of the parameters required for collection by the EDR, such as the time when the front air bag deploys. Time zero is defined as the beginning of the first 20 ms time interval in which the trigger threshold is met during an event. Time zero is used to determine many of the parameters required for collection by the EDR, such as the time of front air bag deployment.

Recording multi-event crashes. A crash may encompass several events. For example, a vehicle may sideswipe a guardrail and then hit a car, or a vehicle may hit one vehicle, then another, and finally a tree. In fact, analysis of crash data from NHTSA's NASS-CDS data system shows that while 54 percent of the crashes involve a single event, 28 percent involve 2 events, and 18 percent involve 3 or more events. Thus, if an EDR captures only a single event as the depiction of a multi-event crash, in nearly one-half of the cases, it could be difficult to determine the event of the crash with which the EDR record was associated.

Current EDRs vary with respect to the number of events they capture. For example, current Ford systems capture single events. GM systems can capture two events, one non-deployment event and one deployment event. These two events can be linked ones under certain circumstances. If they are linked, the amount of time between events is recorded. Current Toyota EDRs can capture up to three events. These can also be linked to a chain of events making up a single crash sequence.

To prevent unassociated events from being captured in the multi-event EDR, we are proposing that the maximum time from the beginning of the first event to the beginning of the third event be limited to 5.0 seconds. To understand the timing between the associated events, we are proposing to require that the number of associated events be included as a data element, and that the time from the first to the second event and the time from the first to the third event also be included as a data element.We are proposing to require that EDRs be capable of capturing up to 3 events in a multi-event crash. For any given event that generates a change in velocity that equals or exceeds the trigger threshold, the EDR would be required to record and possibly capture that event and any subsequent events, up to a total of three, that begin within a 5 second window from time zero of the first event. Subsequent events are events that meet the trigger threshold more than 500 milliseconds after time zero of the immediately preceding event. We note it is very likely that in a crash, the trigger threshold could be met or exceeded many times Thus, we are requiring that when the EDR is currently recording event data, the exceeding of the trigger threshold be disregarded until 500 milliseconds has elapsed.

The pre-event data, such as vehicle speed and engine RPM, need to be recorded continuously. Similarly, pre-event acceleration data need to be recorded continuously. Finally, pre-event statuses, such as safety belt usage, determined at –1.0 second, need a similar treatment. The recording of these data is sometimes referred to as a circular buffer; that is, data are continuously updated as they are generated. When the trigger threshold is met, additional types of data are recorded, including acceleration data and rollover angle.

Capture of EDR data. Once the trigger threshold has been met or exceeded, the data discussed above are recorded by the EDR. The EDR continues to analyze the acceleration signal(s) to determine if a second or third event, determined by the trigger threshold's being equaled or exceeded more than 500 milliseconds after time zero of the immediately preceding event, will occur in a possible multi-event crash. This continues for 5 seconds after time zero of the first event.

A decision is then required to determine if these recorded data should be captured in the EDR's memory bank or discarded in favor of a previously captured data set. This decision is based on the maximum delta-V in the sequence of up to 3 events and air bag deployment status.

The maximum delta-V for a multi-event crash would be defined as the absolute value of the maximum of the individual delta-Vs from each of the events in the crash. Since events in a multi-event crash may occur from the front, side, or rear, we are proposing that the maximum delta-V be based on the magnitude of the value, that is, irrespective of the direction, or sign of the value.

We are proposing that the recorded data be captured in the EDR's memory only if the maximum delta-V for the recorded crash sequence exceeds that of the maximum delta-v associated with the data currently stored in the EDR's memory. We are making this proposal to prevent the capturing of EDR crash data with data from new events that may occur subsequent to the event of greatest interest. In the absence of such a requirement, the trigger threshold might be exceeded when the vehicle is towed from the scene or moved in a salvage yard, thus capturing a new record and erasing data regarding the event of greatest interest.

With regard to air bag deployment status, we are proposing that an event that generates information related to an air bag deployment, either frontal or side bag systems, must be captured by the EDRs and cannot be overwritten.

We note that on current GM systems, the EDR locks the data in memory after a crash that involves an air bag deployment. This results in the air bag control system's needing replacement as part of the vehicle's repair after an air bag deployment. On Ford vehicles, the file is not locked when an air bag deploys. However, it is Ford's current service policy that the control module must be replaced after each deployment event.

In the case of multi-event crashes, some of the pre-crash data will be common to each event. For example, vehicle speed data would be collected for 8 seconds prior to the first event. If the second event occurs 1 second later, an additional sample of speed data would be recorded before the second event. For these cases, only the additional pre-crash data that occur during and between the events would need to be recorded as part of the subsequent event.

To prevent confusion between different multi-event crashes, we are proposing that if a crash includes an event that has a maximum delta-V of sufficient magnitude to warrant capturing the data relating to that event, all previously captured data in the EDR memory must be erased and replaced with that new data. We believe that unless this is done, events that occur days or months apart may be mistakenly interpreted as being part of the same crash.

E. *Privacy*

The recording of information by EDRs raises a number of potential privacy issues. These include the question of who owns the information that has been recorded, the circumstances under which other persons may obtain that information, and the purposes for which those other persons may use that information.

We recognize the importance of these legal issues. The EDR Working Group, too, recognized their importance and devoted a considerable amount of time to discussing them. It also included a chapter on them in its August 2001 final report. Among other things, the

chapter summarizes the positions that various participants in the EDR Working Group took on privacy
issues.

We also recognize the importance of public acceptance of this device, whether voluntarily provided by
vehicle manufacturers or required by the government.
We note that General Motors informed the EDR
Working Group (Docket No. NHTSA-99-5218-9; section 8.3.5) that it believes the risk of private citizens
reacting negatively to the "monitoring" function of the
EDR can be addressed through honest and open communications to customers by means of statements in
owners' manuals informing them that such data are
recorded. That company indicated that the recording
of these data is more likely to be accepted if the data
are used to improve the product or improve the general cause of public safety.

While we believe that continued attention to privacy
issues is important, we observe that, from the standpoint of statutory authority, our role in protecting privacy is a limited one. For example, we do not have
authority over such areas as who owns the information that has been recorded. Some of these areas are
covered by a variety of Federal and State laws not
administered by NHTSA.

Moreover, we believe that our proposed requirements
would not create any privacy problems. We are not
proposing to require the recording of any data containing any personal or location identifiers. In addition, given the extremely short duration of the
recording of the information and the fact that it is only
recorded for crashes, the required information could
not be used to determine hours of service of commercial drivers.

The recorded information would be technical, vehicle-
related information covering a very brief period that
begins a few seconds before a crash and ends a few
seconds afterwards. Many of these same data are routinely collected during crash investigations, but are
based on estimations and reconstruction instead of
direct data. For example, investigators currently estimate vehicle speed based on a variety of factors such
as damage to the vehicle. The proposal would simply
help ensure a more accurate determination of these

factors by providing direct measurements of vehicle operation during a crash event.

To help address possible concerns about public knowledge about EDRs, we are proposing to require manufacturers of vehicles equipped with EDRs to include a standardized statement in the owner's manual indicating that the vehicles are equipped with an EDR and that the data collected in EDRs is used to improve safety. The proposed statement would read as follows:

This vehicle is equipped with an event data recorder. In the event of a crash, this device records data related to vehicle dynamics and safety systems for a short period of time, typically 30 seconds or less. These data can help provide a better understanding of the circumstances in which crashes and injuries occur and lead to the designing of safer vehicles. This device does not collect or store personal information.

Moreover, while access to data in EDRs is generally a matter of state law, we believe that access is and will continue to be possible in only limited situations. While the proposal would require public access to information on the protocol for downloading EDR data, this will not result in public access to EDR data. The interfaces for downloading EDR data will most likely be in a vehicle's passenger compartment. The interface locations will not be accessible to individuals unless they have access to the passenger compartment.

Further, in our own use of information from EDRs, we are careful to protect privacy. As part of our crash investigations, including those that utilize EDRs, we often obtain personal information. In handling this information, the agency complies with applicable provisions of the Privacy Act of 1974, the Freedom of Information Act (section (b)(6)), and other statutory requirements that limit the disclosure of personal information by Federal agencies. In order to gain access to EDR data to aid our crash investigations, we obtain a release for the data from the owner of the vehicle. We assure the owner that all personally identifiable information will be held confidential.

F. *Leadtime*

We are proposing an effective date of September 1, 2008. This would enable manufacturers to make design changes to their EDRs as they make other design changes to their vehicles, thereby minimizing costs.

G. *Response to Petition from Dr. Martinez*

As discussed earlier, in October 2001, the agency received a petition from Dr. Ricardo Martinez, President of Safety Intelligence Systems Corporation, asking us to "mandate the collection and storage of onboard vehicle crash event data, in a standardized data and content format and in a way that is retrievable from the vehicle after the crash." We are granting the petition in part and denying it in part.

As discussed above, our proposed regulation would specify requirements concerning the collection and storage of onboard vehicle crash event data by EDRs, in a standard data and content format, and in a way that is retrievable from the vehicle after the crash. To that extent, we are granting Dr. Martinez's petition. We are not proposing to mandate EDRs, however, and to that extent we are denying the petition.

We believe that the motor vehicle industry is continuing to move voluntarily in the direction of providing EDRs. As indicated earlier, we estimate that 65 to 90 percent of model year 2004 passenger cars and other light vehicles have some recording capability, and that more than half record such things as crash pulse data.

The trends toward installation of EDRs in greater numbers of motor vehicles, and toward designing EDRs to record greater amounts of crash data, are continuing ones. General Motors (GM) first began installing EDRs in its air bag equipped vehicles in the early 1990's. In 1994, that company began phasing in upgraded EDRs that record crash pulse information. GM upgraded its EDRs again around 1999–2000 to begin recording pre-crash information such as vehicle speed, engine RPM, throttle position, and brake status.

Also around 1999-2000, Ford began equipping the Taurus with EDRs that recorded both longitudinal and lateral acceleration and several parameters associated with the restraint systems, including safety belt use, pretensioner deployment, air bag firing, and others. Also in the past few years, Toyota began installing EDRs in its vehicles.

As of now, GM, Ford and Toyota record what would be considered a large amount of crash data. Honda, BMW and some other vehicle manufacturers record small amounts of crash data.

Given these trends, we do not believe it is necessary for us to propose to require EDRs at this time. Moreover, we believe that as manufacturers provide advanced restraint systems in their vehicles, such as advanced air bags, they will have increased incentives to equip their vehicles with EDRs. Vehicle manufacturers will want to understand the real world performance of the advanced restraint systems they provide. EDRs will provide important data to help them understand that performance.

We believe our focus should be on helping to ensure that when an EDR is provided in a vehicle, it will record appropriate data in a consistent format and will be accessible in a manner that makes it possible for crash investigators and researchers to use them easily.

We note that we believe our proposed regulation would not adversely affect the numbers of EDRs provided in motor vehicles. We recognize that, if a regulation made EDRs costly, it could act as a disincentive to manufacturers' providing EDRs. However, as discussed earlier, vehicle manufacturers have minimized the costs of adding EDR capability by designing the air bag control system to capture into memory data that are already being processed by the vehicle. Similarly, in developing our proposal, we focused on the recording of the most important crash-related data that are already being processed by vehicles, and not using the rulemaking to require such things as additional accelerometers. The additional costs associated with an EDR meeting the proposed requirements, compared with those currently being provided voluntarily by the vehicle manufacturers, would therefore be small.

By June, 2004 the stage was set for the "public" debate over automotive black boxes. In retrospect, seven years after the National Transportation Safety Board (NTSB) safety recommendation, and following three petitions, two working group reports, ongoing global standards initiatives, and four international symposiums the NHTSA issued a Notice of Proposed Rulemaking (NPRM) on June 11, 2004.   There was enough research to indicate the value of the technology.

Nationwide, 100 news stories reported the story. The following is a sample of the headlines:

Dallas Morning News: *Federal proposal wouldn't require vehicle black boxes*

KTRX, Texas: *Your car could have a black box tracking your moves*

The Boston Channel: *Government proposes voluntary back boxes for cars*

Charleston Post Courier: *Vehicle black box optional*

Just-Auto.com: *USA: Feds decline to mandate vehicle 'black boxes'*

Miami Herald: *Vehicle black boxes wouldn't be required under federal proposal*

Detroit Free Press: *AUTO INDUSTRY REPORT: Recorders not required*

Guardian, UK: *Feds say vehicle black boxes not needed for crash data*

New York Times: *US proposes uniform data on car crashes*

Philadelphia Inquirer: *Feds say vehicle black boxes not needed*

A NHTSA Press release put it this way:

> NHTSA Proposes Requirements
> For Voluntarily Installed Event Data Recorders
>
> The National Highway Traffic Safety Administration (NHTSA) today proposed standard requirements for Event Data Recorders (EDR) that manufacturers choose to install in light vehicles. The proposed rule would not require the installation of EDRs.
>
> "EDRs are in most new vehicles and are already providing valuable safety information for our crash investigators and researchers," said NHTSA Administrator Jeffrey W. Runge, M.D. "We hope that eventually this crash information will be available in real time to emergency medical systems and physicians to improve trauma care after a crash."
>
> NHTSA is proposing, beginning in September 2008, to: (1) require that the EDRs voluntarily installed in light vehicles record a minimum set of specified data elements useful for crash investigations; (2) specify requirements for that data; (3) increase the survivability of the EDRs and their data by requiring that they function during and after front, side and rear crash tests; (4) require vehicle manufacturers to make publicly available information that would enable crash investigators to retrieve data from the EDR; and (5) require vehicle manufacturers to include a brief,

standardized statement in the owner's manual indicating that the vehicle is equipped with an EDR and describing the purposes of EDRs.

An EDR is an electronic device that detects a crash and records certain information for several seconds of time before, during and after a crash. For instance, an EDR may record pre-crash data, such as impact speed, forces on the vehicle during the crash, safety belt use and air bag performance and allow activation of an automatic collision notification to emergency medical personnel.

NHTSA first began EDR studies after a 1997 recommendation from the National Transportation Safety Board. The agency's studies of the EDR records of more than 2000 crashes led to today's proposal.

Out of the approximately 200 million light vehicles in the US, NHTSA estimates that 15 percent of the vehicle fleet (30 million cars, pickups, vans, sport utility vehicles and multi-purpose vehicles) are equipped with EDRs that can be easily read, and that between 65 and 90 percent of new light vehicle models will be equipped with EDRs.

NHTSA will accept comments on this notice of proposed rulemaking for the next 60 days. Written comments concerning it should be sent to the DOT Docket Facility, Attn: Docket No. NHTSA 2004-18029, Room PL-401, 400 Seventh St., S.W., Washington, D.C., 20590-0001, or faxed to (202) 493-2251.

Now that NHTSA rulemaking was underway many wondered why it took so long to get this far. Some were troubled by the fact that NHTSA set a deadline of 2008—and that was only for voluntary participation. Before then, in the United States, approximately 200,000 people would die and 10 million more would be seriously injured in motor vehicle crashes. Why wait was the simple question. Undoubtedly, the comment period would bring out more arguments both for and against black boxes. The states were also beginning to create and enforce legislation with California and Pennsylvania leading the way. At the federal level, the vexing question at this point focused on voluntary vs. mandated.

CHAPTER 12

The Road Ahead: Automotive's Second Century

Without question, the development of the automobile was among the most significant events of the 20th century. In a little over one hundred years the automobile has been transformed from a horseless carriage, a curiosity and in many cases, viewed by the public at the time as an annoyance, into one of most important modes of transportation in the world today. From a wooden wagon to a sophisticated vehicle that today has as much computer power and technology on board as the early manned space exploration vehicles. The advent of the automobile has had enormous affects on modern society, shrinking distances and making life in general simpler, more comfortable, and yet at the same time, more complex and sometimes more stressful. The automotive industry has evolved into one of the world's greatest industries; delivering consumers worldwide a broad array of products that offer utility, practicality, luxury, simplicity, enjoyment, comfort, sex appeal, prestige and, in my opinion, most importantly, safety. Over the last several decades alone, we have seen incredible advances in, and demand for, vehicles equipped with devices that create a safer environment for both the driver and passengers. The unfortunate truth, however, is that the automobile remains, despite all the years of safety improvements, one of the leading causes of death and injury in our nation."*

—Mark V. Rosenker, Vice-Chairman of the National Transportation Safety Board (NTSB)

Automotive Industry Trends

It would be hard to find a person living in our society today who isn't on an almost daily basis dependent upon—or at least somehow affected by—the automobile. The personal automobile enables us to live, work and play in ways that were unimaginable a century ago. It would be easy to argue that, indeed, *automobility* is crucial to maintaining the manner of living to which our society has become accustomed. Not only are automobiles essential to our way of life because of how we use them, but our economic standard of living is dependent upon the industry itself.

The fact is, America's automotive industry drives the U.S. economy. No other single industry is responsible for more manufacturing, generates more retail business, or provides more employment in this country. The automotive industry is responsible for 6.6 million jobs nationwide, or about 5% of private sector jobs. More than 3.7% of America's total gross domestic product is generated by the sale and production of new light vehicles. Impressive numbers to be sure, but how many other industries are at the root of 43,000 senseless deaths a year?

We are often left with the impression that the automotive industry is on the cutting edge of safety applications with the design and manufacture of its ever-new vehicles. However, most automotive industry "history" regarding real-world safety improvements has been slow and painful. Government regulation or criticism of industry trends in this regard is often met with resentment and anger. A recent example of the sensitivity between the automotive industry and government regulators came when Senator John McCain opened hearings on the safety of sports-utility vehicles. Reporting on these hearings on February 27, 2003, the *Detroit News* headline was "Automakers Don't Deserve McCain's Disdain: Senator wrongly slurs the auto industry, while ignoring the harm done by federal regulators." The article said Senator McCain was treating the nation's automakers as if they were a bunch of tobacco executives. Furthermore, "The Arizona Republican questioned the veracity of America's automakers and expressed doubts that they could be trusted to produce safe vehicles without federal regulators watching over their shoulders."

"You judge people by their history," McCain said. "Where is their credibility in establishing these voluntary vehicle standards?" Earlier in the week, McCain asked in an interview with the *Detroit News*: "Could this be the same industry that was opposed to seat belts, air bags, 10-mph bumpers, and would not see any increase in CAFE (fuel economy) standards?"

While safety standards have obviously been inadequate to this point, there is now a fundamental re-thinking among industry leaders and engineers about the development of motor vehicles. Once the automobile was considered a mechanical device that included some electronic controls. Now, with improved sensors, processors, and general electronic capabilities, the automobile is becoming more of an electronic rather than a mechanical device. A significant increase in automotive electronic systems, coupled with related demands on power and design, have created a vast array of new engineering opportunities and challenges.

According to the National Academy of Engineering, of the top twenty engineering breakthroughs of the 20th century, five were transportation related:

2. the automobile;
3. the airplane;
11. the interstate highway system;
12. space exploration;
17. petroleum and gas technologies.

The number one engineering breakthrough? Electrification. Electronics have already made a tremendous impact on today's vehicles, but future vehicles will see virtually every major system in the automobile controlled by electronics.

The challenge for the 21st century lies in the successful integration of the 20th century's engineering miracles—electrification, communication, information, and transportation—toward achieving real vehicle safety and highway safety.

Gabriel Leen and Donal Heffernan, researchers at the University of Limerick, Ireland, wrote, "The past four decades have witnessed an exponential increase in the number and sophistication of electronic systems in vehicles. Today, the cost of electronics in luxury vehicles can amount to more than 23 percent of the total manufacturing cost. Analysts estimate that more than 80 percent of all automotive innovation now stems from electronics. To gain an appreciation of the change in the average dollar amount of electronic systems and silicon components—such as transistors, microprocessors, and diodes—in motor vehicles, we need only note that in 1977 the average amount was $110, while in 2001 it had increased to $1,800. The growth of electronic systems has had implications for vehicle engineering. For example, today's high-end vehicles may have more than 4 kilometers of wiring—compared to 45 meters in vehicles manufactured in 1955. In July 1969, Apollo 11 employed a little more than 150 Kbytes of onboard memory to go to the moon and back. Just 30 years later, a family car might use 500 Kbytes to keep the CD player from skipping tracks. The resulting demands on power and design have led to innovations in electronic networks for automobiles. Researchers have focused on developing electronic systems that safely and efficiently replace entire mechanical and hydraulic applications, and increasing power demands have prompted the development of 42-V automotive systems."

Ninety-five percent of all miles in the United States (2.5 trillion miles) are traveled in personal vehicles as opposed to all types of public transportation. People are spending more time in their personal vehicles than ever before. North Americans spend 500 million passenger-hours in their vehicles each week, or 26 billion hours annually. With so much time increasingly spent behind the wheel, the need for addressing safety concerns with solutions that truly work becomes an ever more pressing concern.

The good news is that automobile safety will be enhanced by digital developments as manufacturers shift their focus from under the hood to behind the wheel. By employing the power of digital technology, vehicles will run faster, smoother, and even smarter.

So, what can consumers expect in the automobiles they'll be purchasing and driving in the near future? Consider the following trends:

Electronic technology enhancements in safety will include: forward, side, and rear collision warning systems, adaptive cruise control, night vision, high-intensity discharge lights, front and side air bags, occupant sensing, low tire-pressure sensing, and intelligent center high-mounted brake lights.

The automotive industry is moving from gasoline to hybrid to fuel cell. As the fuel economy increases, the amount of time that a vehicle may be operated will also increase. Event Data Recorder (EDR) technologies will be important in post-crash analyses of these vehicles.

The automotive industry is moving from mechanical connection to "drive-by-wire." The new systems will reduce the weight of a vehicle. They also require direct electronic control of steering, braking, suspension and power-train actuators dependent on the current driving conditions and environmental influences. Event Data Recorder (EDR) technologies will be a valuable tool in verifying the credibility and reliability of these emerging technologies.

The automotive industry is moving from proprietary electrical/hardware/software systems to standardized "architectures." Automotive standards and protocols (*i.e.* vehicle networks and communication protocols) will be replaced by wired and wireless in-vehicle networks and open standards interface technology (AML, *etc.*) This is an important step in assuring access to Event Data Recorder (EDR) data.

The automotive industry is moving towards the adoption of IT standards in the technology of the vehicle (XML, Web Services) This increases the need for standardization. It will accelerate the move from analog to digital and provide a more robust interface to vehicle architecture with a higher bandwidth. These advancements, combined with emerging EDR technologies, will add even more real safety and security features to motor vehicles.

Perhaps the most significant trend is the move within the automotive industry from "on-demand telematics services" to "always-on" vehicle connectivity to the Internet. The question is not "if" EDR technologies will be introduced to vehicles, but "when." There is a need to explore the relationship that EDRs will have to consumer desire for privacy and personal information security in an "always on" connected vehicle.

The trends cited above, combined with Event Data Recorder (EDR) technologies, can serve as a catalyst for a national debate on the efficacy of emerging transportation safety technologies. Everyone should be involved in the debate, because everyone is affected by the solutions.

The overall trend in the automotive industry is towards increasing a number of safety-related electronic systems in vehicles that are directly responsible for active and passive safety.

Balancing Active and Passive Safety

During the first century of the automobile—when the vehicle was viewed as a mechanical device—*safety* developed with the predominant focus on offering the best available crash protection to occupants. This is known as *passive safety* performance. Although there have been tremendous improvements in passive safety over the last few decades, fatality and serious injury rates remain high globally. One reason for this is that increasing traffic densities and total miles driven per year tend to offset the passive safety advances.

To measure passive safety performance, a number of consumer tests have been established, such as the Insurance Institute of Highway Safety (IIHS) tests, European New Car Assessment Program (Euro NCAP), United States New Car Assessment Program (US NCAP), Japan New Car Assessment Program (Japan NCAP) and the Australian New Car Assessment Program (Australian NCAP).

Consumers pay attention to these ratings, and vehicle safety performance considerably exceeds current legal requirements. The current state-of-the-art rating is four-stars. In fact, many vehicles achieve the highest score of five stars. The paradox, though, is that despite the best efforts of industry and government, the public is still not adequately protected from motor vehicle crash injuries and fatalities.

While governments are striving to reduce fatality rates significantly (Europe 50% by 2010, Japan approximately 10% by 2010, and the USA approximately 20% by 2008), these targets are impossible to attain through passive safety measures alone. A new philosophy is emerging, one that seeks to balance active and passive safety with the goal of bringing comprehensive safety measures to a new generation of vehicles. The ultimate safety measure of the future will be an active approach toward helping drivers avoid critical situations in real world traffic altogether.

What is needed in order to make this evolutionary step? Objective scientific data derived from automotive black boxes.

Need for Adequate Data

The United States Department Of Transportation (USDOT) mandate includes providing for the safety of the traveling public. Identifying transportation risks and potential remedies requires the acquisition and utilization of adequate data. The USDOT maintains in excess of 40 programs that capture either safety data or crucial information such as measures of exposure. Despite reductions of injuries and deaths in every major category of transportation, we seem to have reached a point of diminishing gains. When programs reach their performance limits, but we still have increases in injuries and death, something else needs to happen to break the stalemate.

A recent Millennium Paper of the Transportation Research Board (TRB) noted, "The frequency and severity of accidents depends on the amount of travel by mode, how cars and roads are built, and how people behave. The amount of travel, what mode of travel is used, on what roads travel takes place, what vehicles are in use, and—to some extent—how people behave, is determined by our own political, planning, and design decisions. For such decisions to be rational, we need to be able to foresee the safety consequences of contemplated actions. The ability to anticipate the safety consequences of an action constitutes *safety knowledge*. The richer the body of safety knowledge, the larger the scope of rational road safety management."[1]

The article notes that, although it would be difficult to make the case against knowledge-based safety management, in fact much of current transportation planning, regulation, design, and decision-making does not entail (quantitative) considerations of safety consequences. In other words, we don't have the factual knowledge necessary to make our roads and vehicles safer. This knowledge must be gleaned from objective sources and be made practically available to professionals who are capable of applying the knowledge to decision-making.

Another TRB Millennium Paper notes, "When policy issues arise for the U.S. Department of Transportation (DOT), metropolitan planning organizations (MPOs), state DOTs, or the private sector, it is usually already too late to begin data collection.[2] They cannot respond, 'Hold on for a year or so, we'll get right back!' When a policy question arises, data professionals usually can answer in:

> Three minutes, if it's on the shelf;
> Three hours, if a little searching is required;
> Three days, if some manipulation is required;
> Three weeks; if a computer program is involved;
> Three months, if major data processing is required; or
> Three years, if new data collection is required.

In other words, transportation professionals are forced to work with what they have in the database 'cupboard' when a policy issue arises. Therefore, all policy will be made with the exact statistical data set."

The need for accurate, reliable, and readily available data cannot be overstated. These are the essential components of the tool needed to provide the kind of knowledge we need. The Motor Vehicle Event Data Recorder (MVEDR) is that tool. MVEDR technologies will identify, quantify, and minimize travel risks associated with vehicles, highway infrastructure, or human factors.

Asking the Public

I updated an earlier online survey (conducted 11/22/02–02/26/03) of College-Age individuals at Sandhills Community College in Pinehurst, North Carolina about the use of EDRs. The questions posed, along with the answers of the 549 respondents, follow:

Do you think Event Data Recorders (EDRs) have the potential to improve motor vehicle safety? 523 respondents answered, "Yes," while 26 answered "No."

What level of improvement do you think EDRs can make to highway traffic safety? 36 answered, "Little if any," 225 answered, "a moderate amount," and 288 answered "a great improvement."

Does EDR technology have potential safety applications for all classes of motor vehicles? 489 respondents answered, "Yes," 60 answered, "No."

Do you believe different types of EDRs should be used for different vehicle types such as passenger vehicles, light duty trucks, heavy trucks, city transit buses, and school buses? 493 respondents answered, "Yes," 56 answered, "No."

Which is more important to you, privacy or safety? 35 answered, "Privacy," 193 answered "Safety," and 331 answered, "Equally important."

If you had to choose between privacy or safety as a first concern, which would it be? 88 answered "Privacy," 461 answered, "Safety."

This question allowed the respondent to type in comments on what additional safety benefits are likely from continued development. The majority of comments stressed the desire for safer vehicles.

Should the National Highway Traffic Safety Administration continue participating in the development of EDRs? 516 respondents answered, "Yes," 33 respondents gave a "No" answer.

Should the NHTSA mandate standardized EDRs for motor vehicles?
114 answered, "For some," 398 responded, "All vehicles," and 0 respondents
answered, "No."

**Should NHTSA mandate that all EDR codes be public or left private
(proprietary)?** 342 respondents answered "Public," 207 answered, "Proprietary."

U.S. Safety Nose Dives

On November 27, 2003, the *New York Times* ran a front page article titled, "Once
World Leader in Traffic Safety, U.S. Drops to No. 9" by Danny Hakim which
lamented the decline in vehicle and highway safety. Ironically, the article
appeared on Thanksgiving Day, the first day of a long holiday weekend in which
the National Safety Council (NSC) forecast that 544 people would die and 28,380
people would suffer non-fatal injuries from motor vehicle crashes.

Hakim, an automotive industry reporter, noted:

> The United States, long the safest place in the world to
> drive and still much better than average among indus-
> trialized nations, is being surpassed by other
> countries. Even though the nation has steadily low-
> ered its traffic death rates, its ranking has fallen from
> first to ninth over the last 30 years, according to a
> review of global fatality rates adjusted for distances
> traveled. If the United States had kept pace with Aus-
> tralia and Canada, about 2,000 fewer Americans would
> die because of traffic accidents every year; if it had the
> same fatality rate as England, it would save 8,500 lives
> a year.

> Many safety experts cite several reasons the United
> States has fallen in the rankings, despite having vehi-
> cles equipped with safety technology that is at least as
> advanced as, if not more than, any other nation. They
> include lower seat-belt use than other nations; a rise in
> speeding and drunken driving; a big increase in deaths
> among motorcyclists, many of whom do not wear
> helmets; and the proliferation of large sport utility
> vehicles and pickup trucks, which are more dangerous
> to occupants of other vehicles in accidents and roll
> over more frequently.

> "Our fatality rates are lowering, but not to the degree
> they have lowered in other regions of the world," said
> William T. Hollowell, Director of the Office of Applied
> Vehicle Safety Research at the National Highway Traf-
> fic Safety Administration.

> Traffic deaths and injuries are growing as a global
> health issue. The World Health Organization, prepar-

ing a report on the issue, says traffic accidents will
become the world's third-leading cause of death and
disability by 2020, up from ninth today—a toll particularly costly because victims are so often young adults.
Indeed, automobile accidents will be the main subject
of World Health Day next April, supplanting diseases
like H.I.V./AIDS and malaria.

"It's going to be a bigger World Health Day than usual
because of the magnitude of the issue," said Dr.
Etienne Krug, Director of the World Health Organization's department for injuries and violence prevention.

"Because there's very little emphasis on it, and emphasis on other health problems, we don't expect to make
progress on traffic safety, which is why the ranking is
expected to get worse," Mr. Krug said. He is mainly
referring to the developing world, where preventing
traffic injuries lags behind fighting disease.

Industrialized nations like the United States are well
ahead of developing nations like China, where death
rates are not only far higher but also rising.

Transportation Secretary Norman Y. Mineta has laid
out an ambitious target of reducing the nation's traffic
death rate to 1 death per 100 million miles traveled
from 1.5 deaths by 2008. That would translate into
roughly 12,000 fewer deaths per year, given projec-
tions for increased road use. Last year in the United
States, 42,815 people died in traffic accidents, the most
since 1990.

"Here we are losing 43,000 people," Mr. Mineta said. "If
we had that many people die in aviation accidents, we
wouldn't have an airplane flying. People wouldn't put
up with it. They ought not to put up with 43,000
uncles, aunts, mothers, dads, brothers and friends
whose lives are snuffed out by traffic accidents."

Getting to his target would require a radically faster
pace of improvement. As of last year, the death rate in
the United States had fallen to 1.51 deaths per 100 million miles traveled from 1.58 in 1998.

Since 1970, the United States traffic death rate has
fallen from nearly 4.8 deaths per 100 million miles
traveled. By 2000, the rate in Britain had fallen to 1.2
deaths per 100 million miles from 6.1 in 1970. The new
figure is the lowest traffic death rate compiled by the

Organization of Economic Cooperation and Development, which collects a variety of statistics from industrialized countries.

Australia's death rate has fallen from 7.13 in 1971—the country did not estimate distances traveled the previous year—to 1.45 in 2001. Canada's death rate is slightly less.

Other nations have much higher rates. Turkey's was close to 11.74 deaths per 100 million miles in 2001 and the Czech Republic was 5.21. The economic organization's median figure in 2001 was about 2.1 deaths.

The Bush administration is mainly focusing on seatbelt use and drunken driving in the near term because they are two major areas in which the United States lags behind some other leading nations.

Dr. Jeffrey W. Runge, Administrator of the National Highway Traffic Safety Administration, said: "If everybody buckles up, we can save between 7,000 and 9,000 lives a year. That would drop our fatality rate off the table. The only way you get to 1.0 is to deal with these very important human factors." Most traffic safety experts agree that the seat belt remains the world's most effective safety device. The nation's usage rate has risen considerably over the past couple of decades, to nearly 80 percent today. But top safety regulators in Canada and Australia say their use of seat belts is about 10 percentage points higher.

One reason more Canadians and Australians buckle up is so-called primary seat-belt laws that allow the police to stop motorists simply for not wearing a seat belt. Less than half of the states in this country have such laws.

Dr. Runge has been lobbying states to add primary belt laws. A provision in a federal highway financing bill before Congress would divide $600 million among states that either have primary belt laws or reach a 90 percent usage rate.

Drunken driving rates are also on the rise in this country. Last year, almost 18,000 people died in alcohol-related accidents, the most since 1992.

The administration is pushing for broader use of sobriety courts, which emphasize counseling and treatment as well as jail time. And Dr. Runge wants local jurisdictions to designate special prosecutors for drunken driving. Another problem is motorcycle deaths, which have risen more than 50 percent since 1997. Only 20 states require riders to wear helmets, down from 47 in 1975, when federal highway financing was tied to helmet laws.

The most contentious topic in the safety debate is the effect of sport utility vehicles, pickup trucks and minivans. Such vehicles made up about a fifth of new vehicle sales in 1980 but now account for more than half. Studies by the traffic safety agency have shown that light trucks, particularly big sport utilities and pickups, pose considerably more risk to the occupants of cars than other cars do.

Sport utilities are also no safer than cars for their own occupants; traffic statistics show that because of advantages they get from their bulk, they are offset by a greater propensity to roll over.

A new study by two researchers, Marc Ross from the University of Michigan and Tom Wenzel from the Lawrence Berkeley National Laboratory, estimates that 3,500 fewer people would die each year if 60 percent to 80 percent of the light trucks—sport utilities, minivans and pickups—on the road were cars or station wagons.

But the auto industry has disputed such claims. And other studies have attributed thousands of deaths to another cause: fuel economy regulations adopted in the 1970's forced automakers to make lighter passenger cars. But the weights of cars, as well as light trucks, have been rising for the past decade and a half.

Dr. Leonard Evans, a top safety researcher who retired after more than three decades at General Motors, said any potential improvements in vehicle design would be far outweighed by improvements in driver behavior. He believes the regulators and the news media are too focused on blaming vehicles.

"We've got to have much more focus on avoiding rather than surviving crashes," he said.

The administration is aiming at both driver and vehicle.

Dr. Runge pressured the industry to collaborate on an effort to make sport utilities and pickups less dangerous to people in cars. A revamped rollover rating system, due next year, seeks to better inform the public about rollovers, which account for more than 10,000 deaths each year.

Mr. Hollowell, a top researcher at the traffic safety agency, said the death rate in the United States had not fallen further for several reasons unique to the country. "The motorcycle fatalities have gone up, the rollovers have gone up, which is a function of a greater numbers of light trucks and vans, and another aspect, in vehicle to vehicle crashes, is that we have a changing fleet," he said.

Dr. Runge, who early in his tenure took heat from Detroit for critical remarks about sport utilities, said, "We've got the safest vehicles in the world, so when you consider where we fall in the scheme of things, we can't blame the vehicles."

He asserted: "We have a unique fleet in this country and we're addressing that. But we could have the perfect vehicles, and until we address the human factors, we're not going to change our ranking."

In response to this article, the *New York Times* published a Letter to the Editor on December 4, 2003. Charles Komanoff, the coordinator of the pedestrian advocacy group, *Right of Way*, wrote:

The criterion by which the United States is ranked No. 9 in traffic safety, fatalities per million miles driven (news article, Nov. 27), bears little connection to real life. Americans drive more than any other people—twice as much per capita as the equally prosperous Germans, for example—because our sprawl-strewn landscapes require it. That our rate of deaths per mile is relatively low is cold comfort to hyper-mobile families and communities that lose more than 40,000 loved ones a year to traffic crashes. By the more tangible measure of traffic-caused funerals per million miles, the United States scores 27th among 31 countries in an international road accident database.

On March 4, 2004 Canada's *Globe and Record* published an article written by Emile Therien, president of Canada Safety Council, titled *Your Car's Black Box Can Watch Over You:*

Excerpts note:

> Air, rail and marine carriers all have crash data record-
> ers. The "black box" or flight-data recorder on an
> aircraft provides vital information about the last few
> moments before a catastrophe. That material is
> invaluable to investigators who try to determine what
> went wrong and identify ways to avoid another
> tragedy.
>
> A similar device installed in motor vehicles can pro-
> vide data about a road crash. How fast was the vehicle
> moving? Was the driver's foot on the gas or the brake?
> How big a jolt did the occupants suffer? Were they
> wearing seat belts?
>
> Knowing what was happening in a vehicle just prior to
> a crash is of tremendous value to collision—investiga-
> tion experts as they analyze causes and recommend
> preventive measures. Such information also enables
> automobile manufacturers to design safer vehicles.
>
> Canadian legislators should be looking at ways to
> make this information more readily available to those
> who can use it to make driving safer.
>
> Many vehicles now on the road are equipped with a
> module that records the last few moments before a
> crash. General Motors has had the modules in all its
> models since 1999, and Ford since 2000. Less sophisti-
> cated than an airliner's black box, the Event Data
> Recorder (EDR) is part of the air-bag deployment
> system. The GM devices record vehicle speed, engine
> speed, brake application, throttle position and
> whether seatbelts were fastened.
>
> Installing EDRs in fleets, with the knowledge of the
> drivers, has been shown to reduce collisions. A 1992
> study by the European Union found that EDRs
> reduced the collision rate by 28 per cent and costs by
> 40 per cent in police fleets; the drivers knew they were
> being monitored. Most North American drivers, how-
> ever, do not realize that an electronic device may be
> monitoring their driving, despite the fact this is
> explained in their owner's manual.
>
> Police and collision-reconstruction experts are already
> using these devices, with the permission of the owner
> or by means of a court order, but a few issues around
> their use remain unresolved.

In 2001, a speeding Montreal driver smashed into a car, killing a young man. Without skid marks and with only the suspect's testimony about his own actions, there was no way to calculate the car's speed before impact. The EDR showed that the vehicle was travelling 157 kilometres an hour (in a 50-km/h zone), that four seconds before impact the driver floored the gas pedal, and that just before impact he took his foot off the gas but did not brake. Despite the EDR evidence, the driver was acquitted of criminal negligence causing death, and convicted instead on the lesser charge of dangerous driving causing death.

In another case, the data proved a possible suspect was innocent. When a chain-reaction crash on an Ontario highway ended in the death of a child, witnesses blamed a speeding car. The driver of that car gave police permission to download the data on his EDR, which showed he was driving slowly and quite properly.

If a driver blames a crash on vehicle malfunction, the EDR can help confirm or disprove that. EDRs might also help investigators solve some of the numerous "mystery crashes" that occur every year. These are fatal single-vehicle incidents with no witnesses. Perhaps the road was slippery, or the driver fell asleep at the wheel. Perhaps the crash was intentional.

Due to the unique designs used by each manufacturer, there is currently a lack of standardization. As a result, EDR data cannot be easily retrieved at the crash site. This means paramedics cannot yet take advantage of the information, which would help them determine the most suitable treatment based on the actual severity of impact.

Questions must be resolved about who owns EDR data, who can have access to it and for what purposes. There are also concerns about privacy and admissibility in court. The legal community in Canada has expressed the opinion that the data in EDRs are the property of the vehicle owner and access cannot be granted without the owner's consent or unless ordered by a court. Legislators must address these issues soon—because reliable, objective crash data from EDRs are critical to further advancement in the science of traffic safety.

On March 5, 2004 the *New York Times* published an article by Matthew W. Wald, titled *U.S. Presses for Strict Seat Belt Laws.* Both sides of the issue are noted:

> After six years, efforts to persuade states to enact tougher seat belt laws have stalled, prompting the federal government to step up its lobbying to have such bills passed by state legislatures. So far, 20 states have passed primary seat belt laws that let police officers pull over motorists and ticket them for not wearing a seat belt. An additional 29, including Florida, have secondary laws that permit the police to issue a ticket for seat belt violations only if they have first pulled the car over for some other reason. One state, New Hampshire, has no law mandating the use of seat belts by adults, though it does have a requirement for children. Federal officials estimate that if every state adopted the tougher primary law, 1,400 lives would be saved annually, 200 of them in Florida. Federal highway safety officials put Florida at the top of their list for conversion, and on Thursday the Florida House of Representatives voted 80 to 39 to adopt a primary law.
>
> Opponents of the tougher measure say seat belt use should be a matter of personal choice. And they say they are concerned that the tougher law would become a tool for racist police officers to harass minority drivers. Some advocates of strengthening seat belt laws say race arguments are bogus. Representative John Conyers Jr., a Michigan Democrat, observed in a debate in Washington on the issue that a primary seat belt law was not required for the police to harass minority drivers. "There are enough excuses for an offending police officer making stops based on race," Mr. Conyers said.
>
> The federal government's push for adoption of a primary seat belt law made little progress last year, adding only Illinois and Delaware. Massachusetts and Virginia defeated such bills. The use of seat belts nationally is estimated at 79 percent for front-seat passengers, but among states with belt laws, the figures run from 62 percent to 95 percent.
>
> Of the roughly 38,000 people who are killed in vehicles, about 1,400 would survive if the rate for using seat belts rose to 90 percent, according to the Transportation Department. States with primary laws show rates of use about 11 percentage points higher

than states with secondary laws, according to the Transportation Department. Tens of thousands of serious injuries would be eliminated or made much less severe by wide seat belt use.

Floridians use seat belts at a rate of 73 percent, according to the National Highway Traffic Safety Administration, 6 percentage points below the national average, and 12 percentage points below neighboring Georgia, which has a primary law. Despite these statistics, the issue remains a tough sell in some states.

State Representative Jack T. Barraclough of Idaho, after hearing Mr. Mineta speak on the need for stronger seat belt laws at a gathering of state legislators in Washington in December, said, "My heart and my knowledge say this is exactly the right thing to do.

"But we're dealing with independent states in the Rocky Mountains," Mr. Barraclough, a Republican, said. "They hate to have things forced on them."

In 2003, the Legislature in Boise amended Idaho's secondary law to raise the fine for riding or driving unbelted to $10 from $5. "Even that was a struggle," he said.

The federal government is offering extra highway money to states that change to the primary seat belt law. Florida would get an additional $37 million under the budget proposed by President Bush in January.

Mr. Mineta said he did not favor withholding ordinary highway aid. "We don't want to short-change the states," he said.

But some people pushing for a toughening of the laws hope that perhaps the opponents will finally be won over by the thought that they could be saving lives.

"They've been holding it up for 20 years," said State Representative Irving Slosberg, a Democrat, who sponsored the Florida bill. Mr. Slosberg's 14-year-old daughter, Dori, who was not wearing a seat belt, was killed in a traffic accident eight years ago. "They have to get tired of having the blood on their hands," he said.

The need to increase seat belt usage is widely recognized. However, the potential of using black box technologies towards this end has not been fully recognized. In fact, this is where the greatest gains can be achieved. There is a natural link between an MVEDR and seat belt usage. In fact, with MVEDR technologies to turn to for objective, factual data, we can, for instance, eliminate the potential concern of officers using subjective criteria (such as racial profiling) when enforcing seat belt laws. We won't have to wonder who is telling the truth when a motorist claims that because of the color of his skin he was singled out by law enforcement for allegedly not wearing a seat belt. The black box will tell the truth.

So Here We Are

We are living on a planet where the automobile dominates passenger travel, where in the civilized world, at least 75% of distance traveled is by car or light truck (Japan at 63% and Spain 69% being the only exceptions). On our planet of 190-plus nation states, car ownership is rising in every country. Transport is a major part of household expenditure; on average, about 14% of all expenditures, and about 85% of this expenditure is for the purchase and use of cars. Our bills for purchase, maintenance, repair, fuel, and insurance add up, but we pay without much complaint. With few exceptions, we can travel by car in half the time it takes us to get wherever we're going by bus or rail. Cars also afford us more independence than we have with other modes of transportation, allowing us the freedom to live and work quite literally wherever we choose.

There is also a constant media message to buy newer and "better" vehicles, and we continue to buy them even when our old ones are running fine. There are more "cars" (personal vehicles, including SUVs and minivans) than there are adult persons in the United States, and the number of "cars" being driven daily is still rising. There is no doubt that congestion is a growing concern in some areas, but we still seem to be getting wherever we're going without being slowed down too much... maybe because once we get past the congestion, we drive faster to catch up. Even though all these cars are daily taking up more and more space on our roadways, there is no evidence that the level of motorization has reached a limit anywhere, even in the United States. Indeed, one might regard cars as beginning to be more like shoes, where different sets are kept for different occasions.

Cars have become essential to our way of life. They are everywhere. We keep buying them. We buy more than we need. Today there are 230,199,000 registered vehicles in the United States. Right now as you read these lines, over 40 million vehicles are in motion. We cannot imagine life in the 21st century without cars as an integral part of the picture.

Yet nowhere in the world do we have the freedom to truly travel safely in them.

What Will It Take?

Without public outrage, sincere and honest automotive industry involvement, and fierce political will, road crashes will continue to exact a tremendous human and societal toll. The Surgeon General of the United States has noted the epidemic, but acknowledging the problem is only the first step. He must also offer some solutions. The public must demand them.

It will take the voice of the educated consumer, the informed taxpayer, the knowledgeable member of society to break the stalemate in safety. At the end of the road safety is all about you. You are important. Without real world scientific data, the road to safety will continue down many dead ends. Automotive "black boxes" offer the greatest potential for reducing the number and severity of road crashes.

Since the early 1970s, the very private debate over automotive black boxes has swung back and forth like a pendulum, occasionally settling for awhile until the increased kill rate got it moving again. It is not over yet, in fact, for the vast majority of the population who never heard of black boxes the public debate has only just begun.

My hope is that reading *FATAL EXIT* provides you with a feeling of awareness and a sense of revelation. My simple goal was to put this topic on the road map of where we are headed towards enhancing vehicle and highway safety.

This has been very challenging and difficult to do. For over a century we have dismissed reality and deluded ourselves to the state that we readily accept the carnage on the roads as business as usual. That is why we continue to call what happens "accidents."

However, the world must know that the great historical nightmare of our century (death on the roadways) was both predictable and preventable.

The cunning offense of history is that WE DO HISTORY FOR THE FUTURE, it's cunning defense is that at times IT CANNOT SEE ITSELF.

Hopefully as the public debate continues, and hundreds of lives continue to be lost each day we can recognize that the missing link in automotive safety is to find a way past the debate.

To do less is a FATAL EXIT.

The cruelest lies are often told in silence. Black Boxes speak for the victims—they speak the truth.

GLOSSARY OF ACRONYMS AND ABBREVIATIONS

AAA: American Automobile Association

AAM: Alliance of Automobile Manufacturers

AAMVA: American Association of Motor Vehicle Administrators

AASHO: American Association of State Highway Officials

AASHTO: American Association of State Highway and Transportation Officials

ABA: American Bar Association

ABS: Antilock Braking System

ACEP: American College of Emergency Physicans

ACN: Automatic Collision Notification

AIA: American Insurance Associations

AIAM: Association of International Automobile Manufacturers, Inc.

ANSI: American National Standards Institute

AORC: Automotive Occupant Restraints Council

APTA: American Public Transit Association

ATA : American Trucking Association

BI: Bodily Injury

BTS: Bureau of Transportation Statistics

CAN: Controller Area Network

CDR: Crash Data Retrieval Tool

CFR: U.S. Code of Federal Regulations

CIREN: Crash Injury Research and Engineering Network

CODES: Crash Outcome Data Evaluation Systems

ComCARE: Communications for Coordinated Assistance and Response to Emergencies

DataBUS: Electrical system installed in a vehicle that allows vehicle sub-systems to communicate with each other.

Delta-V: Delta (change in) Velocity

DC: Direct Current

DGPS: Differential Global Positioning System

DCX: DaimlerChrsyler Corporation

DDEC IV: Detroit Diesel Electronic Controls IV Electronic Control Module

DERM: Diagnostic and Energy Reserve Module

DLC: Data Link Connector

DLTLCA: Distribution and LTL Carriers Association

DMS: Docket Management System

DMV: Department of Motor Vehicles

DOT: United States Department of Transportation

DRL: Daylight Running Lights

EBV: Equivalency Barrier Velocity

ECM: Electronic Control Module

ECU: Engine Control Unit

EDR: Event Data Recorder

EEPROM: Electronically Erasable Programmable Read Only Memory

EMI: Electromagnetic Interference

EMS: Emergency Medical Services

EPA: Environmental Protection Agency

EPIC: Electronic Privacy Information Center

EPROM: Electronically Programmable Read Only Memory

ESV: Enhanced Safety of Vehicles

FAA: Federal Aviation Administration

FAI: Forensic Accident Investigations

FARS: Fatality Analysis Reporting System

FCC: Federal Communications Commission

FHWA: Federal Highway Administration

FIA: Federation Internationale de l'Automobile

FMCSA: Federal Motor Carrier Safety Administration

FMVSS: Federal Motor Vehicle Safety Standard

FOIA: Freedom of Information Act

FRA: Federal Railroad Administration

FRCP: Federal Rules of Civil Procedure

FRE: Federal Rules of Evidence

FTA: Federal Transit Administration

GES: General Estimates System

GIS: Geographic Information Systems

GPS: Global Positioning System

HOV: High Occupancy Vehicle

IEEE: Institute of Electrical and Electronics Engineers

IEEE-SA: Institute of Electrical and Electronics Engineers Standards Association

IIHS: Insurance Institute Highway Safety

ITS: Intelligent Transportation Systems

ISO: International Standards Organization

ISTEA: Intermodal Surface Transportation Efficiency Act of 1991

IVHS: Intelligent Vehicle Highway System

JPL: Jet Propulsion Laboratory of NASA

MMC: Mitsubishi Motors Japan

MVEDR: Motor Vehicle Event Data Recorder

MVSRAC: Motor Vehicle Safety Research Advisory Committee

NAII: National Association of Independent Insurers

NAESME: National Association of Emergency Medical Physicans

NASS/CDS: National Automotive Sampling System / Crashworthiness Data System

NASA: National Aeronautics and Space Administration

NCSA: National Center for Statistics and Analysis

NCHRP: National Cooperative Highway Research Program

NHTSA: National Highway Traffic Safety Administration

NRC: National Research Council

NSC: National Safety Council

NTSB: National Transportation Safety Board

OBC: On-Board Computers

OBD: On-Board Diagnostics

OBD II or OBDII: On-Board Diagnostics standard effective in vehicles sold after January 1, 1996

OECD: Organization for Economic Co-operation and Development

OEM: Original Equipment Manufacturer

OMB: Office of Management and Budget

OOIDA: Owner-Operator Independent Drivers Association

OTA: Office of Technology Assessment

RP: Recommended Practice

SAC: Standardization Administration of China

SAE: Society of Automotive Engineers

SBMTC: School Bus Manufacturers Technical Council

SCC: Sandhills Community College

SISC: Safety Intelligence Systems Corporation, Inc.

TCA: Truckload Carriers Association

TMA: Truck Manufacturer's Association

TOPTEC: Topical Technical Workshop

TRB: Transportation Research Board

UAB: University of Alabama

UN: United Nations

VTS: Vehicular Technology Society of the IEEE

WHO: World Health Organization

VTTI: Virginia Tech Transportation Institute

Active Safety: Designing a vehicle that will help the driver avoid crashes is known as "active safety.

Air bags: Safety devices installed in vehicles that inflate to protect the driver or passengers in case of a collision.

Alcohol interlock device: An electronic breath testing device connected to the ignition of a vehicle. The driver has to breathe into the device. If the driver's breath alcohol level is above a set limit, the vehicle will not start.

Anti-burst door latch: Door latch in a motor vehicle that is designed not to open under certain conditions in crashes, so preventing vehicle occupants from being ejected.

Automatic enforcement: The enforcement of road traffic rules by means of equipment that records offences without requiring the presence of police officers at the scene, such as speed cameras or radar detectors.

Blood alcohol concentration (BAC): The amount of alcohol present in the bloodstream, usually denoted in grams per decilitre (g/dl). A legal BAC limit refers to the maximum amount of alcohol allowed in the bloodstream that is legally acceptable for a driver on the road. In some countries, the law stipulates an equivalent quantity of alcohol in the air breathed out, in order to facilitate detection of drink-driving.

Breakaway columns: Lighting or telegraph poles, designed to break or collapse on impact.

Breathalyzer: An instrument that measures the relative quantity of alcohol in the air a person breathes out.

Capture: The process of saving recorded data.

Crash: An unusual or unstable event or an occurance in a sequence of events that produces injury, death and/or damage to one or more vehicles involving a motor vehicle in transport.

Crash Pulse: The acceleration-time history of the occupant compartment of a vehicle during a crash. This is represented typically in terms of g's of acceleration plotted against time in milliseconds (1/1000 second). The crash pulse determines the test's severity of the crash: an occupant will undergo greater forces if the crash pulse g's are higher at the peak, or if the duration of the crash pulse is shorter.

Crash Severity: The most severe injury sustained in the crash as recorded on the police accident report (PAR) and consists of: Property Damage Only (no injuries), Minor or Moderate (Evident, but not incapacitating; complaint of injury; or injured, severity unknown), Severe or Fatal (killed or incapacitating).

Change in velocity during a collision (Delta V): In crash reconstructions, the change in velocity occurring as a result of an impact—usually at the centre of gravity of the vehicle—is widely used as the measure of the severity of a collision. At substantial speeds, collisions between cars are almost totally inelastic so there is very little rebound. Thus if a car traveling at 100 km/h strikes a stationary car of the same mass, they will both undergo a change in velocity of 50 km/hr.6V is an important measure of the input severity or energy dosage, that relates to the outcome or injury severity. It is therefore a widely used variable in assessing the characteristics of crashes and the benefits of various countermeasures, such as the use of seat-belts and air bags, and changes in speed limits.

Child restraints: Special seat restraint for children, designed according to age and weight, offering protection in the event of a car crash.

Crash cushions: Energy-absorbing applications that can be attached to barrier terminals and other sharp-ended roadside objects to provide crash protection on impact.

Crash-protective roadsides: Collapsible or breakaway roadside objects or energy-absorbing "cushions" on barriers and rails that reduce the severity of injury on contact.

Crash-protective vehicles: Vehicles designed and equipped to afford interior and exterior protection to occupants inside the vehicle as well as to road users who may be hit in the event of a crash.

Data Definition: A description of the format, structure, and properties of a data element.

Data Dictionary: A collection of entries specifying the name, source, usage and format of each data element used in a motor vehicle system or set of systems.

Data Element: A uniquely named and defined component of a data definition; a data "type" in which data items (actual values) can be placed.

Dual-Stage Event: An event that is a sequence of two single-stage events within a period of time.

Event: A crash or physical occurence that causes the trigger threshold to be met or exceeded after the end of the 500 ms period for recording data regarding the immediately previous event.

Forgiving roadside objects: Objects and structures designed and sited in such a way that they reduce the possibility of a collision and severity of injury in case of a crash as well as accommodating errors made by road users. Examples are collapsible columns, guard fences and rails, and pedestrian refuges.

Guard fences and rails: Rigid, semi-rigid or flexible barriers which are situated at the edge of a carriage way to deflect or contain vehicles, or in the central reserve to prevent a vehicle crossing over and crashing into oncoming traffic.

Hands-free mobile telephones: A mobile telephone device usually fitted to the dashboard of a vehicle that does not require manual operation.

Headway: the distance between two vehicles traveling one in front of the other.

High-mounted brake lights: Brake lights fitted to the rear window of a vehicle so that they are at eye level with the driver of the car behind and can therefore be easily and quickly seen.

High visibility enforcement: Patrolling by the police which is easily seen by passing road users, for example, random alcohol and sobriety checkpoints.

Human capital approach: An approach based on human capital theory that focuses on the centrality of human beings in the production and consumption system. The "human capital approach" model includes both direct and indirect costs to individuals and society as a whole due to road traffic injuries. Such costs include emergency treatment, initial medical costs, rehabilitation costs, long-term care and treatment, insurance administration expenses, legal costs, workplace costs, lost productivity, property damage, travel delay, psycho social impact and loss of functional capacity.

Incident: An event in which the safety of the vehicle or any person is threatened.

Ignition interlock function: A device that prevents the ignition from starting until certain conditions have been met, such as putting on a seat-belt.

Integrity of the passenger compartment: ability of a vehicle's passenger compartment to stay whole and not collapse on impact with another vehicle or object.

Intelligent speed adaptation: A system by which the vehicle "knows" the permitted or recommended maximum speed for a road.

Intelligent vehicle applications: Technologies that include communication systems, route and traffic information systems, systems for autonomous control of the vehicle, and smart air bags.

Low-cost and high-return remedial measures: Low-cost, highly cost-effective engineering measures applied at high-risk sites following systematic crash analysis.

Median barrier: Safety barrier positioned in the centre of the road that divides the carriage way deflects traffic and often has energy-absorbing crash-protective qualities.

Motor Vehicle Event Data Recorder (MVEDR): A device that is installed in a motor vehicle to record technical vehicle and occupant based information for a period of time before, during and after a crash. MVEDRs may record (1) pre-crash vehicle dynamics and system status, (2) driver inputs, (3) vehicle crash signature, (4) restraint usage/deployment status, and (5) certain post-crash data such as the activation of an automatic collision notification (ACN) system. The term MVEDR does not include any type of device that either produces exclusively an audio or video record, or records exclusively the hours-of-service for drivers of commercial motor vehicles subject to Federal or State regulation.

Motorized two-wheelers: A two-wheeled vehicle powered by a motor engine, such as a motorcycle or moped.

Non-motorized transport: Any transport that does not require a motor to generate energy. Included in this term are walking, bicycling, and using animal-drawn or human-drawn carts.

Offset deformable barrier test: A frontal crash test that aims to reproduce real-world conditions of car-to-car frontal crashes. In this test, the front of the striking vehicle partially overlaps a deformable barrier.

On-board electronic stability program: An on-board car safety system that enables the stability of a car to be maintained during critical maneuvering.

"Out of position" occupant: A vehicle driver or passenger who is out of his or her seating position at the time of the crash, for example, a child lying across the rear seat.

Padding: energy-absorbing lining of crash helmets or vehicle interiors, offering protection against crashes.

Passenger air bags: Safety devices installed in vehicles in front of the front-seat passenger, that inflate to protect the passenger in certain collisions.

Passenger compartment intrusion: The collapse or partial collapse of the passenger seating area of a vehicle as a result of impact by another vehicle or object, contributing to greater crash severity and injury.

Passive safety: Any device that automatically provides protection for the occupant of a vehicle, such as seat-belts, padded dashboard, bumpers, laminated windshield, head restraints, collapsible steering columns and air bags.

Physical self-enforcing measures: Road engineering measures—such as road humps, chicanes and rumble strips—that force drivers to reduce or lower speeds, without any additional enforcement or intervention by the police.

Post-Crash: Condition inside/outside a vehicle following a crash.

Pre-Crash: Condition inside/outside a vehicle prior to a crash.

Post-crash automatic collision notification: A manual or automatic emergency notification system installed in a vehicle that can lead emergency rescue services or the police directly to the position of the crash, by means of a satellite-based Global Positioning System.

Random breath testing: Alcohol breath tests administered randomly at roadside checkpoints by the police, without any necessary cause for suspicion.

Recorder: Device or method to detail action and/or reaction. See Motor Vehicle Event Data Recorder.

Red-light cameras: Cameras installed at traffic lights that photograph vehicles going through the junction when the traffic lights are on red.

Reflectors: Materials that reflect light as an aid to visibility. They may also be fitted to non-motorized transport and roadside objects.

Road infrastructure: Road facilities and equipment, including the network, parking spaces, stopping places, draining system, bridges and footpaths.

Roadside furniture: Functional objects by the side of the road, such as lamp posts, telegraph poles and road signs.

Road traffic accident: A collision involving at least one vehicle in motion on a public or private road that results in at least one person being injured or killed.

Road traffic crash: A collision or incident that may or may not lead to injury, occurring on a public road and involving at least one moving vehicle.

Road traffic fatality: A death occurring within 30 days of the road traffic crash.

Road traffic injuries: Fatal or non-fatal injuries incurred as a result of a road traffic crash.

Road user: A person using any part of the road system as a non-motorized or motorized transport user.

Rumble strips: A longitudinal design feature installed on a roadway shoulder near the travel lane. Rumble strips are made of a series of indented or raised elements that alert inattentive drivers through their vibration or sound. They may also be used for speed reduction.

Safety: Includes highway safety and vehicle safety and highway and vehicle safety-related research and development, including research and development relating to vehicle highway and driver characteristics, crash investigations, communications, emergency medical care, and transportation of the injured.

Single-Stage Event: A single-stage event is an event that is caused by a single-trigger or the logical result of two or more triggers.

Safety barriers: Barriers that separate traffic. They can prevent vehicles from leaving the road or else contain vehicles striking them, thus reducing serious injury to occupants of vehicles.

Safety performance standards: Definitions or specifications for equipment or vehicle performance that provide improved safety. They are produced nationally, regionally, or internationally by a variety of standard-producing organizations.

Satellite-positioning system: A communication system that gives an exact reference for a ground position.

Seat-belt: Vehicle occupant restraint, worn to protect an occupant from injury, ejection or forward movement in the event of a crash or sudden deceleration.

Seat-belt anchorages: Points in the vehicle to which seat-belts are attached.

Seat-belt reminder systems: Intelligent visual and audible devices that detect whether or not belts are in use in different seating positions and give out increasingly aggressive warning signals until the belts are used.

Skid-resistant surfacing: Surface material on a road or pavement designed to prevent vehicles skidding or pedestrians slipping.

Sobriety checkpoints: Checkpoints at which drivers are stopped by the police and breath-tested if there is reasonable cause for suspicion that alcohol has been consumed.

Speed bump: A device for controlling vehicle speed, usually a raised form placed across a road. It can be permanent or temporary.

Speed cameras: Cameras at fixed sites or employed by mobile police patrols that take photographs of vehicles exceeding the speed limit. Their purpose is to enforce speed limits.

Speed hump: A convex elevation installed across the road that acts on the dynamics of vehicles in such a way that drivers have to reduce speed to avoid discomfort to themselves or damage to their vehicles.

Sustainable transport: Transport that achieves the primary purpose of movement of people and goods, while simultaneously contributing to achieving environmental, economic and social sustainability.

Traffic management: Planning, coordinating, controlling and organizing traffic to achieve efficiency and effectiveness of the existing road capacity.

Traffic mix: Form and structure of different modes of transport, motorized and non-motorized, that share the same road network.

Transition zones: Road marking or features forming a gateway which marks transition from higher speed to lower speed roads, for example, rumble strips, speed humps, visual warnings in the pavement and roundabouts.

Telltale: A display that indicates the activation of a device, a correct or defective functioning or condition, or a failure to function. Telltales are considered visible when activated.

Trigger: Is either any data parameter that exceeds a predefined threshold, or an external input. A trigger initiates the capture of data.

Under-run guards in trucks: Front, side and rearguards that can be fitted to trucks to prevent cars and other vehicles running under the trucks in a collision. Under-run guards can also provide energy-absorbing points of contact for other vehicles to protect them in the event of a crash.

Unforgiving roadside objects: Objects and structures designed and sited in such a way that they increase the chances of collision and severity of injury in case of a crash. Examples are trees, poles and road signs.

Utility poles: Poles at the roadside with a particular function, such as telegraph poles, road traffic sign poles and lighting poles.

Vehicle-to-vehicle compatibility: Improving the structural interaction between vehicles when they collide.

Vehicle speed limitation device: A device fitted in a vehicle that does not permit speeds in excess of a maximum limit.

Vulnerable road users: Road users most at risk in traffic, such as pedestrians, cyclists and public transport passengers. Children, older people and disabled people may also be included in this category.

Vehicle Class: A term that includes sedans, station wagons, ambulances, buses and trucks, or different categories of vehicles such as light vehicles, medium vehicles and heavy vehicles.

Vehicle Identification Number (VIN): As defined in 49 Code of Federal Regulations (CFR) 567 a set of about 17 alphanumeric characters, assigned to a vehicle at the factory and inscribed on a small metal label attached to the dashboard and visible through the windshield. VIN is a unique identifier for the vehicle and therefore is often found on insurance cards, vehicle registrations, vehicle titles, safety or emission certificates, insurance policies, and bills of sale. The coded information in the VIN describes characteristics of the vehicle such as make, model, manufacturer, and other limited vehicle details.

Crash Outcome Data Evaluation System (CODES) (NHTSA): Injuries resulting from motor vehicle crashes remain a major public health problem. These injuries cause an unnecessary burden of increased taxes and insurance premiums. They can be prevented or reduced, but only if we understand what the severity of these crashes are, and their associated health care costs. Crash data alone do not indicate the injury problem in terms of the medical and financial consequences. By linking crash, vehicle, and behavior characteristics to their specific medical and financial outcomes, we can identify prevention factors. CODES evolved from a congressional mandate to report on the benefits of safety belts and motorcycle helmets. NHTSA has funded Alaska, Arizona, Connecticut, Delaware, Georgia, Hawaii, Iowa, Kentucky, Maine, Maryland, Massachusetts, Minnesota, Missouri, Nebraska, Nevada, New Hampshire, New Mexico, New York, North Dakota, Oklahoma, Pennsylvania, Rhode Island, South Carolina, South Dakota, Tennessee, Utah and Wisconsin to link statewide crash and injury data. Probabilistic linkage techniques make it possible for the states to link large state data files in a phenomenally short amount of time at relatively low cost.

Crashworthiness Data System (CDS) (NHTSA): NHTSA's Crashworthiness Data System (CDS) has detailed data on a representative, random sample of thousands of minor, serious, and fatal crashes. Field research teams located at Primary Sampling Units (PSUs) across the country study about 5,000 crashes a year involving passenger cars, light trucks, vans, and utility vehicles. Trained crash investigators obtain data from crash sites, studying evidence such as skid marks, fluid spills, broken glass, and bent guardrails. They locate the vehicles involved, photograph them, measure the crash damage, and identify interior locations that were struck by the occupants. These researchers follow up on their on-site investigations by interviewing crash victims and reviewing medical records to determine the nature and severity of injuries. Interviews with people in the crash are conducted with discretion and confidentiality. The research teams are interested only in information that will help them understand the nature and consequences of the crashes. Personal information about individuals—names, addresses, license and registration numbers, and even specific crash locations—are not included in any public NASS files.

General Estimates System (GES) (NHTSA): Data for GES come from a nationally representative sample of law enforcement reported motor vehicle crashes of all types, from minor to fatal. The system began operation in 1988, and was created to identify traffic safety problem areas, provide a basis for regulatory and consumer initiatives, and form the basis for cost and benefit analyses of traffic safety initiatives. The information is used to estimate how many motor vehicle crashes of different kinds take place and what happens when they occur. Although various sources suggest that about half the motor vehicle crashes in the country are not reported to law enforcement, the majority of these unreported crashes involve only minor property damage and no significant personal injury. By restricting attention to law enforcement-reported crashes, GES concentrates on those crashes of greatest concern to the highway safety community and the general public. GES data are used in traffic safety analyses by NHTSA as well as other DOT agencies. GES data are also used to answer motor vehicle safety questions from Congress, lawyers, doctors, students, researchers, and the general public.

National Center for Statistics and Analysis (NCSA) (NHTSA): NCSA, an office of the National Highway Traffic Safety Administration (NHTSA), an agency in the United States Department of Transportation is responsible for providing a wide range of analytical and statistical support to NHTSA and the highway safety community at large, in the general areas of: · Human, vehicle, environmental, and roadway characteristics, as they relate to crash frequency and injuries.

National Transportation Library (BTS): The National Transportation Library is a repository of materials from public and private organizations around the country. The Library is intended to facilitate the exchange of information related to transportation. The National Transportation Library is administered by the Bureau of Transportation Statistics in cooperation with the Transportation Administrative Services Center (TASC), the operating administrations and the Office of the Secretary of the U.S. Department of Transportation. TRIS Online, the largest and most comprehensive source of information on published transportation research, is now on the web at *http://ntl.bts.gov/tris/*.

ANSI D.16—Manual on Classification of Motor Vehicle Traffic Accidents: This is ANSI Standard provides detailed instruction on the classification of motor vehicle accidents. It is available on-line and in hard copy from the National Safety Council.

ANSI D.20—Data Element Dictionary for Traffic Records Systems: This ANSI Standard provides detailed guidelines of highway safety data systems including driver license, vehicle registration, traffic crash, etc. It is available in hard-copy form from AAMVA.

Data Elements for Emergency Departments (DEEDS): Standardized data elements and definitions for a minimum data set sponsored by CDC and developed with the input of the major stakeholders of ED data. Uniform pre-hospital EMS Data Elements: Standardized data elements and definitions for a minimum data set developed from a consensus process involving all EMS stakeholders and sponsored by NHTSA.

UB92 (Hospital Data): Uniform billing data set mandated for electronic submission to all third party payers of hospital inpatient, emergency department and other medical services.

CHRONOLOGY

1974	NHTSA Disc Recorder Project
1975	Office of Technology (OTA) Assessment: Automobile Collision Data
1976	Sensing & Diagnostic Module (SDM)
1990	Diagnostic & Energy Reserve Module (DERM)
1992	European Union Drive Project
1994	Johns Hopkins University Automatic Collision Notification (ACN) Study
1996	NHTSA Special Crash Program (SCI)
1997	NASA Jet Propulsion Laboratory (JPL) Air Bag Report to NHTSA
1997	NTSB Safety Recommendation H-97-18
1998	16th Motor Vehicle Safety Research Advisory Committee (MVSRAC) recommends NHTSA Research &Development Event Data Recorder (EDR) Working Group (WG)
1998	October, Initial Meeting of NHTSA R&D EDR WG
1998	November, NHTSA Denies EDR Petition
1999	May, NTSB International Symposium on Transportation Recorders
1999	June, NHTSA Denies 2nd Petition for EDRs
1999	November, NTSB Safety Recommendations: H-99-53 & 54
2000	December 1, *USA Today* article *GM Sued Over Automobile Black Boxes*
2000	April, NTSB Symposium: Transportation Safety & the Law
2000	May, Delphi announces Accident Data Recorder to be featured in Indy 500 race.
2000	June, Initial Meeting of NHTSA R&D Truck & Bus EDR WG
2001	March, NHTSA posts website devoted to EDRs
2001	June, TRB/NCHRP Call for Project 17-24: Use of EDR Technology for Highway Crash Data Analysis
2001	August, NHTSA EDR WG Issues Final Report (100 pages)
2001	October 29, 3rd Petition to NHTSA for EDRs

2001	December, IEEE MVEDR Project 1616 Approved
2002	January 22, 1st IEEE Black Box Standards MVEDR meeting
2002	February 26, 2nd IEEE MVEDR meeting
2002	March 26, 3rd IEEE MVEDR meeting
2002	April 30, 4th IEEE MVEDR meeting
2002	May, NHTSA EDR Final EDR Report for Heavy Vehicles
2002	May 28, 5th IEEE MVEDR meeting
2002	June 24–25, 6th IEEE MVEDR meeting
2002	July 12, NHTSA initiates EDR rulemaking initiative
2002	July 29–30, 7th IEEE MVEDR meeting
2002	September 23–24, 8th IEEE MVEDR meeting
2002	October 11, NHTSA Call for Comments based on three petitions
2002	December 26, Tab Turner comments to NHTSA
2002	December 27, University of Alabama at Birmingham comments to NHTSA
2002	December 3–4, 9th IEEE EDR meeting
2002	*New York Times* front page article on EDRs
2002	Automakers form Vehicle Event Data Interface (VEDI) committee and continue to participate in IEEE P1616 standards activities
2003	January 7, School Bus Manufacturers Technical Council (SBMTC) comments to NHTSA
2003	January 9, Chalmers University of Technology, Goteborg, Sweden comments to NHTSA
2003	January 11–15 Transportation Research Board (TRB) Annual Meeting & EDR Session on 1/13 "Use of EDR Data Recorders in Crash Reconstruction: Past, Present & Future"
2003	January 13, Bendix Commercial Vehicle Systems LLC comments to NHTSA
2003	January 13, Insurance Institute of Highway Safety (IIHS) comments to NHTSA
2003	January 14, Owner-Operator Independent Drivers Association, Inc. comments to NHTSA
2003	January 14, American Trucking Association Distribution & LTL Carriers Conference Truckload Carriers Association (DLTLCA) comments to NHTSA
2003	January 15, Automotive Occupant Restraints Council (AORC) comments to NHTSA
2003	January 27, New Jersey Department of Transportation (NJDOT) comments to NHTSA
2003	February 10–11 10th IEEE MVEDR meeting

2003	February 10, NHTSA Associate Administrator for Safety Performance Standards speech to the Society of Plastics Engineers Global Automotive Safety Conference, Detroit, MI.
2003	February 14, IEEE Motor Vehicle Event Data Recorder (MVEDR) Committee comments to NHTSA
2003	February 20, American Insurance Association (AIA) comments to NHTSA
2003	February 20, Virginia Tech Transportation Institute (VTTI) comments to NHTSA
2003	February 21, National Association of EMS Physicians (NAEMSP) comments to NHTSA
2003	February 24, Anthony A. Huffman comments to NHTSA
2003	February 21, National Association of Independent Insurers (NAII) comments to NHTSA
2003	February 23, Harris Technical Services comments to NHTSA
2003	February 24, Garthe Associates comments to NHTSA
2003	February 27, American Bus Association (ABA) comments to NHTSA
2003	February 27, American Automobile Association (AAA) comments to NHTSA
2003	February 27, Forensic Accident Investigators (FAI) comments to NHTSA
2003	February 28, National Transportation Safety Board (NTSB) comments to NHTSA
2003	February 28, Alliance of Automobile Manufacturers (AAM) comments to NHTSA
2003	February 28, Advocates for Highway and Auto Safety (Advocates) comments to NHTSA
2003	February 28, Association of International Automobile Manufacturers, Inc. (AIAM) comments to NHTSA
2003	February 28, Electronic Privacy Information Center (EPIC) comments to NHTSA
2003	March 3, Art Kellerman comments to NHTSA
2003	March 3, EDR Online Survey from Sandhills Community College, Pinehurst, NC, presented to NHTSA
2003	March 4, American College of Emergency Physicans (ACEP) comments to NHTSA
2003	March 12, Public Citizen comments to NHTSA
2003	March 14, American National Standards Institute (ANSI) article on U.S. House Resolution 1086: Standards Development Organization Advancement Act of 2003 that was introduced on March 5, 2003
2003	April 2, Mitsubishi Motors Research and Development America, Inc. (MMC) comments to NHTSA
2003	April 11, Society of Automotive Engineers (SAE) Vehicle Event Data Interface (VEDI) committee meeting

2003	April 18, IEEE Standards Association denies automakers appeal
2003	April 22, DaimlerChrsyler Corporation representative (DCX) resigns from IEEE P1616 standards initiative
2003	April 24, General Motors Corporation, Inc. representative (GMC) resigns from IEEE P1616 standards initiative
2003	April 24, Society of Automotive Engineers (SAE) Vehicle Event Data Interface (VEDI) committee meeting at SAE International Headquarters, Troy, MI.
2003	April 28, Surgeon General of the United States *Call for Better Data* statement to IEEE P1616
2003	May 5–6 11th IEEE MVEDR meeting
2003	May 7, Consumer's Union comments to NHTSA
2003	May 19–22, 18th International Technical Conference on the Enhanced Safety of Vehicles (ESV) conducted at Nagoya, Japan
2003	June 4–5, SAE Vehicle Recorder TOPTEC Symposium, Alexandria, Virginia
2003	June 10, U.S. House of Representatives passes H.R. 1086: Standards Development Organization Advancement Act
2003	July 21, Article in *Electronic Times* (*EE Times*) "Automakers Face Standards Choice for Black Box Recorders"
2003	July 22, Toyota representative resigns from IEEE P1616 standards initiative
2003	August 11–12 12th IEEE MVEDR meeting
2003	September 4, National Cooperative Highway Research Project (NCHRP) 17-24: Use of Event Data recorder (EDR) Technology for Roadside Crash Data Analysis Intern Project report at the National Academy of Sciences, Washington, DC.
2003	September 9, United Nations General Assembly: Secretary-General Kofi Annan recommendations for road safety
2003	September 22, *Automotive News* article: "Technology: Guardian Angel or Big Brother?"
2003	September 23, California Senate Bill AB 213: Disclosure of EDRs
2003	*New York Times* article on California Bill for Automotive Black Boxes
2003	November 3–4 13th IEEE MVEDR meeting
2003	November 6, U.S. Senate Judiciary Committee unanimously reported its version of the Standards Development Organization Act of 2003 (H.R. 1086)
2003	November 11, Society of Automotive Analysts debate "Black Box Telematics: Who Owns the Data?" in Southfield, MI.
2003	November 19, 9th Circuit U.S. Appeals Court ruling on privacy and emergency roadside service assistance

2003	November 20, U.S. Department of Transportation / National Highway Traffic Safety Administration Safety Performance Meeting / Safety Performance Standards conducted at Romulus, MI.
2003	November 27, *New York Times* front page article: "Once World Leader in Traffic Safety, U.S. Drops to No. 9"
2003	December 19, NHTSA sends EDR rulemaking report to Office Secretary of Transportation (OST)
2004	February 9, A broadly based coalition of safety, medical, insurance, automotive, and law enforcement groups joined with U.S. Senators John Warner (R-VA), Hillary Rodham Clinton (D-NY), Mike DeWine (R-OH), and the National Black Caucus of State Legislators (NBCSL) to call for passage of the National Highway Safety Act of 2003 (S. 1993)
2004	February 11, U.S. Senate rejects measure on seat belt laws. It was defeated 56–42.
2004	February 11, the American Insurance Association (AIA) urges standardized use of event data recorders (EDRs)
2004	February 27, *New York Times* reports "Another Recall Involving Ford, Firestone Tires and SUVs"
2004	March 4, Canada's *Globe and Record* article *"Your Car's Black Box Can Watch Over You"* by Emile Therien of Canada Safety Council.
2004	March 5, *New York Times* article *"U.S. Presses States for Strict Seat Belt Laws"*
2004	March 9, White House Office of Management and Budget receives request from NHTSA to proceed to a Notice of Proposed Rulemaking (NPRM) for Event Data Recorders (EDRs)
2004	March 16, IEEE Project 1616: Motor Vehicle Event Data Recorder draft sent for pre-sponsor ballot review
2004	April 5, IEEE Standards Association begins call for sponsor ballot
2004	April 7, IEEE Standards Association approves Working Group Draft for sponsor ballot
2004	April 7, World Health Day: World Report on Road Traffic Injury Prevention
2004	June 1, IEEE Black Box standards sponsor ballot achieves majority approval rate
2004	June 3–4, National Transportation Safety Board (NTSB) Academy—Highway Vehicle EDR Symposium at George Washington University, Virginia Campus, Ashburn, VA.
2004	June 3, White House Office of Management and Budget (OMB) approves and returns EDR rulemaking to NHTSA
2004	June 14, NHTSA published a Notice of Proposed Rulemaking (NPRM) for Event Data Recorders (ERDs) followed by a comment period ending August 18, 2004
2004	July 13, Pennsylvania House Democratic Party Policy Committee meeting on Event Data Recorders conducted at Monessen, Pennsylvania

2004   July 14, Alliance of Automobile Manufacturers request sixty day extension from NHTSA to respond to NHTSA-18029 Docket, Notice of Proposed Rulemaking

2004   August 2, The National Transportation Safety Board (NTSB) recommended to the National Highway Traffic Safety Administration (NHTSA) that all newly manufactured passenger vehicles include Event Data Recorders (EDRs)

2004   August 30, 112 docket submissions received by NHTSA towards Notice of Proposed Rulemaking (NPRM) for Event Data Recorders (EDRs)

NOTES

ACKNOWLEDGEMENTS

1. I attended meetings regarding event data recorders with the following: Adam Reardon, Adrianne Archer, Alan German, Alex Damman, Ali Naqvi, Ami Gadhia, Andre F. A. Fournier, Andy Mackevicus, Anthony Geller, Barbara E. Wendling, Barry Hare, Bill Williams, Bob Arturi, Bob Doughlas, Bob Ferlis, Bob Norton, Brad Cohen, Brian Everest, Brian Shaklik, Carl Hayden, Charles Gauthier, Charles Holt, Chip Chidester, Christina Mullen, Chris Tinto, Chuck Niessner, Chuck Plaushin, Clay Gabler, Craig Taylor, Dan Angell, Dan D'Angelo, Dan May, Dan Floyd, Daniel P. Fuglewicz, Dave Snyder, David Bauch, David Clark, David Eiswerth, David McKendry, David Phelps, David Willis, Deborah M. Freund, Dennis Bodson, Dennis Kramer, Dayi Denj, Don F. Anderson, Don Gilman, Doug Gabauer, Doug McKelvey, Douglas Gurin, Douglas Read, Duane Perrin, Ed Jetter, Edward O'Hara, Edward Rashba, Edward Ricci, Eric Ogilvic, Gabriclle Baymc, Garold Yurko, Gerald Stewart, Greg Niemiec, Greg Shaw, Guillermo Zepeda, Hal Beecraft, Hare Patnaik, Helen Fagerlind, Henry Jasny, Hideki Hada, Hiroshi Tsuda, Jack Haviland, Jack Pokrzywa, Jack Ribbon, Jack Volk, James Onder, Janice Bachman, Jeff Hagarty, Jeff Morin, Jeff Scaman, Jeffrey S. Augestein, Jennifer Ogle, Jerry L. Cage, Jeya Padmanaban, Jim Keller, Joan Claybrook, Joe Marsh, John Bradley, John Brophy, John Carney, John C. Steiner, John Hinch, John Hus, John Mackey, John Neal, John Yurtin, Joseph Lloyd, Joseph Mickey, Kathy Piersall, Keith A. Cota, Kevin George, Klaus Hitzeroth, Ken Dodson, Kenneth R. Stack, King K. Mak, Kris Bolte, Larry Kuhn, Larry Williams, Laura MacCleery, Lurae Stuart, Liz Garthe, Lou Lombardo, Lori Summers, Malcolm Ray, Mack O. Christensen, Marcus Behrendt, Martin W. Hargrave, Mary Ellen Tucker, Mary Rennie, Mary Russell, Matt Wald, Michael Bracki, Michael Cammisa, Mike Alwais, Mike Spitzley, Mike White, Miranjam Kulkarmi, Nancy Roberson, Norm Littler, Park Wu, Patricia Gerdon, Patrick Halley, Patrick Lujan, Paul Kostek, Paul Menig, Paul Tremont, Prabal Dutta, Ralph Hitchock, Raul Arbelaez, Regina Dillard, Ricardo Martinez, Rich Paulson, Richard Cox, Richard Pain, Richard Pandolfi, Richard Smolenski, Richard R. Peter, Rick Synder, Robert Cameron, Robert McElroy, Robert Pinto, Robert Thompson, Robert Yakushi, Roger L. Boyell, Rod Nash, Ruben Payen, Sally Greenberg, Sandra Miller, Sarah McComb, Scott Kidd, Scott McClellan, Scott Schmidt, Sharon Vaughn, Sophia Rayner, Stephen Webb, Steve Belden, Steve Ezar, Steven Johnson, Susan Tatiner, Susan Walker, Thomas Eymann, Thomas M. Kurihara, Tom Berringer, Tom Mercer, Tom Peacock, Tom Roston, Tony Hanson, Tony Huffman, Tony Reynolds, Ty Lasky, Vernon Roberts, Vernon Wright, Wanda Curtis, Whit Harris, and Will Schaefer. This is a virtual who' who's list of the important people from the automotive industry, the insurance industry, government, researchers, medical providers, advocates, academia and the public.

INTRODUCTION

1.Department of Transportation / National Highway Traffic Safety Administration / Docket No. NHTSA-02-13546; Notice 1] RIN 2127-A172, "Event Data Recorders" Request for Comments. *Federal Register* / Vol. 67, No. 198 / Friday, October 11, 2002 / Notices. Washington, DC. See note 1 on page 63393. Full document available at *http://dms.dot.gov/ search/document.cfm?documentid=196624&docketid=13546.*

2.See: World Report on Road Traffic Injury Prevention at *http://.www.who.int/world-health-day/2004/en/.*

3.Ibid.

4.B.J. Campbell

5.Ibid

CHAPTER 1: QUESTION EVERYTHING

1. Approximately 300 college students at Sandhills Community College, Pinehurst, North Carolina asked the questions.

CHAPTER 2: NOTHING HAPPENS FOR THE FIRST TIME

1.Bob Dylan songs see: *http//www.bobdylan.com/songs/blowin.html.*

2. The Congressional Office of Technology Assessment closed on September 29, 1995. During its 23-year history, OTA provided Congressional members and committees with objective and authoritative analysis of the complex scientific and technical issues of the late 20th century. See *http://www.wws.princeton.edu/~ota/* for a site dedicated to the legacy of OTA. An Automobile Collision Data Workshop convened January 16 and 17, 1975, at which the requirements for, and various approaches to, better collision data gathering were presented and discussed in depth by experts in all aspects of the problem. Individuals who participated in the Workshop were the following:

Lynn Bradford (National Highway Traffic Safety Administration)

Paul Browinski (AVCO Systems Division)

B.J. Campbell (Highway Safety Research Center–University of North Carolina)

Charles Conlon, Jr. (AVCO Systems Division)

J. Robert Cromack (Southwest Research Institute)

John Edwards (Ford Motor Company)

M.D. Eldridge (National Highway Traffic Safety Administration)

Vincent J. Esposito (National Highway Traffic Safety Administration)

William Fitzgerald (AVCO Systems Division)

John Garrett (Calpsan Corporation)

Howard P. Gates, Jr. (Economics & Science Planning, Inc.)

Lawrence A. Goldmundtz (Economics & Science Planning, Inc.)

Walton Graham (Economics & Science Planning, Inc.)

James Hofferberth (National Highway Traffic Safety Administration)

John F. Hubbard, Jr. (Center for Auto Safety)

Paul R. Josephson (Center for Auto Safety)

Charles Kahane (Calspan Corporation)

Phil Klasky (Teledyne Geotech)

Gene G. Mannella (National Highway Traffic Safety Administration)

Don Mela (National Highway Traffic Safety Administration)

Charles A. Moffatt (National Highway Traffic Safety Administration)

David Morganstein (Center for Auto Safety)

James O'Day (Highway Safety Research Institute–University of Michigan)

Brian O'Neill (Insurance Institute for Highway Safety)

L. M. Patrick (Wayne State University)

Steven J. Peirce (National Highway Traffic Safety Administration)

Louis W. Roberts (Transportation Systems Center, Department of Transportation)

A. J. Slechter (Ford Motor Company)

John Versace (Ford Motor Company)

Richard Wilson (General Motors Safety Research and Development Laboratory)

3. Office of Technology Assessment: Automobile Collision Data: An Assessment of Needs and Methods of Acquisition. Economics & Science Planning, Inc. Washington, D.C. 20036. February 1975. pg. iii.

4. Ibid, pg. iv.

5. Ibid, pg. iv.

6. Ibid, pg. 1.

7. Ibid, pg. 1.

8. Ibid, pg. 1.

9. Ibid, pg. 1.

10. Teel, S.S.; Peirce, S.J.; Lutkefedder, N.W. Automatic Recorder Research—A Summary of Accident Data and Test Results. National Highway Traffic Safety Administration, Washington, D.C. 57 pgs. International Conference on Occupant Protection. 3rd Proceedings. SAE, New York, 1974. Pg 14-70. Report No. SAE 740566.

11. Office of Technology Assessment: Automobile Collision Data: An Assessment of Needs and Methods of Acquisition. Economics & Science Planning, Inc. Washington, D.C. 20036. February 1975. pg.11.

12. History Section of NHTSA EDR website at *http://www-nrd.nhtsa.dot.gov/edr-site/ history.html.*

13. Ibid.

14. Ibid.

15. Ibid.

16. See "NHTSA's Field Operational Test of an Automated Collision Notification (ACN) System at NHTSA Research & Development site: *http://www-nrd.nhtsa.dot.gov/ departments/nrd-01/summaries/ITS_13.html* and Technology Alternatives for an Automated Collision Notification System, DOT HS 808 288, August 1994, The John Hopkins University Applied Physics Laboratory.

17. Ibid.

18. Ibid.

19. NHTSA EDR Website: History Section

20. Transportation Research Board Special Report 248: Shopping for Safety: Providing Consumer Automotive Safety Information, 1996. The National Academies Press, Washington, DC. 20001.

21. Ibid.

22. Ibid.

23. See: NTSB Safety Recommendation Letters at: *http://www.ntsb.gov/Recs/letters/ letters.htm#Highway.*

24. See: NASA Advanced Air Bag Technology Assessment at: *http://csmt.jpl.nasa.gov/air bag/contents.html.*

25. Ibid.

CHAPTER 3: SHIFTING GEARS

1.The following Committee members and alternates were present (*represents alternates):

Christine M. Branche* (Centers for Disease Control)

Susan Cischke) (Chrysler/AAMA)

Kazuo Higuchi (Honda/AIAM)

Robert L. Muellman (Society for Academic Emergency Medicine)

Andrew Burgess* (American College of Surgeons)

Adrian K. Lund (Insurance Institute for Highway Safety)

Kenneth L. Campbell (University of Michigan's Transportation Research Institute)

Bill Eagleson* (Ford Motor Company/AAMA)

Ray Resendes* (Federal Highway Administration)

Dietmar Haenchen (Volkswagen of America/AIAM)

Elaine Petrucelli (Association for the Advancement of Automotive Medicine)

Marcus Martin (University of Virginia, American College of Emergency Physicians)

NHTSA employees attending this meeting included Ricardo Martinez,M.D., Raymond P.Owings, Ph.D, Lousi Brown, John Hinch, Donna Gilmore, Tom Hollowell, Rolf Eppinger, Joe Kanianthra, Riley Garrott, Duane Perrin, August Burgett, Bruce Spinney, Ken Hardie,

Ed Graham, Joe Scott, Lou Molino, Clay Gabler, George Soodoo, Linda McCray, Tom Hollowell, Lou Lombardo, Heidi Coleman, Gerald Stewart, Ellen Hertz, Lori Summers, Riley Garrott, and Lee Franklin. There were approximately 20–25 members of the public present.

2. The following transcript is from a key meeting that helped establish research and development of EDRs. Copies of meeting presentations and transcripts are available in NHTSA Public File 88-0-1.

MR. HINCH:

So, what I'm here for and I'm going to quickly, we're asking your permission—Ray thinks that the NTSB is very important, and I'm asking your permission to start a working group to look at these event data recorders. I'm sure that vehicle people know what event data recorders are, but the medical people—it's a way that we can measure some of the properties of a crash during the crash, you know, for real vehicles that are in real crashes. Right now we do crash testing. We administer—for real crashes and all we have is what happens after the crash. We show up on the scene, the crash has occurred, we'd like to know what the deceleration, what was the speed of the vehicle—all this would be very interesting. With the data electronic systems, all of ours have air bag sensors, which is basically a crash analyzer, so we can take that data from the crash analyzer and store a little bit of that information. Just a little bit of information will go a long way in providing this researcher and department—a way to—this group. My thought is to talk about setting up one of these committees to study the current—It was mentioned the need for real crash data, more details, crash position. This would go along the—you know, stuff that we are doing. We do spend a lot of money here collecting crash information, $20 million a year or so. This is not something we thought up on our own, although we'd like to take the credit. It was last year NTSB had a big public forum. One of the recommendations out of the NTSB meeting was for NTSB Center establishing methodologies for crash recorders, setting up objectives. This working group, we're not trying to mandate the crash recorders, set a rulemaking requirement at all as far as crash recorders will—what we'll do is get together and set some technical requirements. Some benefits, more timely data. As I talked about, you collect data during this, right after the crash, investigators go out and maybe plug the computer, download the information from the crash recorder and immediately have data. It could reduce costs of, you know, crash collection.

MR. CAMPBELL:

The proposal is that someday every new car would have one of these?

MR. HINCH:

As a matter of fact, that is not the proposal. The proposal is to get a working group together. The working group would set up some specifications or some guidelines for manufacturers to use when—we can see the technology evolving. General Motors already has this, other manufacturers have it. They're starting to produce this information, but it's all proprietary and it's difficult for crash investigators to use. I don't know how it's connected with the computers that download the information or what the additional signals mean. So, what we were trying to do is get together the government officials, activity officials in industry to try to talk about how we could maybe standardize what parameters would be measured and how they would be collected by crash investigators, but this is in no way intended to be a mandate effort. Just to give you a feeling, we'll just talk about the Hadden matrix and here is the Hadden matrix. Right now, we do crash investigations. We can collect some post-crash information, but with crash recorders, there's an explosion of information in the area that we're really interested. There are chances to collect pre-crash information and crash information of the quality that we have never had, and these things actually become more and more involved in crashes, as more manufacturers put them on their cars, there's going to be lots and lots of crashes where these crash recorders are being used and information will be—it really fills out the matrix and gives us a lot of information. There's a couple of slides so I think I'll kind of close with this. As you can see, manufacturers—we met with General Motors, Federal Highway, Transportation Research Board, National Transportation Safety Board, NTSB, we met about a month ago. We talked over the concept of wanting to participate in this effort. So, we, as a group, NTSB and these other federal government and safety boards, we're trying to think of a place to develop this work and your committee came to mind as the proper place for this to work. We propose that we formulate a committee so we can get on with this. We'd like to do this in a short time period. We see this effort as being something that we would work diligently and quickly on to try to come to some closure in a quick time period of maybe a year or maybe slightly more, but not a long range—

Thank you.

MR. EAGLESON:

So, basically what you're trying to do is establish some criteria, protocol for data recorders?

MR. HINCH:

Yes, sir, we would—

MR. EAGLESON:

Over the course of the year—

MR. HINCH:

That's correct. We would get together, set up the group, figure out what's feasible, what's likely in the near future and long term, set up from the—research side, look at what crash data are real important and what crash data would be nice in the long term, and then try to marry those together along with technologies for data collection, where we would work with maybe third party suppliers of Delco or Ford Motor parts people, who develop these kind of outside vendor products that we could—that could be marketed to crash investigators, federal government—that allows them to come in and plug in a box into the car and download only the information that the manufacturer wanted to allow that investigator to get on a real time basis. There's a lot of complications in there, whether the car is still functional or not, but that would be the gist of the thing. We'd try to set up, you know, some policy, I guess, for manufacturers, what they think about this device. If they're thinking about installing it in their vehicle in the long-term, in the next three to five years, just some guidelines from this committee on what would be important parameters to try to collect in this type of device.

MR. EAGLESON:

Do you have any thoughts right now about—

MR. HINCH:

Right, well, Rolf was way ahead of me. He has actually sent out a letter and to the committee and got a nomination back from the committee. I haven't done that. I'm here today as the first step in this process to ask permission to formulate this committee to do some, to send out under Joe's signature. We would put this under the Crashworthiness Subcommittee and obviously under Joe's name. Letters go out to members of that subcommittee to ask them for nominations or themselves to work on such a working group. I have 7,500 people who are interested in this area and who are working with these special devices of such data. We would like to get the right people to get the job done quickly.

MR. OWINGS:

This is Ray Owings. (Inaudible.) I don't care about—the mechanisms before the data are there. We can go taking subsets, certain fleets of vehicles that may be—allow this data to be collected. (Inaudible.)—and I think in connection with STAR and programs like that, where you have this—and now you have—and you marry them together, then I'm starting to get better with 350 cadaver like test procedures a year. The best minds in the country looking at the medical results. Technical—real good idea—just the next generation is so much better and—

MR. BURGESS:

Obviously, I'm interested—I would suggest that from the beginning, you have a strong legal presence on the committee, because our downfall is they come in as a John Doe and we start to approach them and the man is unconscious. They have enough time to—resources and any kind of untoward circumstances, they've already contacted—We get blocked a lot. Once they know that there's crash history which would apply back to the driver's behavior patterns, excessive—or whatever was applied, your family attorney advises them not to sign any institutional form—so, this invaluable tool, I think, should be clear from the beginning some kind of strong legal oversight. Once it arrives on the scene, we want to use it like a great tool. If the same amount of engineering and medical expertise goes to covering the legal bases, you approach this—we've had to on the inside in a lot of our videotaping of resuscitations and stuff like that, where we assigned that quality assurance test which makes it non discoverable. I don't think this will fit that, but we do the dance around these issues, and they're—issues. It would invalidate—I don't know—but 20 percent, or 30 percent or more of this collected data. When you just get the perfect case, the ones that tell a story, by marrying the medical data and the crash data, it's going to be blocked.

MR. HINCH:

I think you're on point there. We have—everybody we talked to about this says, well, what about privacy issues? Where are you going to get the data, who owns the data? Does the data belong to the owner? You know, who's going to allow you to read the data? Those are all concerns that we have actually—you met Heidi Coleman this morning. She's our general law lawyer and she's going to be working with us on this committee. We're going to have strong legal help right from the get go. We're not going to wait to the end and say, whoops, what about that data? There may be, you know, a backbreaker. It could actually turn out to be a backbreaker, that only the data we'll be able to collect will be data that's owned by the government or, you know, fleets, rental car fleets, you know, that data may be available to us. Government vehicles, crashed cars. So, it may be a real backbreaker, but we're going to work that real hard. My opinion here was, don't give up. It's there.

MR. BURGESS:

I'm thinking impact, let's go find fleets where we can put it into. We—some of the quality control in some kind of program that's been put in place. The program itself was the industry's cooperation, you know, in doing certain sort of a Q&A of their own mitigation. I mean, that's the—and obviously insure the family and sticking to our words. Attorneys come after us, as you know, on a regular basis. It's nothing we've been able to—

MR. HINCH:

We have—of all our databases and you heard Riley talking this morning when he talked about the ABS program and they wanted to get the bins for these crashes and the Legal Department has had a strong objection. The engineering people that think of any other possibility and then I'll think again whether I'll give it to you. They still haven't decided. We say that there's no possible way that—they still say no, you can't have it. So, you can see that we those issues. We've looked at them, we'll look at other agencies, collect this kind of data and see if their thoughts can be applied here. Yes, Ken?

MR. CAMPBELL:

Ken Campbell from the University of Michigan. Certainly a crash recorder and better aid—is an excellent idea. I think everyone would like it, but I really think I have to introduce a little bit of reality here, because I think when anybody starts thinking about the first case with one of these recorders that comes to one of our crash centers, you're dreaming about something which is at least 15 years away from today. I mean, someone needs to sit down and do a back up calculation about what fraction of the car fleet you expect to have these in, when is the first year that that car fleet is going to be out in service, and how long is it going to take for a portion to build up, and then what percentage of the crashes are you getting in CIREN? Are you getting 350 out of 12 million? Just do a backup of the envelope calculation, but don't expect to see this first CIREN case any time in the next ten years.

MR. HINCH:

But, that's not a reason not to go with this.

MR. CAMPBELL:

No, no. But, you're over—

(Multiple voices.)

MR. HINCH:

We may not be overselling it as much as you think. GM has already introduced these systems in 1998 models and they plan to implement the fleet totally in the next couple of years, so all the GM vehicles, which make up a large percentage of the fleet, starting in '99 and 2000, are going to have these. So, we're going to have, you know, a lot of vehicles. I agree that's only a quarter percent of the fleet and the chances of those being involved in a CIREN are actually one in a gazillion, probably.

MR. KANIANTHRA:

Suppose we start out with rental fleet cars?

MR. CAMPBELL:

I don't object. I'm just saying, when you start with automobiles, start with the rental fleet, and you say 350 cases in a year with complete crises—you're just a little bit ahead of—

(Multiple voices.)

MR. OWINGS:

I overstated my case.

MR. HINCH:

However, I have to tell you that the benefits are so great that I have to believe in that, because I really think it's just a quantum leap in our ability to establish cause and effect.

MS. CISCHKE:

Sue Cischke from Chrysler. I just have a couple comments to make. Some of the information we do collect today, in terms of vehicle speed and some other things that are measured. A lot of the things that you have on your chart, as far as—and some of the other positioning of the vehicle, that's sensors and things that we don't currently have and wouldn't be collecting that data. So, to get everything that you need, and it is going to mean added expense that someone would be incurring if we ended up putting them on vehicles. The other issue on the litigation side of it, when this was discussed at the NTSB forum, there was more experience in Europe with collecting information regarding seat belts being buckled and not buckled and there was a lot of issue just with the legal nature of that in Europe, let alone in the United States. It's too bad Adrian had to leave, because I think the insurance companies were the ones that were really struggling with that information. So, before we get too far down the road, I think you're really going to have to do a reality check because even the rental fleets are driven by people who will eventually have lawsuits, so you'll have to figure out what kind of data you'll ever be able to use from that.

MR. HINCH:

I understand.

MS. CISCHKE:

However, there seems to be no compelling reason not to have a working group to at least look at the feasibility of doing this in the next decade or whenever.

MR. KANIANTHRA:

I'm saying, even if it is a small step, if we can validate our mass database, that gives a lot of comfort—

MS. PETRUCELLI:

So, even if your feasibility working group says it's probably not realistic, at least it's—

MR. OWINGS:

I wouldn't think so. There's no reason to take systems that are in existence today and make sure that any time the government is involved in the test, whether it's a crash test or some sort of standards test, that we don't make sure that we know how these systems are turned on, and we don't bother to go collect the data and we don't start establishing the relationship between what our dummies say and what this data says, then one learns to run by walking first. I believe the—benefits are so great, we really have to go off and explore, but all those hurdles that are out there, they're there and they're real and whatever. Nowadays, the technology is there. A lot of the data is—you know, computer—that are there, it's a matter of capturing the data—I think it's time that we do. I think they considered it, I think it was back in the '70s, whether it was really an added cost, would be much different. At least we think we should do it. Two, we think this is a reasonably good forum because the data that we get from them is probably used by— what I'd really like to do is go ahead and say we approve and let's go see what can happen. I haven't talked about it, but I hope that we get together again. We've talked about it a lot today, and we'd probably have a different way of presenting some of our results—If there isn't any objection, I'd go ahead and—I guess I should tell you that when you're doing that, at least some of them represent big organizations and I assume that means that you would help us support this working body.

MS. CISCHKE:

I would just like to say that I think you need to define, you know, what the group is going to look at in terms of feasibility, because it's much different if you're designing some kind of system to be used on a test fleet versus a production application, realistic in terms of trying to store as much data as you need. This should really be a feasibility study to say, does it make sense to go and trim around that, because this is a big job.

MR. OWINGS:

Well, at least in one case that we know of, the manufacturer has—and we can start using it in some limited way—Now, general, functional requirements, what things are desirable, what refined steps we'll need, how we use it in crashworthiness, that's when the crash occurs, how we get enough data, some insight into how much data and what to have before a crash. All these, I think, we want to explore.

MR. HAENCHEN:

Dietmer Haenchen, Volkswagen. I think it's a good idea to investigate it. I think the main problems will be legal issues. If it can be resolved—

MR. OWINGS:

Any other issues? Yes? (Inaudible.)

MR. OWINGS:

We're doing this ourselves to some extent with the ACN program that Dr. Martinez talked about. Ken's absolutely right. If you look at the number of vehicle days you need to start getting any significant crash, it's really—so, I think we have to look at all those programs and start bringing that information in. And, we take what's already being paid for in the automobiles and make use of it, in terms of very little additional funding, so we can have this—going into vehicles that—15 million miles of vehicle travel. You'd need a lot of vehicle days to see that, which is, I think, important here.

MR. CAMPBELL:

It isn't a good idea—

MALE VOICE:

Doug—with SAE. As a result of SAE presence last week, I think with you, Ray, and Joe, the recorder certainly lends itself to a service as combined with—together. So, I think if this council is looking at the possibility of putting the working group together, we can certainly provide a service to bring those people together and discuss the issues.

MR. OWINGS:

That's a real possibility, but I really think that I would like to use the structure we have here to first get back to the—get committees. Yes, we have discussed that. I think it's something we should do. Anyone else?

3. A series of meetings, symposiums, standard activities, etc. followed.

4. See: The 16th International Technical Conference on the Enhanced Safety of Vehicles (ESV) Proceedings—Windsor, Ontario, Canada, May 31–June 4, 1998 at *http://www-nrd.nhtsa.dot.gov/departments/nrd-01/esv/esv.html*.

5. See: USDOT / Docket Management System (DMS) NHTSA-99-5218-1 at *http://dms.dot.gov*.

6. See: Federal Register 63 FR 60270 (Nov. 9, 1998)

7. See: USDOT / Docket Management System (DMS) NHTSA-99-5218-2 at *http://dms.dot.gov*.

8. See: Proceedings of International Symposium on Transportation Recorders at *http://www.ntsb.gov/events/symp_rec.htm*.

9. Ibid.

10. See: Federal Register 64 FR 29616 (June 2, 1999)

11. See: USDOT / Docket Management System (DMS) NHTSA-99-5218-3 at *http://dms.dot.gov*.

12. Ibid, NHTSA-99-5218-4

13. See: NTSB Safety Recommendation Letters at: *http://www.ntsb.gov/Recs/letters/letters.htm#Highway*.

14. Ibid.

15. Ibid.

16. See: USDOT / Docket Management System (DMS) NHTSA-99-5218-3 at *http://dms.dot.gov*.

17. See: Vetronix Corporation website at: *http://www.vetronix.com/*.

18. See: Proceedings of Transportation Safety & the Law at *http://www.ntsb.gov/events/symp_rec.htm*.

19. See: USDOT / Docket Management System (DMS) NHTSA-99-

20. Ibid.

21. Ibid.

22. See: IEEE Vehicular Technology Society (VTS) Conference information at *http://vtssociety.org/*.

23. See: The 17th International Technical Conference on the Enhanced Safety of Vehicles (ESV) Proceedings—Amsterdam, The Netherlands, June 4–7, 2001 at *http://www-nrd.nhtsa.dot.gov/departments/nrd-01/esv/esv.html*.

24. "Summary of Findings by the NHTSA EDR Working Group," August 2001, Final Report, DOT-HS-043, *http://www-nrd.nhtsa.dot.gov/pdf/nrd-10/EDR/WkGrp0801.pdf*.

25. Ibid.

26. Letter from IEEE-SA to Dennis Bodson (sponsor) and Tom Kowalick (Chair) of IEEE P1616 dated December 10, 2001.

27. See Institute of Electrical and Electronics website at *http://www.vtsociety.org/*.

28. See Project 1616 website at *http://grouper.ieee.org/groups/1616/home.htm*.

29. Ibid.

30. Ibid.

31. See April Press Release at *http://grouper.ieee.org/groups/1616/new_page_5.htm*.

32. See January 2002 final minutes at *http://grouper.ieee.org/groups/1616/new_page_4.htm*.

33. Ibid—see February 2002 final minutes.

34. Ibid—see March, April, May, June 2002 final minutes

35. Ibid—see July 2002 final minutes.

36. Ibid—see September 2002 final minutes.

CHAPTER 4: NHTSA CALL FOR COMMENTS

1.Department of Transportation / National Highway Traffic Safety Administration [Docket No. NHTSA-02-13546; Notice 1] RIN 2127-A172, "Event Data Recorders" Request for Comments. Federal Register / Vol. 67, No. 198 / Friday, October 11, 2002 / Notices. Washington, DC.

CHAPTER 5: THINGS ARE FURTHER IN THE DARK

1. See IEEE MVEDR website for meeting minutes at: *http://grouper.ieee.org/groups/1616/new_page4.htm*.

2. Committee on the Future of Personal Transport Vehicles in China, National Research Council, National Academy of Engineering, Chinese Academy of Engineering. See: *http://www.nap,edu/catalog/10491.html*.

3. Ibid.

4. Department of Transportation / National Highway Traffic Safety Administration [Docket No. NHTSA-02-13546; Notice 1] RIN 2127-A172, "Event Data Recorders" Request for Comments. Federal Register / Vol. 67, No. 198 / Friday, October 11, 2002 / Notices. Washington, DC.

5."Federal Motor Vehicle Safety Standards; Event Data Recorders: Docket No. NHTSA-2002-13546-82," NHTSA, 2002.

6."Federal Motor Vehicle Safety Standards; Event Data Recorders: Docket No. NHTSA-2002-13546-10," NHTSA, 2002

7. Turner and Associates website at *http://www.tturner.com/*.

8. Ibid

9."Federal Motor Vehicle Safety Standards; Event Data Recorders: Docket No. NHTSA-2002-13546-10," NHTSA, 2002.

10."Federal Motor Vehicle Safety Standards; Event Data Recorders: Docket No. NHTSA-2002-13546-11," NHTSA, 2002.

11. UAB at Birmingham website at *http://main.uab.edu/*.

12."Federal Motor Vehicle Safety Standards; Event Data Recorders: Docket No. NHTSA-2002-13546-11," NHTSA, 2002.

13. New York Times article from ITS website at www.itsa.org/ITSNEWS.NSF/O0e1c68aae85ee741d85256c9f0002558f?OpenDocument

14. Ralph Nader in Unsafe at Any Speed, pgs 42-43.

15. Ibid, pg. 189

CHAPTER 6: BLOWIN THE HORN

1."Federal Motor Vehicle Safety Standards; Event Data Recorders: Docket No. NHTSA-2002-13546-16," NHTSA, 2002. 2 pgs.

2."Federal Motor Vehicle Safety Standards; Event Data Recorders: Docket No. NHTSA-2002-13546-18," NHTSA, 2002. 7 pgs.

3."Federal Motor Vehicle Safety Standards; Event Data Recorders: Docket No. NHTSA-2002-13546-22," NHTSA, 2002. 6 pgs.

4."Federal Motor Vehicle Safety Standards; Event Data Recorders: Docket No. NHTSA-2002-13546-23," NHTSA, 2002. 3 pgs.

5."Federal Motor Vehicle Safety Standards; Event Data Recorders: Docket No. NHTSA-2002-13546-28," NHTSA, 20026. See: IIHS website at *http://www.hwysafety.org/*.

6. Ibid.

7. Ibid.

8. Ibid.

9."Federal Motor Vehicle Safety Standards; Event Data Recorders: Docket No. NHTSA-2002-13546-30," NHTSA, 2002. 9 pgs.

10. See Transportation Research Board Annual Meeting www.trb.org

11. Ibid

12. "Federal Motor Vehicle Safety Standards; Event Data Recorders: Docket No. NHTSA-2002-13546-32," NHTSA, 2002. 5 pgs.

13. Ibid.

14."Federal Motor Vehicle Safety Standards; Event Data Recorders: Docket No. NHTSA-2002-13546-36," NHTSA, 2002. 3 pgs.

15. Ibid.

CHAPTER 7: CRUISE CONTROL

1. See: IEEE MVEDR website for meeting minutes at: *http://grouper.ieee.org/groups/1616/Feb_10_112003_Minutes_FINAL_EDIT.pdf*.

2. Ibid.

3. E-mail to mvedr@ieee.org received 4/22/03. Letter of Resignation: "Hello, Effective today, I am resigning from the P1616 Committee of IEEE. Given the recent formation of the SAE VEDI Committee, which is developing a standard output format and download protocol for light-duty vehicle EDRs, my management believes my time is better spent participating in this activity, rather than in P1616. regards, Barbara E. Wendling."

4. See: *http://www.nhtsa.dot.gov/nhtsa/announce/speeches/* for Stephen-Kratze-speech.pdf

5. "Federal Motor Vehicle Safety Standards; Event Data Recorders: Docket No. NHTSA-2002-13546-39," NHTSA, 2002. 2 pgs.

6. "Federal Motor Vehicle Safety Standards; Event Data Recorders: Docket No. NHTSA-2002-13546-41," NHTSA, 2002. 2 pgs.

7. "Federal Motor Vehicle Safety Standards; Event Data Recorders: Docket No. NHTSA-2002-13546-45," NHTSA, 2002. 2 pgs.

8. Ibid.

9. "Federal Motor Vehicle Safety Standards; Event Data Recorders: Docket No. NHTSA-2002-13546-46," NHTSA, 2002. 5 pgs.

10. Ibid.

11. "Federal Motor Vehicle Safety Standards; Event Data Recorders: Docket No. NHTSA-2002-13546-48," NHTSA, 2002. 2 pgs.

12. "Federal Motor Vehicle Safety Standards; Event Data Recorders: Docket No. NHTSA-2002-13546-50," NHTSA, 2002. 6 pgs.

13. "Federal Motor Vehicle Safety Standards; Event Data Recorders: Docket No. NHTSA-2002-13546-47," NHTSA, 2002. 7 pg.

14. "Federal Motor Vehicle Safety Standards; Event Data Recorders: Docket No. NHTSA-2002-13546-51," NHTSA, 2002. 36 pgs.

15. "Federal Motor Vehicle Safety Standards; Event Data Recorders: Docket No. NHTSA-2002-13546-55," NHTSA, 2002. 3 pgs.

16. "Federal Motor Vehicle Safety Standards; Event Data Recorders: Docket No. NHTSA-2002-13546-56," NHTSA, 2002. 4 pgs.

17. "Federal Motor Vehicle Safety Standards; Event Data Recorders: Docket No. NHTSA-2002-13546-57," NHTSA, 2002. 6 pgs.

CHAPTER 8: TURN SIGNALS

1. "Federal Motor Vehicle Safety Standards; Event Data Recorders: Docket No. NHTSA-2002-13546-59," NHTSA, 2002. 3 pgs.

2. "Federal Motor Vehicle Safety Standards; Event Data Recorders: Docket No. NHTSA-2002-13546-60," NHTSA, 2002. 2 pgs.

3. "Federal Motor Vehicle Safety Standards; Event Data Recorders: Docket No. NHTSA-2002-13546-62," NHTSA, 2002. 10 pgs.

4. "Federal Motor Vehicle Safety Standards; Event Data Recorders: Docket No. NHTSA-2002-13546-63," NHTSA, 2002. 3 pgs.

5. See: www.epic.org/privacy/edr_comments.pdf

6. Ibid

7. "Federal Motor Vehicle Safety Standards; Event Data Recorders: Docket No. NHTSA-2002-13546-71," NHTSA, 2002. 1 pg.

8. "Federal Motor Vehicle Safety Standards; Event Data Recorders: Docket No. NHTSA-2002-13546-68," NHTSA, 2002. 1 pg.

9. "Federal Motor Vehicle Safety Standards; Event Data Recorders: Docket No. NHTSA-2002-13546-67," NHTSA, 2002. 2 pgs.

10. "Federal Motor Vehicle Safety Standards; Event Data Recorders: Docket No. NHTSA-2002-13546-69," NHTSA, 2002. 150 pgs.

CHAPTER 9: SPEED BUMPS

1. "Federal Motor Vehicle Safety Standards; Event Data Recorders: Docket No. NHTSA-2002-13546-72," NHTSA, 1 pg.

2. Ibid.

3. "Federal Motor Vehicle Safety Standards; Event Data Recorders: Docket No. NHTSA-2002-13546-73," NHTSA, 5 pgs.

4. Ibid

5. "Federal Motor Vehicle Safety Standards; Event Data Recorders: Docket No. NHTSA-2002-13546-75," NHTSA, 9 pgs.

6. Ibid.

7. "Federal Motor Vehicle Safety Standards; Event Data Recorders: Docket No. NHTSA-2002-13546-76," NHTSA, 5 pgs.

8. Ibid.

9. Letter from IEEE-SA Appeals Panel to IEEE P1616 Co-Chairs, April 18, 2003.

10. Ibid

11. Ibid

12. Ibid

13. SAE confirmed minutes: Vehicle Event Data Interface (VEDI) Committee, April 24, 2003 at SAE Automotive Headquarters, 755 W. Big Beaver Road, Suite 1600, Troy, MI 48084-4903, USA.

14. "Federal Motor Vehicle Safety Standards; Event Data Recorders: Docket No. NHTSA-2002-13546-81," NHTSA, 2 pgs.

15. See: IEEE MVEDR website for meeting minutes at: *http://grouper.ieee.org/groups/1616/146_Final_Amended_MAY_MINUTES_08_26_03.pdf*.

16. Ibid

17. Ibid

18. Ibid

19. Ibid

20. "Federal Motor Vehicle Safety Standards; Event Data Recorders: Docket No. NHTSA-2002-13546-79," NHTSA, 2002. 7 pgs.

21. See: The 18th International Technical Conference on the Enhanced Safety of Vehicles (ESV) Proceedings—Nagoya, Japan, May 19-22, 2003. Available at *http://www-nrd.nhtsa.dot.gov/departments/nrd-01/esv/esv.html*.

22. See NTSB website at: *http://www.ntsb.gov/events/symp_vr_toptec/symp_vr_toptec.htm*.

23. See: EE Times Online at *http://www.eetimes.com/issue/mn/OEG20030721S0014* (July 21, 2003).

24. See: IEEE MVEDR website for meeting minutes at: *http://grouper.ieee.org/groups/1616/new_page_4.htm*.

25. See: National Academies of Science, Transportation Research Board (TRB) National Cooperative Highway Research Program Project 17-24 site at *http://www4.nas.edu/trb/crp.nsf/All+Projects/NCHRP+17-24*.

26. New York Times, September 22, 2003.

27. Automotive News, September 22, 2003.

28. See: IEEE MVEDR website for meeting minutes at: *http://grouper.ieee.org/groups/1616/new_page_4.htm*.

29. See: World Health Organization site devoted to World Health Day 2004 at *http://www.who.int/world-health-day/2004/en/*.

30. Ibid.

CHAPTER 10: LEGAL AND PRIVACY ISSUES

1. See, Department of Transportation, National Highway Traffic Safety Administration Event Data Recorders—Request for Comments, 67 Fed. Reg. 63493 (Oct. 11, 2002). Depending on vehicle type, EDRs can provide important crash data including such variables as vehicle speed (in five one-second intervals preceding impact), engine speed (in five one-second intervals preceding impact), brake status, whether seatbelts were engaged, whether air bags were enabled or disabled, and other information critical to crash investigators. Of course, EDRs may, in the future, include additional information as well. Presently, separate devices such as global positioning systems (GPS), can provide data such as vehicle location and speed.

2. Id., at 63493.

3. Matthew L. Wald, The Debate over Event Data Recorders, New York Times, (Sunday, Dec. 29, 2002) (discussing potential controversies surrounding the deployment of EDRs); Dean Narciso, Sensors Tell How Teen Driver Crashed, Columbus Dispatch (Jan. 5, 2002) (discussing EDR benefits versus privacy concerns).

4. Barry Brown, Warning! Your trip may be tracked, MSNBC News (July 10, 2002),

http://www.msnbc.com/news/596601.asp.

5. Bob Van Voris, Black Box Car Idea Opens Can of Worms, Nat'l L. J., June 14, 1999, at A1. Similar types of devices have long been installed in airplanes, trains, and other forms of

(usually public or commercial) transportation. See, *e.g.*, 14 C.F.R. 121.343 (Dec. 10, 1972) & 121.344 (Sept. 12, 1997) (flight data recorders); 14 C.F.R. 121.359 (Jan. 1, 1967) (airplane cockpit voice reorders); 49 C.F.R. 229.5 & 135 (May 26, 1995) (locomotive event recorders).

 6. Barry Brown, Accident Investigators Find Help in 'Black Box' For Cars, Buffalo News, Feb. 4, 2001.

7. OnStar, Wingcast, Qualcomm, and the as yet unnamed AT&T system.

8. Ed Gartson, Ex-NHTSA Chief Works on Auto Data, AP Online, Mar. 6, 2002.

9. Id.

10. American Civil Liberties Union, Are Vehicle "Black Boxes" a Black Hole for Privacy? (June 3, 1999), *http://archive.aclu.org/news/1999/w060399a.html*.

11. Harry Stoffer, Promise and pitfalls seen in black box, 75 Automotive News 5948 (2001).

12. Event Data Recorder Research History, (February 28, 2003) (detailing NHTSA's rejection of petitions to mandate EDRs), *http://www-nrd.nhtsa.dot.gov/edr-site/history.html*.

13. Note that this overview will only consider relevant federal statutory and constitutional law. Individual states may provide greater privacy protections above and beyond the ambit of the Fourth Amendment in their own constitutions and statutes. A general survey of state law, however, is beyond the scope of this paper.

14. U.S. Const. art. I, § 8, cl. 3.

15. Id.

16. See Hodel v. Virginia Surface Mining & Reclamation Ass'n, 452 U.S. 264, 277 (1981) (quoting Perez v. United States, 402 U.S. 146, 150 (1971)).

17. See, Nevada v. Skinner, 884 F.2d 445, 450 (9th Cir. 1989).

18. U.S. 1, 196 (1824).

19. See United States v. Lopez, 514 U.S. 549, 558-59 (1995) (citing United States v. Darby, 312 U.S. 100).

20. 884 F.2d 445, 450 (9th Cir. 1989).

21. Id. At 451.

22. Id.

23. 15 U.S.C.A. § 1381 et seq.; 5 U.S.C.A. § 553.

24. 32 Fed.Reg. 2408, 2415 (1967).

25. 37 Fed.Reg. 3911 (1972).

26. New York v. Class, 475 U.S. 106 (1986) (holding that the police officer's actions in searching the car did not violate the fourth amendment because there is no reasonable expectation of privacy regarding the VIN placement).

27. NHTSA also requires so-called high theft line vehicles to have identification numbers or symbols placed on major parts of certain passenger motor vehicles. 49 C.F.R. 541. Once again, this is to foil theft and to enable authorities to track stolen vehicles and parts.

28. Motor Vehicle Mfrs. Ass'n of U.S., Inc. v. State Farm Mut. Auto. Ins. Co., 463 U.S. 29 (1983) ("Given the effectiveness ascribed to air bag technology by the National Highway

Traffic Safety Administration, the mandate of the Motor Vehicle Safety Act to achieve traffic safety would suggest that the logical response to the faults of detachable seatbelts would be to require the installation of air bags.").

29. For example, DOT's "interlock and buzzer" devices were most unpopular with the public. Congress, responding to public pressure, passed a law that forbade DOT from requiring, or permitting compliance by means of, such devices. Motor Vehicle and School bus Safety Amendments of 1974, § 109, 88 Stat. 1482 (previously codified at 15 U.S.C. § 1410b (b) (1988 ed.)).

30. See, *http://www.nhtsa.dot.gov/people/injury/enforce/Millenium/strategy_45.htm* (discussing strategy for mandating interlock devices); Barnes, Nat, Take a Breather, The Express, Saturday, July 13, 2002 (discussing efficacy of interlock devices).

31. Motorcycle helmet laws are another interesting example of a good idea that has garnered slow, begrudging acceptance. Popular opposition to mandatory helmet laws was difficult to overcome, and met with considerable controversy, even though substantial evidence existed demonstrating their safety. See, generally,

http://www.nhtsa.dot.gov/people/outreach/stateleg/mchelmetUpdateDec2000.htm (discussing efficacy of helmets); Mirkin, Gabe, Riding Bikes or Motorcycles, Helmet use Remains Life-Saving, The Washington Times, September 15, 2002, p. C 8. (arguing against the repeal of helmet laws); Bujol, Jessica, Bikers Battle Helmet Law, Associated Press, State & Local Wire, March 9, 2001 (reporting on opposition to mandatory helmet laws). Hampel, Paul, Lawmaker's 10th Try at Motorcycle Helmet Rule Fails Again; Sponsor of the Bill Blames Strong Anti-Helmet Lobby, St. Louis Post-Dispatch, Tuesday, May 28, 2002 (discussing opposition to mandatory helmet laws).

32. 14 CFR 21.

33. 14 CFR 61.

34. 14 CFR 170.

35. 15 U.S.C. 1395, 1401 & 23 U.S.C. 403.

36. 23 USC 401 (1987).

37. Id.

38. 42 U.S.C §§ 7401 et seq. (1990).

39. EPA Regulations on Automobile Exhaust Systems, Exhaust System Repair Guidelines, *http://exhaust soundclips.com/epa_reg.html* (noting that catalytic converters can neither be removed or tampered with).

40. See generally, 42 USC §§ 7521, 7522, 7541, 7545, et seq.

41. US. Const. Amend. IV.

42. The Fourth Amendment applies both to the federal and the state governments.

43. 5 U.S.C. 552a (2002).

44. Most states contain analogues to this act, but a full review of those statutes is well beyond the purview of this paper.

45. Indeed, DOT and the automakers appear to agree that the car's title-holder owns not only the physical EDR, but the data it collects as well.

46. See, *e.g.*, Daewoo Electronics Co., Ltd. v. U.S.,650 F.Supp. 1003 (1986) (computer files); Rowe Entertainment, Inc. v. William Morris Agency, Inc., 205 F.R.D. 421 (2002) (computer files); Nixon v. Freeman, 670 F.2d 346 (1982) (diary); Tran v. New Rochelle Hosp. Medical Center, 740 N.Y.S.2d 11 (2002) (video surveillance); Congleton v. Shellfish Culture, Inc., 807 So.2d 492 (2002) (video surveillance).

47. See, *e.g.*, Cansler v. Mills, 765 N.E.2d 698 (2002) (holding that mechanic should have been able to testify regarding air bag and sufficient evidence was introduced to show air bag was defective); Harris v. General Motors Corp., 201 F.3d. 800 (2001)(evidence regarding air bag defectiveness should have been admitted); Sipes v. General Motors Corp., 946 S.W.2d 143 (2000) (Automobile manufacturer can be held strictly liable for defect that produces injuries even if the defect did not cause accident); Anderson-Barahona v. Gen. Motors Corp., No. 99A19714 (Ga., Cobb County Cir. Ct. Apr. 7, 2000).

48. Thomas Michael Kowalick, Proactive Use of Highway Recorded Data via an Event Data Recorder (EDR) to Achieve Nationwide Seat Belt Usage in the 90th Percentile by 2002, at *http:www.ntsb.gov/events/symp%5Frec/proceedings/authors/kowalick.htm* (last visited Jan. 26, 2003). Kowalick explains that:

"On the question of whether crash recorder data should be admitted, the main point is whether the recorder is realiable, properly read out, and provides a record of the particular event in question. The data of itself is no depositive of liability, but merely serves as certain evidence of the event. As earlier indicated in this report, there is a good correlation between crash severity a recorder might measure and the extent of the crash deformation to the vehicle in which it was installed; and it would be difficult to refuse evidence on the crash severity magnitude as interpreted from vehicle deformation. Thus if the recorder provides good evidence of the event, it seems appropriate that the evidence should be admitted. It may be possible to restrict through legislation the admissibility of crash recorder evidence, particularly if the recorders are government-owned and the records are retrieved and inspected by government employees. Consider, however, the objective of a very simple and widely used integrating accelerometer that is conveniently and readily read by any police accident investigator without special training. It would appear difficult to prevent testimony by a layman—say a tow-truck operator or an auto mechanic—as to what he saw immediately following the accident. In summary, we believe that (1) the data from a crash recorder would be admissible, if it meets necessary qualifications, in a court of law; 2) the data should be admitted if it is good evidence; (3) it will be difficult to prevent admitting crash recorder data, even by Federal law, if the recorder can be easily read by an untrained person." Id. (citing Office of Technology assessment, Automobile Collision data: An assessment of Needs and Methods of Acquisition (1975)).

49. U. S. Department of Transportation: Federal Motor Carrier Safety Administration, A Report to Congress on Electronic Control Module Technology for Use in recording Vehicle Parameters During a Crash, 16 (Sept. 2001) [hereinafter FMCSA Report]; Proposed RP 1214 (T): Guidelines for Event Data Collection, Storage and retrieval (2001).

51.National Transportation Safety Board Amendments Act of 2000, 49 U.S.C. 1114(c), 1154 (a) (2001).

52. Id. 1114(d), 1154(a).

53. Id. 1114(c)-(d), 1154(a).

54. Id. 1114(c)-(d).

55. Id. 1154(a)(1)(A).

56. Id. 1154(a)(2)-(4)

57. Id. 1154(a)(4)(A)-(B).

58. Donald C. Massey, Discovery of Electronic Data from Motor Carriers—Is Resistance Futile?, 35 Gonz. L. Rev. 145, 173 (2000) (noting that event data collected from train crashes is generally admissible) (citing Stuckey v. Illinois Central R.R. Co., 1998 U.S. Dist. LEXIS 2648 (N.D. Miss. Feb. 10, 1998); National R.R. Passenger Corp. v. H & P, Inc., 949 F. Supp. 1556 (M.D. Ala. 1996)); see also American Trucking Associations Website, Legislative Affairs, Trucking Victory: Truck Recorders Gain Protection Given Airplane Recorders, at

http://www.truckline.com/legislative/101800<uscore>truck<uscore>recorders.html (last visited Jan. 26, 2003) (on file with the Rutgers Computer & Technology Law Journal) (explaining that the National Transportation Safety Board Amendments Act of 2000 does not extend its protection to data recorders).

59. Harris v. General Motors Corp., 201 F.3d 800, 804 (6th Cir. 2000); Batiste v. General Motors Corp., 802 So. 2d 686, 688 (La. Ct. App. 2001).

60. Harris, 201 F.3d at 804.

61. Id. at 804 n.2.

62. Id. (citing Daubert v. Merrell Dow Pharm., Inc., 509 U.S. 579 (1993)) (establishing admissibility test for expert scientific testimony based upon Fed. R. Evid. 702 and a rough framework of criteria focusing on scientific validity, reliability, and relevance).

63. Batiste, 802 So. 2d at 688

64. Id.

65. Donald C. Massey Discovery of Electronic Data from Motor Carriers—Is Resistance Futile? 35 Gonz. L. Rev. 145(2000) (explaining that electronically-stored data is universally admitted into evidence).

66. Associated Press, Insurer's "Black Box" Monitors Drivers, USA Today.com, at

http://www.usatoday.com/life/cyber/tech/review/crg529.htm (Nov. 23, 1999).

67. Id.

68. Rental-car Firm Exceeding the Privacy Limit?, News.com, June 20, 2001, at *http://news.com.com/2100-1040-268747.html?legacy=cnet&tag=tp_pr.*

69. Robert Lemos, Car spy pushes privacy limit, ZDNet News, June 19, 2001, *http://zdnet.com.com/2100-11-5301115.html.*

70. Using GPS To Catch Speeders Found Illegal, Slashdot, July 3, 2001 at *http://slashdot.org/articles/01/07/03/0423218.shtml.*

71. It is important to note that the Fourth Amendment does not act as a restraint on private actors, but only those acting under the color of state law. Thus, the Fourth Amendment does not prevent a private party (like an insurance company) from seizing data recorded by a EDR.

72. Katz v. United States, 389 U.S. 347, 361 (1967) (holding that government's activities in electronically listening to and recording defendant's words spoken into telephone receiver in public telephone booth violated the privacy upon which defendant justifiably

relied while using the telephone booth and thus constituted a 'search and seizure' within Fourth Amendment).

73.186 F.3d 1119, 1126-27 (9th Cir. 1999).

74. Id.

75. U.S. 106 (1986).

76. McIver, 186 F.3d at 1127. Cf. Osburn v. Nevada, 44 P.3d 523 (2002) (holding that officers' placement of a monitoring device to defendant's vehicle without first obtaining a search warrant was not an unreasonable search under the Nevada Constitution) with Oregon v. Campbell, 759 P.2d 1040 (1988) (holding that officers' placement of a tracking device to defendant's vehicle without first obtaining a search warrant constituted an unreasonable search under the Oregon Constitution). These cases demonstrate the differing standards under their respective state constitutions.

77. 533 U.S. 27 (2001).

78. New York v. Class, 475 U.S. 106, 114-115 (1986) (holding that when police officers reached into interior of automobile to remove certain papers obscuring the vehicle identification number it was a search but was sufficiently unintrusive to be constitutionally permissible, thereby justifying officer's seizure of weapon found protruding from underneath driver's seat).

79. See California v. Acevedo, 500 U.S. 565 (1991) (holding that police may search a container located within an automobile, and need not hold the container pending issuance of a warrant, even though they lack probable cause to search the vehicle as a whole; it is enough that they have probable cause to believe the container itself holds contraband or evidence.)

80. Acevedo, 500 U.S. at 574.

81. Katz, 389 U.S. at 361, Cf. United States v. Jacobsen, 466 U.S. 109, 120 n.17 (1984) ("A container which can support a reasonable expectation of privacy may not be searched, even on probable cause, without a warrant.")

82. Id.

83. Michigan v. Clifford, 464 U.S. 287 (1984). (Where fire-damaged home was uninhabitable when fire investigator arrived, where personal belongings remained in home, and where owners had arranged to have house secured against intrusion while they were gone, owners retained reasonable privacy interest in fire-damaged residence and fire investigations were subject to warrant requirement).

84. United States v. Oswald, 783 F.2d 663, 666 (6th Cir.1986) (Defendant abandoned property and did not attempt to retrieve in a reasonable time thus did not violate rights against unreasonable search and seizure).

85. It is important to distinguish searches for causal evidence from searches for evidence of criminal conduct. Causal evidence is subject only to the restraints of an administrative warrant, where the search for criminal evidence requires a criminal obtainable only on a showing of probable cause to believe that relevant evidence will be found on the place searched. Michigan v. Clifford, 464 U.S. 287, 294 (1984).

86. New York v. Class, 475 U.S. 106 (1986).

87. Katz, 389 U.S. at 361.

88. A pen register is a surveillance device that captures the phone numbers dialed on outgoing telephone calls. See, 18 USC 3121.

89. Smith v. Maryland, 442 U.S. 735 (1979) (The installation and use of the pen register was not a search within the meaning of the Fourth Amendment, and no warrant was required) (pen register is a surveillance device that captures the phone numbers dialed on outgoing telephone calls).

90. Id. at 743–744. The standard exceptions to this, of course, include information provided to attorneys, treating physicians, or religious counselors. Such information is normally accorded "privileged" status and is not usable in the even of a criminal prosecution.

91. Oliver v. United States, 466 U.S. 170, 182–183 (1984).

92. Id. at 178.

93. Event Data Recorders: Summary of Findings by the NHTSA EDR Working Group, Final Report, Aug. 2001, at 8.3, No. NHTSA-99-5218, available at *http://nrd.nhtsa.dot.gov/edr-site/uploads/edrs-summaryoffindings.pdf* (last visited Jan. 26, 2003) (recording position of National Highway Traffic Safety Administration regarding EDR data ownership) and 8.3.2 (recording position of Federal Highway Administration regarding EDR data ownership).

94. Id.

95. Id. at 8.3.2.

96. U.S. Department of Transportation: Federal Motor Carrier Safety Administration, A Report to Congress on Electronic Control Module Technology for Use in Recording Vehicle Parameters During a Crash, 23 (Sept. 2001) [hereinafter FMCSA Report]; see also American Trucking Associations, Inc., Technology & Maintenance Council, Recommended Practice, Proposed RP 1214(T): Guidelines for Event Data Collection, Storage and Retrieval (2001). While the FMCSA and ATA recommendations pertain to EDR's in trucks, as opposed to non-commercial vehicles, the same data ownership concerns apply.

97. See FMCSA Report, supra note 68.

98. The Supreme Court's decision in Kyllo v. United States, 533 U.S. 27 (2001) does not undercut this analysis. In Kyllo, the Court had to decide whether the government's use of a thermal imaging device constituted a "search" for Fourth Amendment purposes. Although the Court concluded that it did, it did not address the question whether the heat signature left by the defendant's home was "owned" by him. The Court simply held that the government could not, without a warrant, obtain the home's thermal data.

99. Electronic Communications Privacy Act of 1986, Pub. L. No. 99-508, 101-303, 100. Stat. 1848 (codified at 18 U.S.C. 1367, 2510-21, 2701-10, 3121-26).

100. *E (M.D. La.), aff'd, 87808 Edwards v. Bardwell, 632 F. Supp. 584, 586-F.2d 54 (5th Cir. 1986);*Congressional Findings Act of 1968, Pub. L. No. 90-351, 801, 82 Stat. 211-12 (1968) (observing that wire communications form an interstate network susceptible to substantial eavesdropping and interception of wire, electronic, and oral communications; the purpose of the ECPA is to protect the privacy of such communications).

101. The ECPA defines wire and oral communications as follows:

(1) "wire communication" means any aural transfer made in whole or in part through the use of facilities for the transmission of communications by the aid of wire, cable, or other

like connection between the point of origin and the point of reception ... furnished or operated by any person engaged in providing or operating such facilities for the transmission of interstate or foreign communications ....

"oral communication" means any oral communication uttered by a person exhibiting an expectation that such communication is not subject to interception under circumstances justifying such expectation, but such term does not include any electronic communication. *8 U.S.C. 2510(1)—(2)* (1995).

102. See, *e.g.*, United States v. Hall 488 F.2d 193 (9th Cir. 1973) (holding that a communication was protected as a "wire" communication if one party was on a cellular car phone and the other on a land-based line.; United States v. Smith, 978 F.2d 171, 180 (5th Cir. 1992) (holding that the defendant failed to prove that his expectation of privacy in a cordless phone conversation was reasonable under the Fourth Amendment; the court opined in dicta, however, that a more technologically advanced cordless phone may acquire a societal recognition of a reasonable expectation of privacy sufficient for Fourth Amendment protection), cert.denied, 113 S. Ct. 1620 (1993), Tyler v. Berodt, 877 F.2d 705 (8th Cir. 1989), cert.denied, 493 U.S. 1022 (1990). In Tyler, a cordless phone conversation to an unknown receiver was intercepted by one of Tyler's neighbors. Id. at 705. Information heard during the interception subsequently provided the basis for the ensuing criminal charges against Tyler. Id. at 706. Although Tyler sued his neighbor for civil violations of the ECPA, the United States Court of Appeals for the Eighth Circuit affirmed the summary judgment against Tyler. *Id. at 707;* see also *United States v. Hoffa, 436 F.2d 1243, 1247 (7th Cir. 1970)1992)* (communicating via a cellular car phone provides no reasonable expectation of privacy), cert. denied, 400 U.S. 1000 (1971);United States v. Carr, 805 F. Supp. 1266 (E.D.N.C. (holding that there is no reasonable expectation of privacy in communications via at least one cordless phone and either a cordless or land-based line).

103. *18 U.S.C. § 2510*(12) (defining "electronic communication" as, with certain exceptions, "any transfer of signs, signals, writing, images, sounds, data, or intelligence of any nature transmitted in whole or in part by a wire, radio, electromagnetic, photoelectronic or photo optical system that affects interstate or foreign commerce")

104. *18 U.S.C. § 2510*(17) (defining "electronic storage" as "(A) any temporary, intermediate storage of a wire or electronic communication incidental to the electronic transmission thereof; and (B) any storage of such communication by an electronic communication service for purposes of backup protection of such communication").

105. If the communication has been in electronic storage for 180 days or less, the government must obtain a warrant. See 18 U.S.C. § 2703(a) (2001).

106. *18 U.S.C. § 2510*(15) (2001) (defining ECS as "any service which provides to users thereof the ability to send or receive wire or electronic communications").

107. *18 U.S.C. §§ 2703(*a), 2703(b)(1)(B)(ii), 2703(d) (2001).

108. U.S. v. Humphrey, 279 F.3d 372 (2002) (Video Surveillance); U.S. v. Lightfoot, 6 Fed.Appx. 181 (2001) (Video Surveillance); U.S. v. Walton,217 F.3d 443 (2000) (Video Surveillance); Moyer v. Com, 531 S.E.2d 580 (2000) (Diary); U.S. v. Angevine, 281 F.3d 1130 (2002) (Computer files); U.S. v. Jewell, 16 Fed.Appx. 295 (2001) (Computer Files).

109. United States v. Kreimes, 649 F.2d 1185 (5th Cir. 1981) (holding that police officer was justified in stopping defendant's truck and the warrantless search of luggage found in truck was justified because officer had probable cause to believe that an armed fugitive was at large).

110. U.S. Const. amend. IV.

111. *Florida v. White*, 526 U.S. 559, 564 (1999).

112. *Carrol v. U.S.*, 267 U.S. 132 (1925)(federal agents stopped a car they had probable cause to believe contained illegal liquor and immediately subjected it to a warrantless search).

113. *United States v. Mendoza-Burciaga*, 981 F.2d 192 (Cir. 1992).

114. *Michigan v. Tyler*, 436 U.S. 499 (1978).

115. *Michigan v. Clifford*, 464 U.S. 287 (1984).

116. *Tyler*, 436 U.S. at 501.

117. Id. at 502.

118. Id. at 510.

119. Id. at 511.

120. Id. at 509.

121. *Clifford*, 464 U.S. at 289-91.

122. Id. at 296.

123. *Clifford*, 464 U.S. at 291.

124. *U.S. v. Finnigin*, 113 F.3d 1182 (10th Cir. 1997).

125. Id. at 1185.

126. Id.

127. *Massachusetts v. Mamacos*, 409 Mass 635 (1991).

128. Id.

129. *Michigan v. Tyler*, 436 U.S. 499, 509 (1978).

130. Id.

131. See *United States v. Ross*, 456 U.S. 798, 809-10 (1982); *United States v. Buchner*, 7 F.3d 1149, 1154 (5th Cir.1993), cert. denied, 510 U.S. 1207 (1994); *United States v. Kelly*, 961 F.2d 524, 527 (5th Cir.1992).

132. *United States v. Muniz-Melchor*, 894 F.2d 1430, 1438 (5th Cir.), cert. denied, 495 U.S. 923 (1990).

133. *Ross*, 456 U.S. at 825.

134. *South Dakota v. Opperman*, 428 U.S. 364 (1976).

135. *Cady v. Dombrowski*, 413 U.S.433, 440 (1974).

136. *Opperman,* 428 U.S. at 368.

137. *Opperman,* 428 U.S. at 368.

138. *U.S. v. Ross,* 456 U.S. 798 (1982).

139. *Id.*

140. *Cady v. Dombrowski*, 413 U.S. 433, 440 (1974).

141. *South Dakota v. Opperman*, 428 U.S. 364 (1976).

142. *U.S. v. Ross*, 456 U.S. 798 (1982).

143. *Ferguson v. City of Charleston*, 532 U.S. 67 (2001) (A S.C. hospital selectively tested pregnant women seeking prenatal care and turned their drug test results over to the police who then arrested a number of the women. The court held this to violate the fourth amendment and not within the special needs exception).

144. *Earls ex rel. Earls v. Board of Educ. of Tecumseh Public School Dist.*, 242 F.3d 1264 (10th Cir. 2001) (Existence of drug problem at public high school constituted "special need," for purpose of determining whether school's suspicionless drug testing of students participating in competitive extracurricular activities was reasonable).

145. See *Edmond v. Goldsmith*, 183 F.3d 659 (7th Cir. 1999) ("Randomized or comprehensive searches that have survived the Fourth Amendment are not concerned with catching crooks, but rather with securing the safety or efficiency of the activity in which people who are searched are engaged for example, owners and proprietors (such as state as owner of public roads) have a right to take reasonable measures to protect safety and efficiency of their operations.")

146. The "special needs" exception may not apply absent a specific statute on point. The Sixth Circuit has held that the government may initiate seizure of property without prior hearing under certain very limited circumstances; in each case, seizure has been directly necessary to secure important governmental or general public interest, there must have been special need for very prompt action, and person initiating seizure must have been government official responsible for determining, under standards of narrowly drawn statute, that it was necessary and justified in the particular instance *First Federal Sav. Bank and Trust v. Ryan*, 927 F.2d 1345 (6th Cir. 1991) (emphasis added).

147. See, *e.g.*, *Pennsylvania v. Rhodes. Crim. Div. Doclet No.* 746701 (Montgomery County, Court of Common Pleas) (2002) (defendant pleaded guilty in case where EDR data was admitted to show that when the accident occurred, he was driving in excess of 100 miles per hour); *California v. Beeler,* case. No. SCD158974 (San Diego Sup. Crt.) (2002) (EDR data admitted to show that defendant was traveling at excessive speeds in a 45-mph zone—defendant convicted of felony and pleaded guilty to manslaughter).

148. Noah Bierman, Black box gives crash details: Broward traffic-deaths case among first of its kind, Miami Herald, Tuesday, May 6, 2003, *http://www.miami.com/mld/ miamiherald/5793635.htm.*

149. Case No. 01 CR 967 (1st Judicial District Court, Div. 3, Jefferson County) (2003).

150. See, *e.g.*, *Stanton v. Nat'l R.R. Passenger Corp.,* 849 F. Supp. 1524, 1528 (M.D. Ala. 1994); Dennis Donnelly, Black Box Technology in the Courtroom, 38 APR Trial 41 (April, 2002).

151. 533 U.S. 27 (2001).

152. U.S. Const. Amend. V.

153. See, generally, Allen, Kuhns & Stuntz, Constitutional Criminal Procedure (3rd Ed. 1995); LaFave, Israel & King, Criminal Procedure (3rd Ed. 2000).

154. *384 U.S.* 757 (1966).

155. *Id.*, at 767.

156. *Malloy v. Hogan*, 378 U.S. 1 (1964).

157. See, *e.g.*, *Hoffman v. United States*, 341 U.S. 479 (1951) (articulating the standards under which the privilege may be invoked).

158. This applies strictly to federal courts. State courts, of course, have their own rules of evidence; although, as a practical matter the state rules are generally quite similar to those used in federal courts.

159. Donald C. Massey, Proposed On-Board Recorders for Motor Carriers: Fostering Safer Highways or Unfairly Tilting the Litigation Playing Field?, 24 S. Ill. U. L. J. 453 (2000).

160. *Fed. R. Civ. P.* 26(b)(1).

161. *Fed. R. Civ. P. 34.* The Advisory Committee on Federal Rule 34 specifically considered evolving technology, and particularly, computer generated documents. It stated: [t]he inclusive description of documents is revised to accord with changing technology. It makes clear that Rule 34 applies to electronic data compilations from which information can be obtained only with the use of detection devices, and that when the data can as a practical matter be made usable by the discovering party only through respondent's devices, respondent may be required to use his devices to translate the devices to translate the data into usable form. In many instances this means that respondent will have to supply a print-out of computer data. *Fed. R. Civ. P. 34,* 1970 Advisory Committee Note.

162. *Fed. R. Civ. P.* 1001(1). Rule 1001 provides that a "writing consists of letters, words or numbers, or their equivalent, set down by ... magnetic impulse, mechanical or electrical recording, or other form of data compilation." Id. Moreover, an original of a writing "is the writing or recording itself or any counterpart intended to have the same effect by a person executing or issuing it ...." Id.

163. *Id.*1

164. *Fed. R. Civ. P.* 26(b)(1).

165. See generally Fed. R. Evid. 401.

166. *Fed. R. Evid.* 702.

167. *Frye v. United States*, 54 App.D.C. 46, 47 (1923).

168. Id. It is important to note that some states continue to use variations of the *Frye* test. In *Bachman v. Gen. Motors Corp.,* 332 Ill. App.3d 760 (Ill. 2002), the Appellate Court of Illinois applied the Frye test in permitting the admission at trial of evidence gathered from an air bag sensor. The appellate court concluded that the process of recording and downloading the air-bag sensor data proved no bar to admissibility under *Frye*.

169. *Daubert v. Merrell Dow Pharmaceuticals, Inc.,* 509 U.S. 579 (1993).

170. *Id.*

171. *Daubert*, 509 U.S. 579 (1993).

172. Id. It is important to note that many states still use *Frye*-type tests.

173. See *Fed. R. Evid.* 702 (Advisory Committee's Note to 2000 Amendment).

174. *Daubert*, 509 U.S. at 593-94.

175. See generally *Daubert,* 509 U.S. 579 (1993).

176. See *Fed. R. Evid.* 702.

177. See *Daubert*, 509 U.S. at 594 (describing how the varying factors applied by judges were neither exclusive nor dispositive).

178. See generally *Fed. R. Evid.* 702 (The 2000 Amendment provides that an expert may testify when such testimony is derived from sufficient data.). See, *e.g.*, *Heller v. Shaw Indus., Inc.*, 167 F.3d 146, 160 (3d Cir. 1999) (describing how expert testimony cannot be excluded simply because the expert uses one test rather than another, when both tests are accepted in the field and both reach reliable results); *Ruiz-Troche v. Pepsi Cola*, 161 F.3d 77, 85 (1st Cir. 1988) (holding that "Daubert neither requires nor empowers trial courts to determine which of several competing scientific theories has the best provenance").

179. 201 F.3d. 800 (2001).

The so-called physical facts rule disallows testimony "which is opposed to the laws of nature, or which is clearly in conflict with principles of science, [and] is of no probative value …." Lovas v. Gen. Motors Corp., 212 F.2d 805, 808 (6th Cir. 1954). Such testimony, "which is positively contradicted by the physical facts cannot be given probative value by the court." Ibid.

180. *Id.*, at 804.

181. *Daubert*, 509 U.S. 579 (1993).

182. Id.

CHAPTER 11: PROS AND CONS AND CUSTOMERS OF SAFETY DATA

1. I conducted several research studies at Sandhills Community College, Pinehurst, North Carolina, from 1997 to 2004.

CHAPTER 12: THE ROAD AHEAD

1. National Academy of Sciences / Transportation Research Board (TRB) Millenium Papers website at *http://www4.trb.org/trb/onlinepubs.nsf/web/ millenium_papers?OpenDocument.* See *http://gulliver.tr.org/publications/millennium/ 00147.pdf* for "Making Safety Management Knowledge Based."

2. Ibid, see *http://gulliver.trb.org/publications/millenium/00027.pdf* for "Data, Data, data—Where's the Data?"

BIBLIOGRAPHY

2004

"AIA Urges Co-operative Effort Between AAA/Insurers to Use Standardized Event Data Recorders in Vehicles. "*Insurance Advocate*, 2/16/2004, Vol. 115 Issue 6, p16, 2/5pp.

Associated Press. "Officials deny Baldacci rode without belt." *Press Herald*, Portland, ME. March 2004.

"Black Boxes Unlocks Car Crash Mysteries." *State Legislatures*, February 2004, Vol. 30 Issue 2, p7, 1/3pp.

Fischetti, Mark. Data Driven *Scientific American*, Vol 290 Issue 2, p 90, 1 p, 1 graph, 2 c; February 2004.

Haines, Lester. "Air bag grasses up killer." *The Register.co.uk*. March 2004.

Kelly, Cathy. "Black box tells us plenty." *Toronto Star* (Canada) 24 April 2004.

National Transportation Safety Board (NTSB) & Society of Automotive Engineers (SAE) Highway Vehicle Event Data recorder Symposium Proceedings. NTSB Academy at George Washington University, Virginia Campus, Ashburn, Virginia. Topics include: state-of-the-art in technology, ongoing research and validation, privacy and legal issues, end users, the insurance industry perspective and EDR standards committees. Draft 352 pp. Final report available at *http://www.sae.org*.

Pitcher, G. "Silent Witness." *New Electronics*. ISSN 00479624, pp. 18–20, Findlay Publications Limited, Franks Hall, Horton Kirby, Kent DA4 9LL, England. January 27, 2004.

Thanh Ha, Tu. "Smart air bag proves man was speeding." *Globe and Mail*,Toronto, Canada, March 2004.

"U.S. Proposes Uniform Data on Car Crashes." *New York Times*. Jun 11, 2004. p. C4.

2003

American Bar Association. "Hot Legal Issues and Prominent Figures Featured at ABA Annual Meeting in San Francisco." News Release. 24 Jul 2003. See: *http:www.pressi.com/ file://A:bibliography%203%20reference%202.htm*.

Augenstein, J. ; Perdeck, E. ; Stratton, J. ; Digges, K. ; Bahouth, G. ; Baur, P. ; Borchers, N. 2003. Methodology for the Development and Validation of Injury Predicting Algorithms. *Proceedings of the 18th International Technical Conference of the Enhanced Safety of Vehicles* (ESV) Conference, May 19–22, 2003 at Nagoya, Japan. Paper Number 467, 13 pp.

Ayers, Ian, Nalebuff, Barry; "Black Box For Cars." Aug 2003. *Forbes.com* See: *http;// forbes.com/home/free_forbes/2003/0811/084.html.*

Bellion, P. "Event Data Recorders: What Do They Tell Us?" Monash University–Institute of Transport Studies. 10 pp. Monash University, Victoria, Australia.

"Black Boxes." *Arizonia Republic Online.* 12 Jul. 2003."

Black Boxes Can Protect Against Litigation." NAFA Fleet Executive. Nationwide Mutual Insurance Company, V. 17, No. 8 (August 2003), pp. 10–12. Source of document: Northwestern University Transportation Library. 2003.

Bonilla, Denise. "Behind the Wheel ; A Key Witness at Crash Scenes : the Black Box ; Helping their Investigators Reconstruct Accidents, Data Recorders are not just for Jets Anymore." *Los Angeles Times* 24 June 2003. p. B.2.

Butler, Matthew; "Black Box is Standard Vehicle Technology." The Times Plus [Monroe, Wi.] 25 Jul. 2003. *http:file//A:%bibliography%203.htm.*

Campo-Flores, Arian. "Car Talk" *Newsweek* [New York]. 26 May 2003. Vol. 141, Issue 21. pg.10

Chambliss, John. "Devices have Critics—Auto Recording Boxes Change Routine Cases." *Online Ledger* 23 July 2003.

Champion, H. ; Augenstein, J. ; Blatt, A. ; Cushing, B. ; Digges, K. ; Hunt, R. ; Lombardo, L. ; Siegel, J. 2003. Reducing Highway Deaths And Disabilities with Automatic Wireless Transmission of Ser. *Proceedings of the 18th International Technical Conference of the Enhanced Safety of Vehicles (ESV) Conference,*May 19–22, 2003 at Nagoya, Japan. National Highway Traffic Safety Administration, Washington, DC. Paper Number 406, 15 pgs.

Dolan, Bill; Melvinville Cop Black Box to Be Used in Probe of Fatal Motorcycle Crash— Witness Reports Spur Investigations to Determine Police Car Speed in Accident. Times Online. 4 Aug 2003. *http://www.thetimesonline.com/articles/2003/08/05/news/ region_and_state.*

Echegaray, Chris. "Lake Avenue crash prompts call for black boxes." *Telegram & Gazette.* Worcester, Massachusetts: September 3, 2003. p. A.10.

Frey, Joe; Black Boxes Will Aide in Auto Collision Analysis *http://www.info.insure.com/ auto/collision/blackbox1000.htm.*

Gabler, H. ; Hampton, C. ; Roston, T. 2003. Estimating Crash Severity : Can Event Data Recorders Replace Accident Reconstruction ? *Proceedings of the 18th International Technical Conference of the Enhanced Safety of Vehicles (ESV) Conference,* May 19–22,2003 at Nagoya, Japan. Rowan University. USNHTSA. Paper Number 490, 12 pgs.

General Accounting Office Report: "Highway Safety Research Continues on a Variety of Factors that Contribute to Motor Vehicle Crashes." Report No. GAO-03-436, HS-043 537. 55 pages, March 2003. Available from General Accounting Office, 441 G Street, NW, Washington, DC 20548.

Gilman, Don; Automotive Black Box Data Recovery Systems. 5 Sept. 2003.

http://www.tarorgin.com/art/Dgilman

Hall, James E. "Viewpoint." *Aviation Week & Space Technology*, 1/6/2003, Vol 158 Issue 1, p54, 1p.

Jenkins, Chris. "Petty's Crash at Bristol hardest recorded by NASCAR black box." *USA Today.* 03/31/2003.

Katz, David M. "Privacy in the private sector: use of the automotive industry's event data recorder and cable industry's interactive television in collecting personal data." *Rutgers Computer & Technology Law Journal*, Spring 2003 v 29 i1 p163.

Kullgren, A. ; Krafft, M. ; Lie, A. ; Tingvall, C. 2003. Combining Crash Recorder and Paired Comparisons Technique : Injury Risk Functions In. *Proceedings of the 18th Internatinal Technical Conference of the Enhanced Safety of Vehicles (ESV) Comfernece*, May 19–22, 2003 at Nagoya, Japan. Folksam Research (Sweden) / Monash University Sccident Research Centre, Victoria (Australia). Paper Number 404, 8 pgs.

Lawrence, J.M.; Wilkinson, C.C.; Heinrichs, B.E.; Siegmund, G.P. 2003. The Accuracy of Pre-crash Speed Captured by Data Recorders. MacInnis Engineering Associates, Rich-mond, British Columbia (Canada) 5pp. *Accident Reconstruction* 2003. Warrendale, SAE, 2003, p211–215. Report No. 2003-01-0889. UMTRI-96371 A08

Leyden, John. "Air Bag Black box Nails Killer Driver." *Register* [United Kingdom]

Linder, A,; Avery, M.; Krafft, M.; Kullgren, A. 2003. Change of Velocity and Crash Pulse Characteristics in Rear Impacts: Real-World Data. *Proceedings of the 18th International Technical Conference of the Enhanced Safety of Vehicles (ESV) Conference*, May 19–22, 2003 at Nagoya, Japan. The Motor Insurance Research Centre, Thatchum,(United King-dom)/ Folksam Research, Sweden. Paper Number 285, 9 pgs.

Makowski, Felicia; "Vehicular Vision—Your Car May Be Tracking the Way You Drive." ABC NEWS. 16 July 2003.

http://abcnews.go.com/sections/wnt/DailyNews/blackboxes020716.html

Mohl, Bruce; "Imagine a Black Box For Your Car—It's There." *Boston Globe* 3 Aug. 2003. Section C1.

Murray, Charles J. "Automakers face standards choice for black box recorders." *Electronic Engineering Times*, 7/21/2003 Issue 1279, p4, 1p, 1c.

NAII Encouraging NHTSA to Push Envelope in Development of EDRs for Automobiles. *Insurance Advocate*, 3/3/2003, Vol 114 Issue 9, p22, 2p.

Newman, Richard. "No Place to Hide: Technology tells about your Driving Habits—and who may have caused that Accident." *US NEWS & World Report* 14 July 2003.Vol. 135, Issue 1. p. 32.

"NTSB, in Conjunction with Automotive Engineers, Set Symposium on Transportation Vehicle Recorders." *Insurance Advocate*, 5/5/2003, Vol 114 Issue 17, p30, 1/3pp.

Pochna, Peter. "Auto Manufacturers increasingly Installing Event Recorders in Automo-biles." *Record* [New Jersey]. 20 Jan. 2003.

Sala, D.; Wang, J.T. 2003 Continuously Predicting Crash Severity. *Proceedings of the 18th International Conference of the Enhanced Safety of Vehicles (ESV) Conference*, May 19–22, 2003 at Nagoya, Japan. General Motors Corporation (USA).Paper Number 314, 8 pgs.

Schoeneburg, R.; Baumann, K.; Justen, R.; 2003. Pre-Safe—The Next Step in the Enhancement of Vehicle Safety. *Proceedings of the 18th International Technical Conference of the Enhanced Safety of Vehicles (ESV) Conference*, May 19–22, 2003 at Nagoya, Japan. DaimlerChrysler (Germany). Paper Number 410, 8 pgs.

Sharp, Deborah. "Autos' Black box Data Turning up in Courtrooms." *USA Today* 16 May 2003, p. 3A.

Smith, Jeff. "Snitch or Savior? Black Box for Cars Stores Crash Data: Most Drivers unaware Tracking Device Could be Under their Hoods." *Rocky Mountain News* [Colorado]. 28 June 2003. p. 1C.

"Trial of Student Charged in Student's Death Postponed." *The Clarion Ledger.* 8 Jul. 2003. *http://clarionledger.com/news/o307/08/m16.html.*

"Vetronix and Injury Sciences Announce Extension of Strategic Partnership for Crash Data retrieval System." *Business Wire* [Ft. lauderdale, Fla.] 23 Jul. 2003.

Wald, Matthew l. "Privacy law in California Shields Drivers." *New York Times*, 9/23/2003, Vol. 152 Issue 52615, pA18, Op.

Wales, Elspeth. "Black Boxes Record Taxi Accidents—Special Report: Smart Garages." *Australian* 15 April 2003, pg.T08.

Wilso, Kevin A. "Something to Think About." *AutoWeek*, 6/9/2003, Vol 53 Issue 23, p10, 2/3p, 1 c.

Wright, Jeanne. " Wheels; Black Boxes can Monitor Teen Drivers; Parents can Install the Devices to help keep Tabs on the Kids Actions when they are behind the Wheel." *Los Angeles Times.* [California] 29 Jan. 2003 pg. G.1.

Zitter, Josh; Deely, Liam: "Racing Toward Safer Cars—Ford Uses Crash Data Recorder in Race Cars to Build Safer Cars." *TECH TV.*

2002

Architecture Development Team, Iteris, Inc.; Lockheed Martin. Standards Requirements Document. Washington DC: Federal Highway Administration: April 2002.

"Building a Black Box." *Fleet Owner*, June 1, 2002.

Chan, C. 2002. On the Detection of Vehicular Crashes: System Characteristics and Architecture. Partners for Advanced Transit and Highways, Berkeley, Calif. 14 pp. *IEEE Transactions on Vehicular Technology.* Vol. 51, No. 1, Jan.2002, pp. 180–193. UMTRI-62788

Clark, D. E. and Cushing, B. M. "Predicted Effect of Automatic Crash Notification on Traffic Mortality". *Accident Analysis & Prevention.* Volume 34, Issue 4, July 2002, Pages 507–513.

Commercial Vehicle Information Systems and Networks (CVISN) Guide to Electronic Screening. Prepared by staff of the Johns Hopkins University Applied Physics Laboratory (JHU/APL). Washington, DC: Federal Motor Carrier Safety Administration, 2002. Baseline

Version 1.0. POR-99-7193. Full text of current CVISN documents may be downloaded from this site: *http://www.jhuapl.edu/cvisn/Documents/ Document_Nav_Frame_Page.shtml*.

Commercial Vehicle Information Systems and Networks (CVISN) Guide to Safety Information Exchange. Prepared by staff of the Johns Hopkins University Applied Physics Laboratory (JHU/APL). Washington, DC: Federal Motor Carrier Safety Administration, 2002. Full text of current CVISN documents may be downloaded from this site: *http:// www.jhuapl.edu/cvisn/Documents/Document_Nav_Frame_Page.shtml*.

Content, Thomas; Miller, Stanley A. "Sensors Could Sketch out Crash—At Least two Cars had Data Recorders that Investigators can tap for Clues." *Milwaukee Journal Sentinel* [Wisconsin]. 15 Oct. 2002.p. 01.

Evangelista, Benny. "Car-Crash Recorders are Moving from Airliners to Autos." *San Francisco Chronicle* 2 Sept. 2002, p.E1.

Fay, R.; Robinette, R.; Deering, D.; Scott, J. 2002. Using Event Data in Collision Reconstruction. Fay Engineering Corporation, Denver, Colo. 13 pp. *Accident Reconstruction* 2002. Warrendale, SAE, 2002, pp. 1–13. Report No. SAE 2002-01-0535 UMTRI-95830 A01

Flavelle, Dana. "Black box takes shape." *Toronto Star* (Canada) 25 March 2002.

FY 2003 Performance Plan. Washington, DC: National Highway Traffic Safety Administration, 2002. Full text at: *http://www.nhtsa.dot.gov/nhtsa/whatis/planning/perf-plans/2003/ Index.html*.

Garner, Dwight. The Car Black Box. *New York Times Magazine*, 15 December 2002. Vol 152, Issue 52333, p. 69, 1 p.

Gates, K; Rayner, G. "Preliminary Results on the Use of Video Event Data Recorders as Part of a Driver Safety Training Program." *9th World Conference on Intelligent Transport Systems* (Chicago, Illinois) sponsored by ITS America, ITS Japan, ERTICO (Intelligent Transport Systems and Services—Europe). 2002.

General Motors Corporation Delphi–Delco Electronic Systems. Automotive Collision Avoidance Field Operational Test Warning Cue Implementation. Summary Report. Washington, DC: National Highway Traffic Safety Administration, 2002. Report No. DTNH22-99-H-07019. Full text at: *http://www-nrd.nhtsa.dot.gov/pdf/nrd-12/acas/ACAS-WarningCue-2002-05.pdf*.

Golob, T. F. and Regan, A. C. "Trucking Industry Adoption of Information Technology: a Multivariate Discrete Choice Model." Transportation Research Part C: Emerging Technologies. Volume 10, Issue 3, June 2002, pp. 205–228.

Hu, P., Boundy B., Truett, T., Chang, E. and Gordon, S. Cross-Cutting Studies and State-of-the-Practice Reviews: Archive and Use of ITS-Generated Data. Washington, DC: Federal Highway Administration, 2002.

Full text at: *http://www.itsdocs.fhwa.dot.gov/%5CJPODOCS%5CREPTS_TE%5C/ table_of_contents.htm*.

Kraft, M.; Kullgren, A.; Ydenius, A.; Tingvall, C. Influence of crash pulse characteristics on whiplash associated disorders in rear impacts: crash recording in real life crashes. Folksam Research, Stockholm (Sweden) / Sweden, National Road Administration. 9 pp.

Lawrence, J. M.; Wilkinson, C. C.; King, D. J.; Heinrichs, B.E.; Siegmund, G.P. The Accuracy and Sensitivity of Event Data Recorders in Low-speed Collisions. MacInnis Engineering Associates. 11 pp. *Advances in Safety Test Methodology,* Warrendale, SAE, 2002, pp. 1–11. Report No. SAE 2002-01-0679. UMTRI-95613 A01.

Li, H; Ogle, J; and Bachman, W. "Evaluating Driver Behavior and Safety with GPS Event Recorders and GIS." *Proceedings of GIS-T 2002: Melting Down to the Stove Pipes.* 18 pages. Available from American Association of State Highway & Transportation Office, 444 North Capitol Street, NW, Suite 249, Washington, DC 20001.

Machrone, Bill.; More Security, More Privacy. *PC Magazine,* Vol 21 Issue 14 p65

MacWilliams, Joel. "Investigating Real-World Crashes: An Interview with UMTRI's Crash Investigator, Joel MacWilliams." *UMTRI Research Review.* Vol. 33, Number 1, pp. 7–9, ISSN: 07397100. Available at University of Michigan Transportation research Institute, 2901 Baxter Road, Ann Arbor, MI., 48109. January 2002.

McIntosh, Jil. "Big Brother in a Box: Safety vs. Privacy." *Toronto Star* 16 Mar. 2002. pg. G28.

McKinnon, Julie. "Little Black Box Hidden in GMC, other Vehicles." *Blade* [Toledo, Ohio]. 13 Oct. 2002. p. A01.

McNamara, Mary. "Drive Time; Growl if you Think Teenage Monitoring has gone too far." *Los Angeles Times* . 17 July 2002. p.E.2.

Media Release. Anderson Launches Satellite Navigation Policy. Canberra: Commonwealth of Australia Department of Transport and Regional Services, 28 August 2002. Full text of the media release, with abstracts of case studies at: *http://www.ministers.dot-ars.gov.au/ja/releases/2002/august/A106_2002.htm#attachment.* **Note:** *There is an associated policy document,* Positioning For The Future—Australia's Satellite Navigation Strategic Policy. Canberra: Commonwealth of Australia Department of Transport and Regional Services, 2002. Full text at: *http://www.agcc.gov.au/positioning_for_the_future.pdf.*

Murray, Charles J. "Engineering Join in the push for Automotive Black Boxes." *Electronic Engineering Times* 22 April 2002. Issue 1215, p.1.

Mussa, R. N. and Upchurch, J. E. "Simulator Evaluation of Incident Detection by Transponder-equipped Vehicles." *Transportation.* Vol. 29 (3): 287–305, August 2002.

"NHTSA Wants Comments on Event data recorders." *Transport Topics.* Source of Document: Transport Topics, No 3508 (October 21, 2002), p. 31, Northwestern University Transportation Library. 2002.

National Transportation Safety Board Safety Recommendation, H-02-35. NTSB Report No. HS-043 509, Publication date December 19, 2002. 5 pages. Available from National Transportation Safety Board, 490 L'Enfant Plaza, SW, Washington, DC 20594.

Poellabauer, C. and Schwan, K. Power-Aware Video Decoding Using Real-time Event Handlers. Session on Wireless and Mobile Networks. Performance. Proceedings of the Fifth ACM international workshop on Wireless Mobile Multimedia. 2002. Atlanta, Georgia: ACM, 2002. Pages: 72–79. ISBN: 1-58113-474-6.

Real Time Information—Sources and Applications. London, U.K.: The Department for Transport. 2002. Full text at: *http://www.dft.gov.uk/research/rti/pdf/rti.pdf.*

2002 Safety in Numbers Conference Compendium. Georgetown University, Conference Center, Washington, DC: Bureau of Transportation Statistics, U.S. Department of Transportation, 2002. Link to full text of all conference papers and presentations: *http://www.bts.gov/sdi/conferences/2002_01_09/index.html.*

Safety Report. Analysis of Intrastate Trucking Operations. Washington, DC: National Transportation Safety Board, 2002. NTSB/SR-02/01. NTIS PB2002-917001. Notation 7448. Full text at: *http://www.ntsb.gov/publictn/2002/SR0201.pdf.*

Technical Specifications for the Digital Tachograph. Commission Regulation (EC) No 1360/2002 of 13 June 2002. Official Journal of the European Communities. August 5, 2002. Link for a gateway to the full text of the technical specifications: *http://europa.eu.int/comm/transport/themes/land/english/lt_8_en.html.*

Notes: Readers may choose from text versions in several languages, including English. The complete file of the specifications is quite large, and may take some time to download.

Wald, Matthew. "Automakers Block Crash Data Recorders that Could Save Lives, Critics Say." *New York Times*. 29 Dec. 2002. p. 1.24.

White, Joseph B. "How Much Information Should Cars Have? Black Box Technology Accelerates onto Roadways." *Wall Street Journal* 20 March 2002 Vol 239 Issue 55, p. B.5.F.

Ydenius, A. 2002. Influence of Crash Pulse Duration on Injury Risk in Frontal Impacts Based on Real-Life Crashes. Folksam Research, Stockholm (Sweden) 12 p. International IRCOBI Conference on the Biomechanics of Impacts. Proceedings. Munich, Germany, 2002. UMTRI-96237 A12.

2001

A report to Congress on electronic control module technology for use in recording vehicle parameters during a crash. Federal Motor Carrier Safety Administration, Washington, DC. 30 pgs. see: *http://grouper.ieee.org/groups/1616/544thRpt2Congr.pdf.*

Abstract Booklet: 17th International Conference on the Enhanced safety of Vehicles, Amsterdam, The Netherlands. June 4–7, 2001. National Highway Traffic Safety Administration (NHTSA) Report HS-809 243. Held from June 4, 2001 to June 7, 2001. Available from NHTSA, 400 7th Street, SW, Washington, DC 20590.

Arai, Y.; Nishimoto, T.; Ezak, Y; Yoshmoto, K. June, 2001. Accidents and Near-Misses Analysis by Using Video Drive-Recorders in a Fleet Test. *Proceedings of the 17th International Technical Conference on the Enhanced Safety of Vehicles (ESV) Conference*, June 4–7, 2001 Amsterdam, The Netherlands. National Highway Traffic Safety Administration, Washington, DC. HS 809 220, June 2001.

Arai, Y.; Nishimoto, T.; Nakatani, I.; Yoshimoto, K. 2001. Accident Analysis Using Drive-Recorders in a Fleet Test. 6 pp. *JARI Research Journal*, Vol. 23, No. 9, Sept. 2001. pp. 467–471. UMTRI-62470.

Augenstein, J.; Digges, K.; Ogata, S.; Perdeck, E.; Stratton, J.; 2001.Development and Validation of the Urgency Algorithm to Predict Compelling Injuries. Lehman, William, Injury Research Center, Miami, University, Coral Gables, School of Medicine, Fla. 6 pp. ESV; *17th International Technical Conference on the Enhanced Safety of Vehicles, 2001*. Report No. 352. UMTRI-96752 C53.

Croft, Arthur. "Sensing Diagnostic Module (SDM): The Modern Motor Vehicle's Black Box." *Dynamic Chiropractic.* 22, Oct. 2001. Vol. 19, Issue 22 p. 6.

Cullen, David. "Black Boxes Get Big boost." *Fleet Owner.* Oct. 2001. Vol. 96, Issue 10. p. 20.

Cameron, M.; Narayan, S.; Newstead, S.; Ernvall, T., Laine, V.; Langwieder, K. June, 2001. Comparative Analysis of Several Vehicle Safety Rating Systems. *Proceedings of the 17th International Technical Conference on the Enhanced Safety of Vehicles (ESV) Conference,* June 4–7, 2001 at Amsterdam, The Netherlands. National Highway Traffic Safety Administration, Washington, DC. DOT HS 809 220, June 2001. Paper Number 68, 12 pgs.

Carra, J.S.; Stern, S. D. June, 2001. Large Truck Crash Data Collection. *Proceedings of the 17th International Technical Conference on the Enhanced Safety of Vehicles (ESV) Conference,* June 4–7, 2001 at Amsterdam, The Netherlands. National Highway Traffic Safety Administration, Washington, DC. DOT HS 809 220, June 2001. Paper Number 209, 3 pgs.

Chidester, A.; Hinch, J.; Mercer, T.C.; Schultz, K.S. Recording automotive crash event data. National Highway Traffic Safety Administration, Washington, DC/General Motors Corporation, Detroit, MI. 14 pgs.

Chidester, A.C.D; Hinch, J; Roston, T.A. June, 2001. Real World Experiences with Event Data Recorders. *Proceedings of the 17th International Technical Conference on the Enhanced Safety of Vehicles (ESV) Conference,* June 4–7, 2001 at Amsterdam, The Netherlands. National Highway Traffic Safety Administration, Washington, DC. DOT HS 809 220, June 2001. Paper Number 247, 11 pgs.

Chidester, A.C.D.; Isenberg, R. A. June, 2001. Final Report—The Pedestrian Crash Data Study. *Proceedings of the 17th International Technical Conference on the Enhanced Safety of Vehicles (ESV) Conference,* June 4–7, 2001 Amsterdam, The Netherlands. National Highway Traffic Safety Administration, Washington, DC. DOT HS 809 220, June 2001. Paper Number 248, 12 pgs.

Chidester, A.C.D.; Roston, T.A. June, 2001. Air Bag Crash Investigations. *Proceedings of the 17th International Technical Conference on the Enhanced Safety of Vehicles (ESV) Conference,* June 4–7, 2001 at Amsterdam, The Netherlands. National Highway Traffic Safety Administration, Washington, DC. DOT HS 809 220, June 2001. Paper Number 246, 12 pgs.

Correia, J.T.; Iliadis, K.A.; McCarron, E.S.; Smolej, M.A. June, 2001. Utilizing Data from Automotive Event Data Recorders. Hastings, Boulding, Correia Consulting Engineers. *Proceedings of the Canadian Multidisciplinary Road Safety Conference XII;* June 10–13, 2001; London, Ontario. 16 pgs.

Donnelly, B. R.; Galganski, R. A.; Lawrence, R. D.; Blatt, A. 2001. Crash reconstruction and injury-mechanism analysis using event data recorder technology. Veridian Engineering, Buffalo, N.Y. 21 pp. *Association for the Advancement of Automotive Medicine.* 45th annual conference. Proceedings. Barrington, Ill., Association for the Advancement of Automotive Medicine, 2001, pp. 331–351. UMTRI-95176 A22.

Donnely, B.;Galganski, R.A.; Donnelly, B.R.; Blatt, A.; Lombardo, L.V. June, 2001. Crash Visualization Using Real-World Acceleration Data. *Proceedings of the 17th International Technical Conference on the Enhanced Safety of Vehicles (ESV) Conference,* June 4–7, 2001 at Amsterdam, The Netherlands. National Highway Traffic Safety Administration, Washington, DC. DOT HS 809 220, June 2001. Paper Number 357, 10 pp.

Event Data Recorders: summary of findings. National Highway Traffic Safety Administration, Research and Development, Event Data Recorder Working Group, Washington, DC. 101 pp. UMTRI-96641 2001.

Event Data Recorders: summary of the findings by the NHTSA EDR Working Group. Final report. National Highway Traffic Safety Administration, Motor Vehicle Safety Research Advisory Committee, Event Data Recorder Working Group, Washington, DC. 101 pp. UMTRI-95054. Subjects: Accident Investigations; Vehicle Crashworthiness; Data Acquisition Methods; Data Analysis; Recorders; Crash Recorders. 2001.

Gabler, H.C.; DeFuria, J.; Schmalzel, J. L. June 2001. Automated Crash Notification Via the Wireless Web: System Design and Validation. *Proceedings of the 17th International Technical Conference on the Enhanced Safety of Vehicles (ESV) Conference*, June 4–7, at Amsterdam, The Netherlands. National Highway Traffic Safety Administration, Washington, DC. DOT HS 809 220, June 2001. Paper Number 71, 5 pp.

Galganski, R.A; Donnelly, B.R; Blatt, A; Lombardo, L.V. *Proceedings of 17th International Technical Conference on the Enhanced Safety of Vehicles*, held Amsterdam, The Netherlands, 4–7 June 2001. CD ROM. DOT 809220 See: Crash Visualization Using Real-World Acceleration Data. 10 pp. 2001.

Garthe, E. A.; Mango, N. K. 2001. Conflicting Uses of Data From Private Vehicle Data Systems. Garthe Associates, Marblehead, Mass. 15 pp. Intelligent Vehicle Initiative (IVI): Technology and Navigation Systems. Warrendale: SAE, 2001, pp. 79–93. Report No. SAE 2001-01-0804. UMTRI-94222 A10.

German, A.; Comeau, J.L; Monk, B.; McClafferty, K.; Tiessen, P.F; Chan, J. June, 2001. The Use of Event Data Recorders in the Analysis of Real-World Crashes, *Proceedings of the Canadian Multidisciplinary Road Safety Conference XII; June 10–13, 2001*; London, Ontario. 15 pp.

Hansen, Brian. "Black Boxes." *CQ Researcher*, 10/26/2001, Vol 11 Issue 37, p 890, 2 pp.

Hendrie, D.; Lyle, G. June, 2001. Safety Benefits of Improvements in Vehicle Design Since the Introduction of the ANCAP Crash Test Program. *Proceedings of the 17th International Technical Conference on the Enhanced Safety of Vehicles (ESV) Conference*, June 4–7, 2001 at Amsterdam, The Netherlands. National Highway Traffic Safety Administration, Washington, DC. DOT HS 809 220, June 2001. Paper Number 259, 10 pp.

Hill, J.; Thomas, P.; Smith, M.; Byard, N.; Rillie, I. June, 2001. The Methodology of On the Spot Accident Investigations in the UK. *Proceedings of the 17th International Technical Conference on the Enhanced Safety of Vehicles (ESV) Conference*, June 4–7, 2001 at Amsterdam, The Netherlands. National Highway Traffic Safety Administration, Washington, DC. DOT HS 809 220, June 2001. Paper Number 350, 10 pp.

Hook, P. 2001. "Skunk in the Trunk? Journey and Collision Data Recorders: Asset or Liability?" *Traffic Technology International* 2001. (2001).

Kowalick, T. M. May, 2001. Pros and Cons of Emerging Event Data Recorder (EDR) Technologies in the Highway Mode. *Proceedings of the Institute of Electrical and Electronic Engineers (IEEE) VTS 53rd Vehicular Technology Conference*, May 6–9, 2001 at Rhodes, Greece. IEEE catalog number 01CH37202C, ISBN: 0-7803-6730-8. 10 pp.

Kowalick. T. M. June, 2001. Real-World Perceptions of Emerging Event Data Recorder (EDR) Technologies. *Proceedings of the 17th International Technical Conference on the*

Enhanced safety of Vehicles (ESV) Conference, June 4–7, 2001 at Amsterdam, The Netherlands. National Highway Traffic Safety Administration, Washington, DC. DOT HS 809 220, June 2001. Paper Number 146, 8 pp.

Krafft, M.; Kullgren, A.; Lie, A.; Tinggvall, C. June, 2001. Injury Risk Functions for Individual Car Models. *Proceedings of the 17th International Technical Conference on the Enhanced Safety of Vehicles (ESV) Conference,* June 4–7, 2001 at Amsterdam, The Netherlands. National Highway Traffic Safety Administration, Washington, DC. DOT HS 809 220, June 2001. Paper Number 168, 8 pp.

Krafft, M.; Kullgren, A.; Ydenius, A.; Tingvall, C. June, 2001. The Correlation Between Crash Pulse Characteristics and Duration of Symptoms to the Neck—Crash Recording in Real Life Rear Impacts. *Proceedings of the 17th International Technical Conference on the Enhanced Safety of Vehicles (ESV) Conference,* June 4–7, 2001 at Amsterdam, The Netherlands. National Highway Traffic Safety Administration, Washington, DC. DOT HS 809 220, June 2001. Paper Number 174, 7 pp.

Laine, V.; Ernvall, T.; Cameron, M.; Newstead, S. June, 2001. Agressivity Variables and Their Sensitivity in Car Agressivity Ratings. *Proceedings of the 17th International Technical Conference on the Enhanced Safety of Vehicles (ESV) Conference,* June 4–7, 2001 at Amsterdam, The Netherlands. National Highway Traffic Safety Administration, Washington, DC. DOT HS 809 220, June 2001. Paper Number 190, 10 pp.

Linder, A.; Avery, M.; Krafft, M.; Kullgren, A.; Swensson, M.Y. June, 2001. Acceleration Pulses and Crash Severity in Low Velocity Rear Impacts—Real World Data and Barrier Tests. *Proceedings of the 17th International Technical Conference on the Enhanced safety of Vehicles (ESV) Conference,* June 4–7, 2001 at Amsterdam, The Netherlands. National Highway Traffic Safety Administration, Washington, DC. DOT HS 809 220, June 2001. Paper Number 216, 10 pp.

Langwieder, K.; Fildes, B.; Ernvall, T; Cameron, M. June, 2001. Quality Criteria for Crashworthiness Assessment from Real-World Crashes. *Proceedings of the 17th International Technical Conference on the Enhanced safety of Vehicles (ESV) Conference,* June 4–7, 2001 at Amsterdam, The Netherlands. National Highway Traffic Safety Administration, Washington, DC. DOT HS 809 220, June 2001. Paper Number 389, 15 pp.

Masler, Ross. "Emerging Technologies." *New York Law Journal* 3 June 2003.

Mateja, Jim. "Technology can put Cars in a Telling Situation." *Chicago Tribune,* 6 July 2001.

Mooi, H.G.; Galliano, F. June, 2001. Dutch In-Depth Accident Investigation: First Experiences and Analysis Results for Motorcycles and Mopeds. *Proceedings of the 17th International Technical Conference on the Enhanced safety of Vehicles (ESV) Conference,* June 4–7, 2001 at Amsterdam, The Netherlands. National Highway Traffic Safety Administration, Washington, DC. DOT HS 809 220, June 2001. Paper Number 236, 10 pp.

Needham, Peter. "Safety and Efficiency Using Telematics Systems." *European Conference on Vehicle Electronic Systems* (2001: Coventry, England). Vehicle Electronic Systems 2001. VDO Lienzle, UK, Ltd. Available from University of California., Berkeley: Inst Transp Studies Lib Refer to: *http://www.lib.berkeley.edu/ITSL/services.html.*

Needham, P.L. "Collision Prevention: The Role of an Accident Data Recorder (ADR). *Proceedings of the International Conference on Advanced Driver Assistance Systems (ADAS*

2001), held Birmingham, UK, September 2001. ISBN: 0-85296-743-8, Institution of Electrical Engineers (IEE), PO Box 96, Stevenage, Herts SGI 2SD United Kindgom, 2001.

Niessner, Charles W. "Use of Event Data Recorder (EDR) Technology for Roadside Crash Data Analysis Funding Information Announcement." Project 17-24 Funded by National Cooperative Highway Research Program (NCHRP) Washington, DC. Notice date December 31, 2001.

Prasad, A.; Performance of Selected Event Data Recorders. National Highway Traffic Safety Administration. Washington, DC. 20 pp. Report number HS-043 448. Full text at: *http://www-nrd.nhtsa.dot.gov/pdf/nrd-10/EDR/EDR-round-robin-Report.pdf.*

Rayner, G. "I-Witness Black Box Recorder." ITS-IDEA Program Project Final Report. ITS-IDEA Project 84, Final Report. p. 19. Transportation Research Board, 500 Fifth Street, NW, Washington, DC. 2001. Available from: National Technical Information Service, 5285 Port Royal Road, Springfield, VA. 22161.

Rosenbluth, W. June 2001. Investigation and Interpretation of Black Box Data in Automobiles: A Guide to the Concepts and Formats of Computer Data in Vehicle Safety and Control Systems. Jointly published by American Society for Testing and Materials (ASTM) West Conshohocken, PA, and Society of Automotive Engineers (SAE).

Selingo, Jeffrey. "It's the Cars, Not the Tires, That Squeal." *New York Times,* 10/25/2001, Vol 151 Issue 51917, pF1, Op, 1 diagram, 3bw.

Sporner, A.; Kramlick, T. June, 2001. Motorcycle Braking and It's Influence on Severity of Injury. *Proceedings of the 17th International Technical Conference on the Enhanced safety of Vehicles (ESV) Conference,* June 4–7, 2001 at Amsterdam, The Netherlands. National Highway Traffic Safety Administration, Washington, DC. DOT HS 809 220, June 2001. Paper Number 303, 7 pp.

Stewart, Gerald. R. June, 2001. The Role of Innovation and Statistical Methodology in Safety Assessment Projects. *Proceedings of the 17th International Technical Conference on the Enhanced Safety of Vehicles (ESV) Conference,* June 4–7, 2001 at Amsterdam, The Netherlands. National Highway Traffic Safety Administration, Washington, DC. DOT HS 809 220, June 2001. p. 412, 7 pp.

Stoffer, Harry. "Promise and Pitfalls Seen in Black Box." *Automotive News* 17 Sept. 2001Vol. 76, Issue 5948.

Thompson, K.M.; Graham, J.D.; Zeeler, J.W. June, 2001. Risk-Benefit Analysis Methods for Vehicle Safety Devices. *Proceedings of the 17th International Technical Conference on the Enhanced Safety of Vehicles (ESV) Conference,* June 4–7, 2001 at Amsterdam, The Netherlands. National Highway Traffic Safety Administration, Washington, DC. DOT HS 809 220, June 2001. Paper Number 340, 7 pp.

Tingvall, C.; Kullgren, A.; Ydenius, A.; Krafft, M.; 2001. The Correlation between Crash Pulse Characteristics and the Duration of Symptoms to the Neck—Crash Recording in Real-Life Rear Impacts. Folksam Research, Stockholm (Sweden) Swedish National Road Administration, Stockholm. 7 pp. ESV; 17th International Technical Conference on the Enhanced Safety of Vehicles. 2001. Report No. 174. UMTRI-95752 A40.

Ueyama, M. June, 2001. Driver Characteristic Using Driving Monitoring Recorder. *Proceedings of the 17th International Technical Conference on the Enhanced Safety of Vehicles (ESV) Conference,* June 4–7, Amsterdam, The Netherlands. National Highway Traffic

Safety Administration, Washington, DC. DOT HS 809 220, June 2001. Paper Number 426, 10 pp.

Ydenius, A.; Kullgren, A.; 2001. Injury Risk Functions in Frontal Impacts Using Recorded Crash Pulses. Folksam Research, Stockholm, (Sweden) Karolinska Institutet, Stockholm (Sweden) 12 pp. *International IRCOBI Conference on the Biomechanics of Impacts. 2001. Proceedings.* Isle of Man (United Kingdom), 10–12 Oct. 2001, pp. 27-38. UMTRI-95099 A03.

2000

Blatt, A; Donnelly, B; Galganski, R. "The Use of Automated Crash Notification and Black Box Technology to Characterize Crash Severity and Occupant Injury Severity in Real World Crashes." ISATA-Duesseldorf Trade Fair, Epson House, 10C East Street, Epsom, Surrey KT17 1HH United Kingdom. ISBN: 1-902856-10-4, pp. 221–227. 2000.

Goebelbeck, JM. "Crash Data Retrieval Kit Recovers Reconstruction Data from G.M. Black Boxes." Accident Investigation Quarterly. pp. 42–46, ISSN: 10826521. Available from Accident Investigation Quarterly, PO Box 234, Waldorf, MD 20604-0234. 2000.

Goebelbecker, J. M.; Ferrone, C. 2000. Utilizing Electronic Control Module Data in Accident Reconstruction. Triodyne Consulting Engineers, Niles, Ill. 7 pp. Accident Reconstruction: Analysis, Simulation, and Visualization. Warrendale, SAE, 2000, pp. 83–89. Report No. SAE-2000-01-0466. UMTRI-93282 A07.

Grush, Ernie. Ford Motor Company; Research Opportunities With Automotive Crash Recorders, available at: *http://www.ntsb.gov/events/2000/symp_legal/default.htm.*

Hook, P. "Skunk in the Trunk." *Traffic Technology International*, pp. 21–23, ISSN: 1356-9252. UK & International Press, Abinger House, Church Street, Dorking, Surrey RH4 1DF United Kindgom, 2000.

Kahn, J. "High Technology in the Transportation Industry: Is the New Data We Gather Worth All the Costs?" *Transportation Law Journal*. Vol. 28 No. 1 pp. 89–106, ISSN: 0049450X. Available from Transportation Law Journal, Editor-in-Chief, 7039 E 18th Avenue, Denver, CO 80220. 2000.

Kaiser, Rob. "General Motors Moves Beyond Air Bags with Automobile Event Data Recorders." *Chicago Tribune* 23 Feb. 2000.

Krafft, M.; Kullgren, A.; Tingvall, C.; Bostroem, O.; Fredriksson, R. 2000. How Crash Severity in Rear Impacts Influences Short and Long-term Consequences to the Neck. Folksam Research and Development, Stockholm (Sweden)/ Monash University, Accident Research Centre, Clayton, Victoria (Australia)/ Autoliv AB, Vaargaarda (Sweden) 9 pp. Accident Analysis and Prevention, Vol. 32, No. 2, Mar 2000, pp. 187–195. UMTRI-61502.

Kullgren, A.; Krafft, M.; Nygren, AA.; Tingvall, C. 2000. Neck Injuries in Frontal Impacts: Influence of Crash Pulse Characteristics on Injury Risk. Folksam Research and Development, Stockholm (Sweden)/ Karolinska Institutet, Department of Clinical Neuroscience and Family Medicine, Stockholm (Sweden)/ Monash University, Accident Research Centre, Clayton, Victoria (Australia) 9 pp. Accident Analysis and Prevention, Vol. 32, No. 2, Mar 2000, pp. 197–205. UMTRI-61503.

Marsh J. 2000. Ford's New Taurus and Sable; The Safety Network; pp. 4–5; November, 2000.

Massey, Donald. "Proposed on-board Recorders for Motor Carriers: Fostering Safer Highways or Unfairly Tilting the Litigation Playing Field?" *Southern Illinois University Law Journal* Spring 2000. Vol. 24, No. 3.

O'Connell, Dominic. "British System Allows Cars to Provide Information to Drivers." *Sunday Business* [Britain]. 8 Oct. 2000.

Opdam, Pat. "Safe Colors: Much Has Been Made of Black Box Recorders for Motor Vehicles." *Traffic Technology International.* pp. 11–12, 2000. Available at *http:www.lib.berkeley.edu/ITSL/services.html.*

Orr-Munro, T. "Black Boxes to be Fitted to Force Vehicles." *Police Review*, p12 ISSN: 0309-1414 Jane's Information Group, London, United Kingdom. 2000.

Record of the U.S. DOT/National Highway Traffic Safety Administration (NHTSA) Event Data Recorder (EDR) Working Group, Docket NHTSA-00-7699, available at *http://dms.dot.gov.*

Rosenfeld, J.V.; McDermott, F.T.; Laidlaw, J.D.; Cordner,S.M.; Tremayne, A. B. 2000. The Preventability of Death in Road Traffic Fatalities with Head Injury in Victoria, Australia. Royal Children's Hospital, Parkville, Victoria (Australia)/ Melbourne University, Department of Pediatrics (Australia)/ Melbourne University, Department of Surgery (Australia)/ Monash University, Department of Surgery, Melbourne (Australia) / Alfred Hospital, Department of Neurosurgery, Prahan, Victoria (Australia) / Victorian Institute of Forensic Medicine, Southbank (Australia) 8 pp. Journal of Clinical Neuroscience, Vol. 7, No. 6, Nov. 2000, pp. 507–514. Sponsor: Transport Accident Commission, Melbourne, Victoria (Australia); Victoria Department of Human Services (Australia) UMTRI-94852.

Sabow, G. 2000. (IVU Inst). "Driving Data Recorders (FDS) and Young Drivers." Around the World in Two and a Half Days: Lessons from the UK Proceedings (2000).

Symposia Records of the National Transportation Safety Board (NTSB) Transportation Safety and the Law, April 25–26, Washington, DC.

The NTSB held a symposium on issues related to improving transportation safety and the available information in the 21st century. The proceedings can be viewed in their entirety at: *http://www.ntsb.gov/events/2000/symp_legal/default.htm.*

To, H; Choudhry, O.; April, 2000. Mayday Plus Operational Test Evaluation Report. Minnesota Department of Transportation.

Whitten, Daniel L. "NHTSA Looks Into Black Box Technologies and Costs." Transport Topics, no. 3406 (November 6, 2000), p. 54. Source of document: Northwestern University Transportation Library, 2000.

Whitten, Daniel L. "Mandate for On-Board Hours Recorders Pleases No One." Transport Topics, No. 3379 (May 1, 2000), p. 6. Source of document: Northwestern University University Transportation Library, 2000.

Thomas, Andrew. "Car Black Box Privacy Threat." *Register* [United Kingdom].

Stoffer, Harry. "Safety Experts Cautiously Open Automotive Black Box." *Automotive News*, Vol. 74, No. 5873 (May 8, 2000), p. 44.

Wolfson, H. "Coming Soon: Buicks with Black Boxes." *ADTSEA News and Views*. Vol. 6 No. 3. Available from Indiana University of Pennsylvania, Highway Safety Center, Indiana, PA 15705-1092. 2000.

Wouters, P. I. J.; Bos, J. M. J. 2000. Traffic Accident Reduction by Monitoring Driver Behavior with In-Car Data Recorders. Institute for Road Safety Research SWOV, Leidschendam (Netherlands) 8 pp. *Accident Analysis and Prevention*, Vol. 32, No. 5, Sept 2000, pp. 643–650. UMTRI-61880.

1999

deKroes, J. L. "Self Surveillance and Other Tools for Safe Driving." *Proceedings of the Second World Congress on Safety of Transportation: Imbalance Between Growth and Safety?"* The European Commission, Ministry of Transport, Public Works and Water Management, The Netherlands held February 18, 1998 to February 20, 1998. Pagination 548–551.ISBN: 9040718520. Available from Delft University Press, Postbus 98, 2600 MG Delft, The Netherlands. 1999.

Eisenstein, Paul A. "Telematics at a Glance." *Automotive Industries*. Vol. 179, no 11, pp. a6–a12, 1999.

Record of the U.S. DOT/National Highway Traffic Safety Administration (NHTSA) Event Data Recorder (EDR) Working Group, Docket NHTSA-99-5218, available at *http://dms.dot.gov*.

Federal Register, 64 FR 29616 (June 2, 1999) available at *http://www.access.gpo.gov/su_docs/aces/aces140.html*.

Johnson, K; Pool, M. "The Applicability of the Aircraft's Black Box to Other Modes of Transport." *Proceedings of the Second World Congress on Safety of Transportation: Imbalance Between Growth and Safety?"* The European Commission, Ministry of Transport, Public Works and Water Management, The Netherlands held February 18, 1998 to February 20, 1998. pp. 561–565. Available from Delft University Press, Postbus 98, 2600 MG Delft, The Netherlands. 1999.

National Transportation Safety Board (NTSB) International Symposium on Transportation Recorders. May 3–5, 1999, Washington, DC. Goal: To Share Knowledge and Experience Gained from the Use of Recorded Information to Improve Transportation Safety and Efficiency.

See the following papers: (16)

1.Brooks, Jeffrey L. (*Smiths Industries Flight Data/Cockpit Voice Recorders.*) Available at *http://www.ntsb.gov/events/symp_rec/proceedings/authors/brooks.htm* and *http://www.ntsb.gov/events/symp_rec/proceedings/authors/brooks.pdf*.

2Carroll, Joseph A.; Fennell, Michael D. (*An Autonomous Data Recorder for Field Testing.*) Available at *http://www.ntsb.gov/events/symp_rec/proceedings/authors/carroll.htm* and *http://www.ntsb.gov/events/symp_rec/proceedings/authors/carroll.pdf*.

3. Champion, Howard; Augenstein, J.S.; Cushing, B.; Digges, K.H.; Hunt, R.; Larkin, R.; Malliaris, A.C.; Sacco, W.J.; Siegel, J.H. (*Reducing Highway Deaths and Disabilities with Automatic Wireless Transmission of Serious Injury Probability Ratings from Crash Recorders to Emergency Medical Services Providers.*) Available at *http://www.ntsb.gov/events/*

symp_rec/proceedings/authors/champion.htm and *http://www.ntsb.gov/events/symp_rec/ proceedings/authors/champion.pdf.*

4. Chidester, Augustus; Hinch, John; Mercer, Thomas C.; Schultz, Keith S. (*Recording Automotive Crash Event Data.*) Available at *http://www.ntsb.gov/events/symp_rec/proceedings/authors/chidester.pdf.*

5. Dobranetski, Ed; Case, Dave. (*Proactive Use of Recorded Data for Accident Prevention.*) Available at *http://www.ntsb.gov/events/symp_rec/proceedings/authors/dobranetski.htm* and *http://www.ntsb.gov/events/symp_rec/proceedings/authors/dobranetski.pdf.*

6. Dole, Les. (*On-Board Recorders: The "Black Boxes" of the Trucking Industry.*) Available at *http://www.ntsb.gov/events/symp_rec/proceedings/authors/dole.htm* and *http:// www.ntsb.gov/events/symp_rec/proceedings/authors/dole.pdf.*

7. Durkin, Matthew. (*Digital Audio Recorders Life Savers, Educators, and Vindicators.*) Available at *http://www.ntsb.gov/events/symp_rec/proceedings/authors/durkin.htm* and *http://www.ntsb.gov/events/symp_rec/proceedings/authors/durkin.pdf.*

8. Evans, Gregory L. (*Transportation Event Recorder Data: Balancing Federal Public Policy and Privacy Rights.*) Available at *http://www.ntsb.gov/events/symp_rec/proceedings/ authors/evans.htm* and *http://www.ntsb.gov/events/symp_rec/proceedings/authors/ evans.pdf.*

9. Fenwick, Lindsay. (*Security of Recorded Information.*) Available at *http://www.ntsb.gov/ events/symp_rec/proceedings/authors/fenwick.htm* and *http://www.ntsb.gov/events/ symp_rec/proceedings/authors/fenwick.pdf.*

10. Horne, Mike. (*Future Video Accident Recorder.*) Available at *http://www.ntsb.gov/events/ symp_rec/proceedings/authors/horne.htm* and *http://www.ntsb.gov/events/symp_rec/pro- ceedings/authors/horne.pdf.*

11. Kowalick, Thomas Michael. (*Proactive Use of Highway Recorded Data Via an Event Data Recorder (EDR) to Achieve Nationwide Seat Belt Usage in the 90th Percentile by 2002.*) Available at *http://www.ntsb.gov/events/symp_rec/proceedings/authors/kowalick.htm and ck.pdf.*

12. Lehmann, Dr. Gerhard; Reynolds, Tony. (*The Contribution of Onboard Recording Sys- tems to Road Safety and Accident Analysis.*) Available at *http://www.ntsb.gov/events/ symp_rec/proceedings/authors/lehmann.htm* and *http://www.ntsb.gov/events/symp_rec/ proceedings/authors/lehmann.pdf.*

13. Menig, Paul; Coverdill, Cary.(*Transportation Recorders on Commercial Vehicles.*) Available at *http://www.ntsb.gov/events/symp_rec/proceedings/authors/menig.htm* and *http://www.ntsb.gov/events/symp_rec/proceedings/authors/menig.pdf.*

14. Scaman, R. Jeffrey. (*The Benefits of Vehicle-Mounted Video Recording Systems.*) Available at *http://www.ntsb.gov/events/symp_rec/proceedings/authors/scaman.htm* and *http://www.ntsb.gov/events/symp_rec/proceedings/authors/scaman.pdf.*

15. Thomas, Neill L.; Freund, Deborah M. (*On-Board Recording for Commercial Motor Vehicles and Drivers: Microscopic and Macroscopic Approaches.*) Available at *http:// www.ntsb.gov/events/symp_rec/proceedings/authors/thomas.htm* and *http://www.ntsb.gov/ events/symp_rec/proceedings/authors/thomas.pdf.*

16. Thompson, Michael H. (*A Vision of Future Crash Survivable Recording Systems.*) Available at *http://www.ntsb.gov/events/symp_rec/proceedings/authors/thompson.htm* and *http://www.ntsb.gov/events/symp_rec/proceedings/authors/thompson.pdf*.

Posters (4)

Posters are available in HTML (default) or PPT format. Graphics have been included, whenever possible, in the HTML version, but PPT will have the higher-quality image, and requires a PowerPoint viewer.

1. Bolte, Kristin; Jackson, Lawrence; Roberts, Vernon; McComb, Sarah. (*Accident Reconstruction/Simulation with Event Recorders.*) Available at *http://www.ntsb.gov/events/symp_rec/proceedings/posters/bolte/bolte.htm* and *http://www.ntsb.gov/events/symp_rec/proceedings/posters/bolte.ppt*.

2. Kowalick, Thomas Michael. (*Seat Belt Event Data Recorder (SB-EDR).*) Available at *http://www.ntsb.gov/events/symp_rec/proceedings/posters/kowalick/sld001.htm* and *http://www.ntsb.gov/events/symp_rec/proceedings/posters/kowalick.ppt*.

3. Mackey, John J.; Brogan, Christopher J.; Bates, Edward; Ingalls, Stephen; Howlett, Jack. (*Mobile Accident Camera.*) Available at *http://www.ntsb.gov/events/symp_rec/proceedings/posters/mackey/sld001.htm* and *http://www.ntsb.gov/events/symp_rec/proceedings/posters/mackey.ppt*.

4. Scaman, R. Jeffrey. (*The Benefits of Vehicle-Mounted Video Recording Systems.*) Available at *http://www.ntsb.gov/events/symp_rec/proceedings/posters/scaman/sld001.htm* and *http://www.ntsb.gov/events/symp_rec/proceedings/posters/scaman.ppt*.

The proceedings from the NTSB symposium can be viewed in their entirety at: *http://www.ntsb.gov/events/symp_rec/symp_rec.htm*.

Kowalick, T. M. June, 1999. Perceptions of College Students Regarding Utilization of Transportation Recorders in the Highway Mode, Sandhills Community College, Pinehurst, NC, 651 pgs. Available at *http://normandy.sandhills.cc.nc.us/research/recorders.pdf*.

Kullgren, A. 1999. Crash-Pulse Recorders in Real-Life Accidents: Influence of Change of Velocity and Mean and Peak Acceleration on Injury Risk in Frontal Impacts. Folksam Research Foundation, Stockholm (Sweden) Karolinska Hospital, Department of Clinical Neuroscience, Stockholm (Sweden) 8 pp. Crash Prevention and Injury Control, Vol. 1, No. 2, Oct 1999, pp. 113–120. UMTRI-61230.

Kullgren, A. 1999. (Folksam Res, Sweden, Sweden Thompson, R. (Chalmers Univ Technol, and Sweden Krafft, T. M. (Folksam Res. "The Effect of Crash Pulse Shape on AIS1 Neck Injuries in Frontal Impacts." *Proceedings of the 1999 IRCOBI Conference on the Biomechanics of Impact*, September 23–24, 1999, Sitges, Spain. 1999. pp231–42: 18 Refs.

Popov, A. A.; Cole, D. J.; Cebon, D.; Winkler, C. B. 1999. Energy Loss in Truck Tyres and Suspensions. Michigan University, Ann Arbor, Transportation Research Institute, Engineering Research Division. 12 pp. Sponsor: Engineering and Physical Sciences Research Council (United Kingdom); Dunlop Tyre and Rubber, Birmingham (England); Cambridge Vehicle Dynamics Consortium. UMTRI-93076.

Roszbach, R.; Heidstra, J.; Wouters, P. I. J. 1999. Data Recorders in Voertuigen; [Data Recorders in Vehicles] Netherlands, Rijkswaterstaat, Delft. 61 pp. Sponsor: Institute for

Road Safety Research SWOV, Leidschendam (Netherlands) Report No. R-99-26. UMTRI-93452.

Siiri, Bill. "Crash—Data Recorders for Your Car." *Electronics Now* March, 1999. Vol. 70, Issue 3 pp. 25.

Ydenius, A.; Kullgren, A. 1999. Pulse Shapes and Injury Risks in Collisions with Roadside Objects: Result from Real-Life Impacts with Recorded Crash Pulses. Folksam Research Foundation, Stockholm (Sweden) 8 pp. *International IRCOBI Conference on the Biomechanics of Impacts.* 1999. Proceedings. Bron (France), 1999. pp. 435–442. UMTRI-92961 A26.

1998

Federal Register, 63 FR 60222270 (November 9, 1998) available at *http://www.access.gp.*

Kullgren, A.; Ydenius, A.; Tingvall, C. 1998. Frontal Impacts with Small Partial Overlap: Real Life Data from Crash Recorders. Folksam Research (Sweden) Karolinska Institutet, Department of Clinical Neuroscience and Family Medicine, Stockholm (Sweden) Swedish National Road Administration. 10 pp. *International Technical Conference on Experimental Safety Vehicles. Sixteenth. Proceedings.* Volume I. Washington, DC, NHTSA, 1998. pp. 259–268. Report No. 98-S1-O-13. UMTRI-92420 A38.

Krafft, M.; Kullgren, A.; Tingvall, C. 1998. Crash Pulse Recorder in Rear Impacts—Real Life Data. Folksam Research Foundation, Stockholm (Sweden)/ Karolinska Institutet, Stockholm (Sweden) Statens Vaegoch Trafikinstitut, Linkoeping (Sweden) 7 pp. *International Technical Conference on Experimental Safety Vehicles. Sixteenth. Proceedings.* Volume II. Washington, DC, NHTSA, 1998. pp. 1256–1262. Report No. 98-S6-O-10. UMTRI-92421 A50.

Matsumoto, K. 1998. Trends and Priorities in Motor Vehicle Safety for the 21st century: Japan. Japan Ministry of International Trade and Industry, Tokyo. 3 pp. *International Technical Conference on Experimental Safety Vehicles. Sixteenth. Proceedings.* Volume I. Washington, DC, NHTSA, 1998. pp. 85–87. UMTRI-92420 A15.

Melvin, J. W.; Baron, K. J.; Little, W. C.; Gideon, T. W.; Pierce, J. 1998. Biomechanical Analysis of Indy Race Car Crashes. General Motors Corporation, Detroit, Mich./ Kestrel Advisors, Inc. 20 pp. *Stapp car crash conference. Forty-second. Proceedings.* Warrendale, SAE, 1998. pp. 247–266. Report No. SAE 983161. UMTRI-91882 A17.

Phen, Dowdy, Ebbeler, Kim, Moore, and VanZandt; Advanced Air Bag Technology Assessment; JPL Publication 98-3; April 1998. This report can be found on the NASA Jet Propulsion Laboratory web site: *http://csmt.jpl.nasa.gov/air bag/contents.html.*

Steeples, B. "Spy Within." *Testing Technology International,* No. 2, p55–60 ISSN: 1461-8966 September 1998, UK & International Press, Dorking, Surrey, UK.

Ueyama, M.; Ogawa, S.; Chikasue, H.; Muramatu, K. 1998. Relationship Between Driving Behavior and Traffic Accidents—Accident Data Recorder and Driving Monitor Recorder. National Research Institute of Police Science, Tokyo (Japan)/ Yazaki Meter Corporation (Japan) 8 pp. *International Technical Conference on Experimental Safety Vehicles. Sixteenth. Proceedings.* Volume I. Washington, DC, NHTSA, 1998. pp. 402–409. Report No. 98-S1-O-06. UMTRI-92420 A53.

Wright, P. G. 1998. The Role of Motorsport Safety. Federation Internationale de l'Automobile (England) 6 pp. *International Technical Conference on Experimental Safety Vehicles. Sixteenth. Proceedings.* Volume II. Washington, DC, NHTSA, 1998. pp. 1263–1268. Report No. 98-S6-O-12. UMTRI-92421 A51.

1997

Andersson, U.; Koch, M.; Norin, H. 1997. The Volvo Digital Accident Research Recorder (DARR) Converting Accident DARR-Pulses Into Different Impact Severity Measures. Volvo Car Corporation, Automotive Safety Centre, Goeteborg (Sweden) 20 pp. *International IRCOBI Conference on the Biomechanics of Impact. 1997. Proceedings.* Hannover, IRCOBI, 1997. pp. 301–320. UMTRI-92418 A19.

"Colloquium on Monitoring of Driver and Vehicle Performance" *Digest (Institution of Electrical Engineers); No* 1997, no. 122. (1997).

Berg, F.; Alexander, M. 1997. Uwe, Bergisch Gladbach Bundesanstalt Fursstrassenwesen, and Berichte Der Bundesanstalt Fur Strassenwesen. Fahrzeugtechnik. "Accident Data Recorders as a Source of Information for Accident Research in the Pre-Crash Phase" HEFT (1997).

Byrne, R. H.; Pletta, J. B.; Case, R. P.; Klarer, P. R.; Campbell, K. L.; Blower, D. 1997. Commercial Vehicle Incident Monitors. Sandia National Laboratories, Albuquerque, N.M./ Michigan University, Ann Arbor, Transportation Research Institute, Center for National Truck Statistics. 243 pp. Sponsor: Federal Highway Administration, Office of Motor Carriers, Washington, DC. UMTRI-91197.

Faith, N. Crash: The Limits of Car Safety. Boxtree, 25 Eccleston Place, London, SW1W 9NF United Kindgom. ISBN: 0-7522-1192-7, 186 pp. 1997.

Wouters, P.I.J. 1997. (SWOV, Netherlands, and Netherlands BOS JMJ) The Impact of Driver Monitoring With Vehicle Data Recorders on Accident Occurrence: Methodology and Results of a Field Trial in Belgium and The Netherlands. (R-97-8) 64 pgs; 9 Refs.

1996

Korner, J. 1996. (Volvo Car Corp, Sweden. "The Safety Philosophy Guiding Car Design." *Proceedings of the Fifth World Congress of the International Road Safety Organization— Marketing Traffic Safety,* 3–6 October 1994, Cape Town, Republic of South Africa. *1996.* pp. 319–26: 10 Refs.

Lehmann, G. 1996. The Features of the Accident Data Recorder and its Contribution to Road Safety. VDO Kienzle GmbH, Villingen-Schwenningen (Germany) 4 pp. *International Technical Conference on Enhanced Safety of Vehicles. Fifteenth Proceedings.* Volume 2. Washington, DC, National Highway Traffic Safety Administration, 1996. pp. 1565–1568. Report No. 96-S9-W-34. UMTRI-91346 A54.

Melvin, J. W.; Baron, K. J.; Little, W. C.; Pierce, J.; Trammell, T. R. 1996. Investigation of Indy Car Crashes Using Impact Recorders. General Motors Corporation, Research and Development Center, Warren, Mich./ General Motors Corporation, Motorsports, Warren, Mich./ Championship Automobile Racing Teams. 17 pp. *Motorsports Engineering Conference Proceedings.* 1996. Volume 1: Vehicle Design Issues. Warrendale, SAE, 1996. pp. 127–143. Report No. SAE 962522. UMTRI-89565 A02.

The 7th Westminister Lecture on Transport Safety. "A Holistic View of Automotive Safety." 1996 17pp. (1996).

Ueyama, M.; Beppu, S.; Koura, M. 1996. Automatic Recording System and Traffic Accidents at Uncontrolled Intersections. National Research Institute of Police Science, Tokyo (Japan)/ Mitsubishi Electric Corporation (Japan) 11 pp. *International Technical Conference on Enhanced Safety of Vehicles. Fifteenth Proceedings*. Volume 2. Washington, DC, National Highway Traffic Safety Administration, 1996. pp. 1476–1486. Report No. 96-S9-O-17. UMTRI-91346 A44.

1995

Cambourakis, G. "Black Box for Surface Vehicles: Argos Anaylsis and Software Development." *IEEE International Symposium on Industrial Electronics*, (1995: Athens, Greece) ISIE '95: Proceedings Vol. 2. pp. 564–568.

Fincham, W.F; Kast. A.; Lambourn, R.F. 1995. *The Use of a High Resolution Accident Data Recorder in the Field*; Paper No. 950351; SAE.

Korner, J. "Car safety and Its Future Potential." *International Forum on Road Safety Research*, Bangkok, Thailand, October 25–27, 1995 (VTI KONFERENS) Vol.5a:1, pp. 61–73 ISSN 1104-7267. Statens Vaeg—Och Transport Forsknings Institut, Linkoeping, Sweden 1995.

Russell, E. "Black Boxes for Driver-Aided Vehicles." *Automotive Engineer*, Vol. 19, No. 6, pp. 30–31 ISSN: 0307-6490, Mechanical Engineering Publications Limited, Edmunds, United Kingdom, 1995.

1994

Finchman, W; Fowkes, M. "SAMOVAR Uses a Specialized Vehicle Data Recorder to Aid Traffic Reconstruction." *Proceedings of the First World Congress on Applications of Transport Telematics and Intelligent Vehicle-Highway Systems*, November 30–3 December 1994, Paris France, Vol. 4, P1975-82 ISBN: 0-89006-825-9, ENTRICO, Brussels, Belgium 1994.

Kullgren, A.; Lie, A.; Tingvall, C. 1994. The Use of Crash Recorders in Studying Real-Life Accidents. Chalmers Tekniska Hoegskola, Goeteborg, Sweden. 7 pp. International Technical Conference on Enhanced Safety of Vehicles. Fourteenth. Proceedings, Volume 1 Washington, DC, National Highway Traffic Safety Administration, 1994 pp. 856–862. UMTRI-88120 A79.

Norin, H.; Koch, M.; Magnusson, H. 1994. Estimating Crash Severity in Frontal Collisions Using the Volvo Digital Accident Research Recorder (DARR). Volvo Car Corporation, Goeteborg, Sweden. 7 pp. ISATA International Symposium on Automotive Technology and Automation, 27th. *Proceedings for the Dedicated Conference on Road and Vehicle Safety*. Croydon, Automotive Automation Ltd., 1994. pp. 409–415. Report No. 94SF024. UMTRI-87370 A28.

Williams, M.; Hoekstra, E. 1994. Comparison of Five On-Head, Eye-Movement Recording Systems. Final report. Michigan University, Ann Arbor, Transportation Research Institute. 88 pp. Sponsor: Michigan University, Ann Arbor, IVHS Industrial Advisory Board. Report No. UMTRI-94-11. UMTRI-87344.

Van Der Sluis, J. "Electronics in Heavy Vehicles." Stichting Wetenschappelijk Onderzoek Verkeersveiligheid SMOV, Leidschendam, Netherlands 1994.

1993

Aldman, B.; Kullgren, A.; Lie, A.; Tingvall, C. 1993. Crash Pulse Recorder (CPR)—Development and Evaluation of a Low Cost Device for Measuring Crash Pulse and Delta-V. Folksam Research and Development, Stockholm, Sweden/ Chalmers Tekniska Hoegskola, Goeteborg, Sweden. 5 pp. *International Technical Conference on Experimental Safety Vehicles. Thirteenth. Proceedings.* Volume I. Washington, DC, NHTSA, 1993. pp. 188–192. UMTRI-85231 A19.

Lambourn R. F., 1993. 525 School Street SW, Suite 410, Washington, DC 20024 USA Institute of Transportation Engineers. "Road Accident Investigation as a Branch of Forensic Science." Conference Title: Compendium of Technical Papers, ITE, 63rd Annual Meeting Location: The Hague, Netherlands. Sponsored by: Institute of Transportation Engineers. Held: 19930919-19930922. 1993, no. 09. pp. 438–442 (1993): 21 Refs.

1992

Cheng, C. H.; Nachtsheim, C. J.; Benson, P. G. 1992. Statistical Methods for Optimally Locating Automatic Traffic Recorders. Ohio State University, Columbus/ Minnesota University, Minneapolis. 132 pp. Sponsor: Transportation Department, Washington, DC.; Mountain-Plains Consortium. Report No. MPC 92-14. UMTRI-84774.

1991

Salomonsson, O.; Koch, M. 1991. Crash Recorder for Safety System Studies and as a Consumer's Product. Mannesmann Kienzle, Germany/ Volvo Car Corporation, Goeteborg, Sweden. 13 pp. Frontal Crash Safety Technologies for the 90's. Warrendale, SAE, 1991. pp. 21–33. Report No. SAE 910656. UMTRI-80924 A03.

1990

Texas Department of Transportation, 125 East 11th Street Austin TX 78701-2483. "National Traffic Data Acquisition Technologies Conference, Austin, Texas, August 26–30, 1990. PROCEEDINGS." *Conference Title: National Traffic Data Acquisition Technologies Conference* Location: Austin, Texas. Sponsored by: American Society for Testing and Materials; Texas A&M University; University of TX; and Federal Highway Administration. Held: 19900826-19900830. 1990, no. 08. pp. 432 (1990): Phots., Figs., Tabs., Refs.

1989

Adiv, A.; Ervin, R. D. 1989. Examination of Features Proposed for Improving Truck Safety. Final report. Michigan University, Ann Arbor, Transportation Research Institute. 95 pp. Sponsor: Michigan Department of Transportation, Lansing. Report No. UMTRI-89-2. UMTRI-78350.

1988

Panik, F. 1988. Future Aspects in Automotive Electronics. Daimler-Benz, AG, Stuttgart, Germany FR. 54 pp. UMTRI-79073.

Tumbas, N.S; Smith, R.A. 1988. Measuring Protocol for Quantifying Vehicle Damage from an Energy Basis Point of View; SAE 880072.

1987

Panik, F.; Hamm, L.; Reister; Voy (1987) Einfluss der Elektronik auf den Automobilverkehr der Zunkunft; Influence of Electronics on Automobile Traffic of the Future. Daimler-Benz, AG, Stuttgart, Germany FR. 40 pp. UMTRI-79072.

Wilson, F. R. 1987. Measurement of Collision Avoidance Times. *1987 Annual Conference Proceedings: Roads and Transportation Association of Canada*. B41–B61 (14 Refs.) Roads and Transportion Association of Canada, Ottawa, Ontario, Canada.

1986

Volkmar, H.; Koch, S.; Weber, R. 1986. Erhebung und analyse von Pkw-Fahrleistungsdaten mit Hilfe eines mobilen Datenerfassungssystems.; Acquiring and Analyzing Passenger Car Performance Data Using a Mobile Data Acquisition System. Infratest Sozialforschung, Germany FR/ Mannesman Kienzle, Germany FR. 76 pp. Sponsor: Forschungsvereinigung Automobiltechnik e.V., Frankfurt, Germany FR. Report No. 61. UMTRI-76304.

1985

Held, T. H. 1985. The Potential Use of Optical Videodiscs in Automotive Navigational Systems: a Prototype System. MetaMedia Systems, Inc., Germantown, MD. 3 pp. Brown, I. D., Goldsmith, R., Coombes, K., and Sinclair, M. A., eds. Ergonomics International 85. Philadelphia, Taylor and Francis, 1985. pp. 433–435. Report No. E5/3. UMTRI-74960.

1984

Theodorsson, P. "A Black Box for Every Vehicle?" *Journal of Traffic Medicine*, Vol. 12, No. 2, p23–26 ISSN: 03455564 International Association of Accident & Traffic Medicine, Stockholm Sweden. 1984.

Winkler, C. B.; Campbell, J. D.; Hagan, M. R. 1984. Vehicle Motion Measurement Technology. Final report. Michigan University, Ann Arbor, Transportation Research Institute. 63 pp. Sponsor: General Motors Corporation, Proving Ground Section, Milford, Mich. Report No. UMTRI-84-20. UMTRI-71951.

1983

Bacon, G. C; Petty, S.P.F. "Specification for On-Board Crash Test Data Acquisition System: Report of the Task Force On-Board Crash Test Data Acquisition System." Society of Environmental Engineers Journal, Vol. 22–3, No. 98, pp. 14–17 ISSN: 0374356X. Modino Press Limited, London, England, 1983.

1982

Baker, W. T. 1982. Photologging. Federal Highway Administration, Washington, DC. 44 pp. National Cooperative Highway Research Program Synthesis of Highway Practice, No. 94, Nov 1982. Sponsor: American Association of State Highway and Transportation Officials, Washington, DC. UMTRI-55285.

Fraser. P. J. 1982. The ARRB Road Users Data Acquisition System (RUDAS) Australian Road Research Board, Vermont South. 21 pp. Report No. ATM No. 14. UMTRI-47931.

1981

Blauvelt, A. A.; Klein, R. H.; Peters, R. A. 1981. Instrumentation for Measuring Pavement-Vehicle Interaction. Volume III: Kennedy Co. Operation and Maintenance Manual, Formatter and Digital Tape Transport. Final report. Systems Technology, Inc., Hawthorne, Calif. 226 pp. Sponsor: Federal Highway Administration, Structures and Applied Mechanics Division, Washington, DC. Report No. TM-1109-1/ FHWA-RD-80-077. UMTRI-46632.

Blauvelt, A. A.; Klein, R. H.; Peters, R. A. 1981. Instrumentation for Measuring Pavement-Vehicle Interaction. Volume II: Digalog Systems Operation and Maintenance Manual, Data Acquisition System, model DLI 203. Final report. Systems Technology, Inc., Hawthorne, Calif. 98 pp. Sponsor: Federal Highway Administration, Structures and Applied Mechanics Division, Washington, DC. Report No. TM-1109-1/ FHWA-RD-80-076. UMTRI-46631.

Blauvelt, A. A.; Klien, R. H.; Peters, R. A. 1981. Instrumentation for Measuring Pavement-Vehicle Interaction. Volume I: System Description, Operation, Calibration and Maintenance Manual. Final report. Systems Technology, Inc., Hawthorne, Calif. 84 pp. Sponsor: Federal Highway Administration, Structures and Applied Mechanics Division, Washington, DC. Report No. TM-1109-1/ FHWA-RD-80-075. UMTRI-46630.

Bowden, T. J.; Reichert, J. K.; Landolt, J. P. 1981. The Data Acquisition System at the DCIEM Impact Studies Facility. Defence and Civil Institute of Environmental Medicine, Downsview, Ontario, Canada. 8 pp. Report No. SAE 810812. UMTRI-46023.

Bowersock, R. G.; Dupree, J. F.; Bock, D. T. 1981. A Microcomputer-Based On-Vehicle Data Acquisition System. Ford Motor Company, Dearborn, Mich. 11 pp. Report No. SAE 810811. UMTRI-46024.

Fouts, P. G.; Griggs, G. A.; Holdren, E. J. 1981. Digital Recording of Vehicle Crash Data. Chrysler Corporation, Highland Park, Mich. 13 pp. Report No. SAE 810810. UMTRI-46006.

Klaber, K. 1981. Advanced Automotive Crash Recorder Design Development and Test Analysis. National Highway Traffic Safety Administration, Washington, DC. 10 pp. Report No. SAE 810809. UMTRI-46008.

Reichert, J. K.; Landolt, J. P. 1981. Digital and Analog Filters for Processing Impact Test Data. Defence and Civil Institute of Environmental Medicine, Downsview, Ontario, Canada. 11 pp. Report No. SAE 810813. UMTRI-46022.

Scanlon, M. "The Future is Now." *Ward's Auto World*, Vol. 17, No. 11, Report No. HS-032 735, p44. Ward's Communications, Incorporated, Detroit Michigan 1981.

Thatcher, C. D. 1981. Advanced Recorder Design and Development. Final report. Dynamic Science, Inc., Phoenix, Ariz. 187 pp. Sponsor: National Highway Traffic Safety

Administration, Washington, DC. Report No. 8314-80-213/ DOT/HS 805 914. UMTRI-46293.

1980

Waddell, R. L. "Automotive Electronics: The Black Box Comes of Age." *Ward's Auto World*, Vol. 16, No.11, Report HS-030 895, p 23–26. Ward's Communications, Incorporated, Detroit Michigan 1980.

1979

Green, P. B. MacMillan, M. "A Portable Event Recorder." *Traffic Engineering and Control*, Vol. 20, No. 2, pp. 594–595 ISSN: 00410683. Printerhall Limited, London, December 1979.

O'Neill, B.; Wong, J. 1979. A Laboratory Evaluation of a Low Cost Motor Vehicle Crash Recorder. Insurance Institute for Highway Safety, Washington, DC. 7 pp. Accident Analysis and Prevention, Vol. 11, No. 1, March 1979, pp. 43–49. UMTRI-54119.

Ruschmann, P. A.; Carroll, H. O.; Greyson, M.; Joscelyn, K. B. 1979. An Analysis of the Potential Legal Constraints on the Use of Mechanical Devices to Monitor Driving Restrictions. Final report. Highway Safety Research Institute, Ann Arbor, Mich. 56 pp. Sponsor: National Highway Traffic Safety Administration, Washington, DC. Report No. UM-HSRI-79-65/ DOT/HS 805 523. UMTRI-44938.

Sherwin, J. R.; Kerr, J. D. 1979. Advanced Recorder Design Development. Final Report. Teledyne Geotech, Garland, Tex. 46 pp. Sponsor: National Highway Traffic Safety Administration, Washington, DC. Report No. DOT/HS 805 081. UMTRI-43051.

Wyman, J. H. 1979. Event Recorder as a Turning Movement Indicator. Maine Department of Transportation, Augusta, ME. Report Number: IM-3, 18 pgs (5 photos., Figs).

1978

Backaitis, S. H. 1978. Evaluation of New Instruments for Measurement of Differential Crash Velocity and for Sensing the Threshold of Critical Crash Intensity. National Highway Traffic Safety Administration, Office of Motor Vehicle Programs, Washington, DC. 20 pp. International Congress on Automotive Safety. Fifth. Proceedings. Washington, DC, NHTSA, March 1978. pp. 427–446. UMTRI-40399 A24.

Wolf, R. J. 1978. A Solid-State Digital Data Recorder for Monitoring Automotive Crash Environments. Final report. Kaman Sciences Corporation, Colorado Springs, CO. 73 pp. Sponsor: National Highway Traffic Safety Administration, Washington, DC. Report No. DOT/HS-803 666. UMTRI-41371.

1977

Damkot, D. K.; Geller, H. A.; Whitmore, D. G. 1977. Measuring Driver Performance: Instrumentation, Software, and Application. Vermont University, Burlington. 7 pp. Sponsor: National Institute on Alcohol Abuse and Alcoholism, Rockville, MD. Report No. SAE 770813. UMTRI-38078.

Glen, M.G.M; Powell, D.G. 1977. A Low-Cost, Portable Event-Recording System. *Traffic Engineer Control*. 1977/11. pgs 424–6 (1 photo; 3 figs.; 6 refs.).

Kaye, A. M.; Sandover, J.; Thomas, P. D. 1977. Apparatus for Field Studies of Man at Work. London School of Hygiene and Tropical Medicine, England/ Loughborough University of Technology, Leicestershire, England. 2 pp. *Journal of Physiology*, Vol. 268, No. 1, June 1977, pp. 5P–6P. Sponsor: Medical Research Council, London, England; Transport and Road Research Laboratory, Crowthorne, England. UMTRI-38402.

Richter, V.; Kramer, M. 1977. Digitale Messdatenaufnahme und—verarbeitung bei Fussgaenger—Fahrzeug-Unfallexperimenten; Digital Data Collection and Processing in Pedestrian/Vehicle Accident Experiments. Berlin Technische Universitaet, Institut fuer Landverkehrsmittel, Germany FR. 3 pp. ATZ, 79. Jahrgang, Nr. 11, Nov 1977, pp. 509-510, 513. UMTRI-53643.

Strickland, L. R.; Wood, P. 1977. TRI-MET Automated Fare Billing System. Mitre Corporation, Metrek Division, McLean, Va. 48 pp. Sponsor: Urban Mass Transportation Administration, Washington, DC. Report No. MTR-7582 Rev. 2. UMTRI-40497.

1976

1976. Fundamental Consideration on the Generation of Data for the Relation Between Vehicle Handling and Accident Avoidance with the Aid of Drive Recorders. Revised. International Organization for Standardization, Geneva, Switzerland. 16 pp. Report No. ISO/ TC 22/SC 9 Germany-6. UMTRI-34934.

1976. On-Board Computer Testing. 4 pp. Automotive Engineering, Vol. 84, No. 11, Nov 1976, pp. 30–33. UMTRI-53122.

1976. Static Evaluation of Air Cushion Deployment Effects on the Memory Retention of the Solid-State Digital Recorder System. Final report. Kaman Sciences Corporation, Colorado Springs, Colo. 29 pp. Sponsor: National Highway Traffic Safety Administration, Washington, DC. Report No. K-76-64-U(R)/ DOT/HS 802 040. UMTRI-35857.

Abromavage, J. C.; Beemer, R. L. 1976. A Data Acquisition Method for Dynamic Vehicle Testing. Amerco Technical Center, Phoenix, AZ. 7 pp. Report No. SAE 760789. UMTRI-35914.

Backaitis, S. H.; Trout, E. M.; Wolf, R. J. 1976. The Development and Performance of a Self-Contained Solid-State Digital Crash Recorder for Anthropomorphic Dummies. National Highway Traffic Safety Administration, Washington, DC/Federal Aviation Administration, Washington, DC/Kaman Sciences Corporation, Colorado Springs, CO. 32 pp. Report No. SAE 760013. UMTRI-33750.

Courgage, K.G.; Michalopoulos, P. 1976. Bus Priority System Studies Using Instrumented Buses. Florida University, Gainesville. 35 pp. Sponsor: Florida Department of transportation, Tallahassee; Urban Mass Transportation Administration, Washington, DC. UMTRI-33533.

Hofferberth, J. E. 1976. User Data Needs. National Highway Traffic Safety Administration, Washington, DC. 6 pp. Garrett, J. W., ed. *Motor Vehicle Collision Investigation Symposium.* Volume I: Proceedings. Buffalo, Calspan Corporation, Aug 1976. pp. 143–148. UMTRI-35846 A08.

Enserink, E. 1976. Evaluation of Self-Contained Anthropomorphic Dummy Data Acquisition System. Final report. Dynamic Science, Phoenix, Ariz. 141 pp. Sponsor: National

Highway Traffic Safety Administration, Washington, DC. Report No. 3961-75-178/ DOT/ HS 801 827. UMTRI-33788.

Michalopoulos, P. G. 1976. Bus Priority System Studies. Florida University, Gainesville. 6 pp. Traffic Engineering, Vol. 46, No. 7, July 1976, pp. 46–49, 52, 54. Sponsor: Transportation Department, Washington, DC; Florida Department of Transportation, Tallahassee. UMTRI-52996.

Nelson, G.; 1976. FAIRTRAN: Operation of a Credit Card Transit Fare System. RRC International, Inc., Latham, N.Y. 17 pp. UMTRI-33547.

O'Brien, C.; Paradise, M. G. A. 1976. The Development of a Portable Non-Invasive System for Analyzing Human Movement. Nottingham University, Department of Production Engineering and Production Management, England. 3 pp. *International Ergonomics Association. 6th Congress. Proceedings. Santa Monica*, Human Factors Society, 1976. pp. 390–392. UMTRI-34935 A27.

Wolf, R. J. 1976. A Solid-State Digital Data Recorder for Monitoring Anthropomorphic Dummy Impact Environments. Final report. Kaman Sciences Corporation, Colorado Springs, Colo. 74 pp. Sponsor: National Highway Traffic Safety Administration, Washington, DC. Report No. K-76-28U(R)/ DOT/ HS 801 907. UMTRI-34533.

1975

1975. A New look at Tachs—Use of Sangamo Tachographs for Safety. 3 pp. *Diesel Equipment Superintendent*, Vol. 53, March 1975, pp. 32–34. UMTRI-33183.

1975. A Solid-State Recorder for Monitoring Anthropomorphic Dummy Impact Environments. Operator's manual for KSC recorder model ADO2T12. Preliminary edition. Kaman Sciences Corporation, Colorado Springs, Colo. 24 pp. Report No. K-75-95U(R) UMTRI-33675.

1975. American National Standard Guide for the Selection of Mechanical Devices Used in Monitoring Acceleration Induced by Shock. American National Standards Institute, Inc., New York, N.Y. 23 pp. Sponsor: Society of Packing and Handling Engineers, Chicago, Ill. Report No. ANSI-S9.1-1975. UMTRI-33578.

Appleby, M. R.; Bintz, L. J. 1975. Seat Belt Use—Inducing System Effectiveness. Final report. Automobile Club of Southern California, Automotive Engineering Department, Los Angeles. 45 pp. Sponsor: National Highway Traffic Safety Administration, Office of Driver Performance Research, Washington, DC. Report No. DOT/HS 801 503. UMTRI-32135.

Enke, K. 1975. On the Necessity of Employing Driver Recorders for Investigation of the Relation Between the Dynamic Performance of Passenger Cars and Accident Prevention. Daimler-Benz AG, Stuttgart, Germany. 7 pp. UMTRI-34939.

Enke, K. 1975. The Relation Between Vehicle Handling and Accident Avoidance. Daimler-Benz AG, Stuttgart, Germany. 3 pp. International Technical Conference on Experimental Safety Vehicles. Fifth. Report. Washington, DC, GPO, 1975. pp. 815–817. UMTRI-32385 A58.

Hoffer, W. 1975. How They're Using On-Board Crash Recorders to Probe Puzzling Questions About Car Safety. 3 pp. *Popular Science*, Vol. 207, No. 4, Oct 1975, pp. 94–95, 154. UMTRI-32833.

Kidd, E. A. 1975. A Discussion of Data Gathering Systems. Calspan Corporation, Buffalo, NY. 7 pp. Report No. SAE 750892. UMTRI-32932.

1975. Automobile Collision Data; An Assessment of Needs and Methods of Acquisition. Economics and Science Planning, Inc., Washington, DC. 250 pp. Sponsor: Congress, Office of Technology Assessment, Washington, DC. UMTRI-32144.

Gardner, J. A.; Soliday, S. M.; Williamson, G. A. 1975. Design and Implementation of a System to Record Driver Lateral Positioning. Honeywell, Inc., Minneapolis, Minn./ Midwest Research Institute, Kansas City, MO./ North Carolina State University, Raleigh. 10 pp. Transportation Research Record, No. 538, 1975, pp. 59–68. UMTRI-52600 A01.

Johnson, T. M.; Formenti, D. L.; Gray, R. F.; Peterson, W. C. 1975. Measurement of Motor Vehicle Operation Pertinent to Fuel Economy. General Motors Corporation, Noise and Vibration Laboratory, Milford, ME. 30 pp. Report No. SAE 750003. UMTRI-41986.

Priestas, E. L.; Mulinazzi, T. E. 1975. Traffic Volume Counting Recorders. Maryland University, College Park. 13 pp. American Society of Civil Engineers. Transportation Engineering Journal, Vol. 101, No. TE2, May 1975, pp. 211–223. Sponsor: Maryland State Highway Administration, Brooklandville; West Virginia Department of Highways, Charleston. UMTRI-32857.

Soliday, S. M. 1975. Lane Position Maintenance by Automobile Drivers on Two Types of Highway. North Carolina State University, Raleigh, Department of Industrial Engineering. 9 pp. Ergonomics, Vol. 18, No. 2, March 1975, pp. 175–183. UMTRI-52328.

1974

Baker, M. 1974. Unattended Field Measurement Instrumentation. General Motors Corporation, Proving Ground Section, Milford, ME. 5 pp. Report No. SAE 740940. UMTRI-42070.

Fancher, P. S.; MacAdam, C. C. 1974. Data Documentation for Vehicle Handling. Final report. Highway Safety Research Institute, Ann Arbor, ME. 208 pp. Sponsor: National Highway Traffic Safety Administration, Washington, DC. Report No. UM-HSRI-PF-74-4. UMTRI-30757.

Larsson, L. E.; Rumar, K. 1974. A Versatile Recorder of Visual Point of Regard. Uppsala University, Department of Psychology, Sweden. 19 pp. Sponsor: Trygg-Hansa Insurance Company, Sweden; Swedish Transport Research Delegation. Report No. 162. UMTRI-30513.

Machemehl, R.; Lee, C. E. 1974. Dynamic Traffic Loading of Pavements. Final report. Texas University, Center for Highway Research, Austin. 79 pp. Sponsor: Texas Highway Department, Planning and Research Division, Austin. Report No. (TTI) 160-IF. UMTRI-34835.

O'Neill, J. F. 1974. Multiplexing Takes the Measures of Crashes. Data Control Systems, Inc., Danbury, Conn. 4 pp. *Instruments and Control Systems*, Vol. 47, No. 4, April 1974, pp. 41–44. UMTRI-33005.

Ryder, M. O., Jr. 1974. Development and Evaluation of Automobile Crash Sensors—Executive Summary. Summary Final report. Calspan Corporation, Buffalo, NY. 33 pp. Sponsor: National Highway Traffic Safety Administration, Washington, DC. Report No. CAL ZQ-5351-V-3/ DOT/HS 801 262. UMTRI-30722.

Teel, S. S.; Peirce, S. J.; Lutkefedder, N. W. 1974. Automotive Recorder Research—Disc Recorder Pilot Project. Volume II: Results of Tests and Evaluations. Technical report. National Highway Traffic Safety Administration, Office of Operating Systems Research, Washington, DC. 105 pp. Report No. DOT/HS 801 156. UMTRI-29980.

Teel, S. S.; Peirce, S. J.; Lutkefedder, N. W. 1974. Automotive Recorder Research—A Summary of accident Data and Test Results. National Highway Traffic Safety Administration, Washington, DC. 57 pp. *International Conference on Occupant Protection*. 3rd. Proceedings. SAE, New York, 1974. pp. 14–70. Report No. SAE 740566. UMTRI-30029 A02.

Warner, C. Y.; Free, J. C.; Wilcox, B.; Friedman, D. 1974. An Inexpensive Automobile Crash Recorder. Brigham Young University, Provo, UT/ Minicars, Inc., Goleta, CA 9 pp. *International Conference on Occupant Protection. 3rd. Proceedings*. SAE, New York, 1974. pp. 71–79. Report No. SAE 740567. UMTRI-30029 A03.

Yurchevski, A. A., et al 1974. [Recording of the Vehicle Trajectory During Tests.] 3 p. *Avtomobil'naya Promyshlennost'*, No. 7, July 1974, pp. 21–23. UMTRI-52289.

1973

1973. Automotive Tape Recorder. Volume 2. Development Test Report. Final report. Avco Corporation, Avco Systems Division, Wilmington, Mass. 167 pp. Sponsor: National Highway Traffic Safety Administration, Washington, DC. Report No. AVSD-0135-72-CR/ DOT/ HS 800 806/ DOT/HS 800 953. UMTRI-27724.

1973. Automotive Tape Recorder. Volume 3. Assembly, Inspection and Pre-Calibration. Final report. Avco Corporation, Avco Systems Division, Wilmington, MA 48 pp. Sponsor: National Highway Traffic Safety Administration, Washington, DC. Report No. AVSD-0135-72-CR/ DOT/HS 800 807/ DOT/HS 800 954. UMTRI-27418.

Baker, R. C. 1973. Automotive Tape Recorder. Volume 4. Installation, Maintenance and Removal. Final report. Avco Corporation, Avco Systems Division, Wilmington, MA. 78 pp. Sponsor: National Highway Traffic Safety Administration, Washington, DC. Report No. AVSD-0135-72-CR/ DOT/HS 800 808/ DOT/HS 800 955. UMTRI-27419.

Conlon, C. M., Jr. 1973. Automotive Tape Recorder. Volume 1. Design and Preliminary Development. Final report. Avco Corporation, Avco Systems Division, Wilmington, MA. 163 pp. Sponsor: National Highway Traffic Safety Administration, Washington, DC. Report No. DOT/HS 800 677/ DOT/HS 800 952. UMTRI-19102.

Dunham, T. D.; Scheidt, D. C. 1973. Automotive Disc Recorder Environmental Tests. Final report. Southwest Research Institute, San Antonio, TX. 110 pp. Sponsor: National Highway Traffic Safety Administration, Washington, DC. Report No. 02-3701/ DOT/HS 801 015. UMTRI-28936.

Holmstrom, F. R.; Hopkins, J. B. 1973. Microwave Crash Sensor for Automobiles. Transportation Department, Washington, DC. Published by Patent Office, Washington, DC. 7 pp. Report No. Patent 3,760,415. UMTRI-35566.

Kanaya, O.; Sakai, H.; Inokuchi, N. 1973. A VTR System, Which Records On-the-Spot Accident Scenes. Japan Automobile Research Institute, Inc., Ibaragi. 16 pp. *International Conference on the Biokinetics of Impacts. Proceedings*. Organisme National de Securite Routiere, Laboratoire des Chocs, Lyon-Bron, 25 May 1973. pp. 171–186. UMTRI-28048 A12.

LeFevre, D.; D'Auteuil, R. 1973. Automotive Tape Recorders. Volume 5. Data Processing and Post-Calibration. Final report. Avco Corporation, Avco Systems Division, Wilmington, MA. 43 pp. Sponsor: National Highway Traffic Safety Administration, Washington, DC. Report No. AVSD-0135-72-CR/ DOT/HS 800 809/ DOT/HS 800 956. UMTRI-27420.

Lutkefedder, N. W.; Teel, S. S. 1973. Automotive Recorder Research and its Effects on Future Vehicle Safety. National Highway Traffic Safety Administration, Washington, DC. 21 pp. Vehicle Safety Research Integration Symposium. National Highway Traffic Safety Administration, Washington, DC, 1973. pp. 353–373. UMTRI-29031 A20.

Merik, B.; Gittery, V. H. 1973. A New Detection System for Automotive Headlamp Photometry. General Electric Company, Cleveland, OH. 6 pp. *Illuminating Engineering Society Journal*, Vol. 3, No. 1, Oct 1973, pp. 77–82. UMTRI-51455.

Moscarini, F. 1973. The Italian Technical Presentation—Progress Report for the Experimental Institute for Motor Vehicles (ISAM). Effect of Vibrations by Air and by Solid Bodies on the Human Organism. Alfa Romeo, Institute for Experiments on Automobiles and Motors, Milan, Italy. 5 pp. International Technical Conference on Experimental Safety Vehicles. Fourth. Report. NHTSA, Washington, DC, 1973. pp. 411–415. UMTRI-29313 A48.

Trenka, A. R. 1973. Basic Research in Crashworthiness II—Comparison of Teledyne-Geotech Crash Recorder Data and Accelerometer Data. Interim technical report. Calspan Corporation, Buffalo, NY. 111 pp. Sponsor: National Highway Traffic Safety Administration, Washington, DC. Report No. CAL YB-2987-V-15/ DOT/HS 800 873. UMTRI-29610.

Teel, S. S.; Peirce, S. J.; Lutkefedder, N. W. 1973. Automotive Recorder Research—Disc Recorder Pilot Project. Volume I: Fleet Status and Data System Procedures. Technical report. National Highway Traffic Safety Administration, Office of Operating Systems Research, Washington, DC. 69 pp. Report No. DOT/HS 801 019. UMTRI-28935.

Trenka, A. R. 1973. Basic Research in Crashworthiness II—Instrumentation and Data Handling Techniques. Interim technical report. Calspan Corporation, Buffalo, NY. 217 pp. Sponsor: National Highway Traffic Safety Administration, Washington, DC. Report No. CAL YB-2987-V-5/ DOT/HS 800 865. UMTRI-28071

1972

Cheeseman, M.; Nelson, P. M. 1972. A Data Logging System for the Measurement of Road Traffic Noise. Transport and Road Research Laboratory, Crowthorne, England. 18 pp. Report No. TRRL LR 479. UMTRI-19484.

Hackbarth, E. W. 1972. Production Engineering of Automotive Triaxial Crash Recorder, Model 35500. Final report. Teledyne Geotech, Garland, TX. 46 pp. Sponsor: National Highway Traffic Safety Administration, Washington, DC. Report No. TR 72-5/DOT/HS 800 733. UMTRI-19864.

Hackbarth, E. W. 1972. Production Engineering of Automotive Triaxial Crash Recorder, Model 35500. Final report. Teledyne Geotech, Garland, TX. 103 pp. Sponsor: National Highway Traffic Safety Administration, Washington, DC. Report No. TR 72-5/DOT/HS 800 732. UMTRI-19863.

Hudson, C. L. 1972. Development of a Vehicle Mounted Crash Recorder. Final report. EG&G, Inc., Santa Barbara Division, Goleta, CA. 65 pp. Sponsor: National Highway Traffic

Safety Administration, Washington, DC. Report No. S-564-R/ DOT/HS 800 664. UMTRI-17675.

Lundstrom, L. C. 1972. Progress in Vehicle Safety (through electronics) General Motors Corporation, Environmental Activities Staff, Milford, MI. 21 pp. UMTRI-28233.

Romeo, D. J. 1972. Crash Test Evaluation of Crash Recorder and Inflatable Driver Restraint. Cornell Aeronautical Laboratory, Inc., Buffalo, NY. 53 pp. Sponsor: National Highway Traffic Safety Administration, Washington, DC. Report No. CAL ZM-5207-K-1. UMTRI-27417.

Sewell, R. 1972. A Data Acquisition System for Studies of Driver and Vehicle Performance Parameters in Real Traffic Conditions. National Research Council, National Aeronautical Establishment, Ottawa, Canada. 16 pp. Report No. LTR-ST.533. UMTRI-28425.

Shirk, B. I. 1972. Maryland Takes a New Look at Highway Accident Reporting. Maryland Department of Public Safety and Correctional Services, Data Center, Pikesville. 2 pp. *Police Chief*, Vol. 39, No. 8, Aug 1972, pp. 28–29. UMTRI-50779.

1971

1971. Recorder Aids Blood Alcohol Program. Honeywell, Inc., Industrial Division, Fort Washington, PA. 4 pp. Instrumentation, Vol. 24, No. 1, 1971, pp. 11–14. UMTRI-19295.

Forbes, R. T. 1971. A New F.M. Recording System. Motor Industry Research Association, Lindley, England. 2 pp. M.I.R.A. Bulletin, No. 2, April/June 1971, pp. 8–9. UMTRI-16714.

Ohtake, K. 1971. Development of a New Eye Mark Recorder. NAC Inc., Engineering Section, Yokohama, Japan. 6 pp. *Society of Photo-Optical Instrumentation Engineers Seminar Proceedings*, Vol. 22, 1971, pp. 83–88. UMTRI-27163.

Waszkewitz, B. 1971. Der Fahrtschreiber als Hilfsmittel der Fahrerkontrolle; Driving Diagrams as a Means to Supervise Drivers. 4 pp. Zeitschrift fuer Verkehrssicherheit, 17. Jahrgang 1971, II. Quartal, Heft 2, pp. 120–123. UMTRI-50388.

1970

Adams, J. E.; Collins, C. C. 1970. Implanted Monitors. California University, San Francisco, Medical Center, Division of Neurological Surgery/ Institute of Medical Sciences, San Francisco, CA. 16 pp. Gurdjian, E. S., Lange, W. A., Patrick, L. M., Thomas, L. M., eds., comps., *Impact Injury and Crash Protection*, Charles C. Thomas, 1970, pp. 180–195. UMTRI-12268 A08.

Klasky, P. S. 1970. Development of an Automotive Crash Recorder. Final report. Teledyne Geotech, Garland, TX. 121 p. Sponsor: National Highway Traffic Safety Administration, Washington, DC. Report No. TR 70-37/ DOT/HS 800 547. UMTRI-16215.

Lamorlette, P. 1970. Systeme de collecte digitale et traitement automatique de donnees de circulation par ruban perfore; Digital Collection and Automatic Processing of Traffic Data by Punched Tape System. Societe E.V.R., Paris, France. 9 pp. Trafic Maritime et Fluvial et Trafic Urbain, AFCET, Centre Universitaire Dauphine, Paris, 1970, pp. 3a.27–3a.35. UMTRI-15514 A01.

INDEX

Association of International Automobile
    Manufacturers, Inc. (AIAM) xiii,
    177, 409
ATX Technologies 313
Australia's death rate 381
Australian New Car Assessment Program
    (Australian NCAP) 376
Authority of the Secretary 268
Authority to mandate the installation of
    EDRs in all new vehicles 260
Auto Week 45
Automakers 333, 408
Automatic Collision Notification (ACN)
    systems 40, 64, 178
Automobile manufacturers 96, 97, 301
Automotive Coalition for Traffic
    Safety 311
Automotive Industry 16
Automotive Industry Trends 373
Automotive News v, 45, 410
Automotive Occupant Restraints Council
    (AORC) xiii, 110, 408

B

*Bachman, et al, v. General Motors
    Corp.,* 318
Ballard, Bill 334
Barraclough, Jack T. 387
Batiste v. General Motors Corporation 275
Bendix Commercial Vehicle Systems
    LLC 95, 408
Big Brother syndrome 304
Bingham, Price T. 60, 69
Birnbaum, Marie E. 59, 69
Black box xiv, 59, 60, 69, 81, 82, 128, 129,
    168, 179, 180, 184, 232, 233, 234,
    236, 259, 261, 275, 304, 307, 311
Blowin' in the Wind 20
BMW Group 163, 321
Britain's death rate 380
British Medical Journal 3
Bush administration 381

C

Cady v. Dombrowski 292
California bill 320
California Senate Bill AB 213 410
California v Sanchez 319
California v. Beeler 318
Cameron, Bob 253, 254
Campbell, B. J. 4
Canada R. v. Brander 319
Canada R. v. Daley 319
Canada R. v. Gratton 319
Canada Safety Council 383
Canada's death rate 381

Capeloto, Alexa 43
Car and Driver 311
Car manufacturers claim 263
Carrol v. U.S 288
Catalyst for a national debate 376
Chair of the Alliance of Automakers Event
    Data Recorder Committee 123
Chalmers University of Technology of
    Goteborg, Sweden 93, 408
Chapter 301, Motor Vehicle Safety 59
Chapter VIII, Title 49 of the Code of
    Federal Regulations 34
Charleston Post Courier 371
China's Traffic Management Research
    Institute 78
China's VDR 78
Claybrook, Joan 78, 85
Cockpit Voice Recorders (CVRs) 267
Collect statistical data 253
Colorado State Patrol 186
Colorado v. Cain 295, 318
*Commonwealth of Massachusetts vs.
    Mamacos* 291
Congress 33, 155, 160, 259, 264, 268, 303
Congress's authority to regulate interstate
    transportation 264
Connellsville Daily Courier 320
Conners, Ellen Engleman 333
Consent of the vehicle owner 277
Constitution 264
Constitution's Commerce Clause 264
Constitutional questions 271
Consumer Protection Law 320
Consumer's Union 410
Consumers 59, 249, 250, 251
Conyers Jr., John 386
Court decisions 259
Crain, Keith 45
Crash Data Retrieval (CDR) system 42
Crash Dynamics 94
Crash Injury Research and Engineering
    Network (CIREN) 178
Crash investigations 75
Crash kinematics 195
Crash Outcome Data Evaluation Systems
    (CODES) 136, 195
Crash pulse 153
Crash reconstruction 105
Crash severity 64, 98, 135
Crash Survivable Modules 43
Crash tests 97
Crashworthiness Subcommittee 61
Csere, Csaba 311
Cusinato, Kelly 46
Czech Republic death rate 381